Sustainable Agriculture Concepts and Approaches

NIPA GENX ELECTRONIC RESOURCES & SOLUTIONS P. LTD.
New Delhi-110 034

About the Author

Dr Alok Kumar Patra graduated in Agriculture from Chiplima campus of Odisha University of Agriculture & Technology in 1988, obtained his M.Sc. (Agriculture) in Agronomy from Institute of Agriculture, Visva Bharati, Santiniketan in 1990 and Ph.D. from Bidhan Chandra Krishi Viswavidyalaya, Mohanpur, West Bengal in 1996. He has an MBA degree from IGNOU, New Delhi.

Dr Patra has published more than 90 research papers, 125 popular scientific articles and bulletins, 6 books and 30 book chapters. He has 30 years of experience in research and teaching in UG and PG level on agroforestry and agrometeorology. Presently he is working as Chief Agronomist in the All India Coordinated Research Project on Integrated Farming Systems, OUAT, Bhubaneswar.

Sustainable Agriculture Concepts and Approaches

Alok Kumar Patra
Professor of Agronomy
All India Coordinated Research Project on Integrated Farming Systems
Odisha University of Agriculture & Technology
Bhubaneswar-751003, Odisha

nipa

NIPA GENX ELECTRONIC RESOURCES & SOLUTIONS P. LTD.
New Delhi-110 034

**NIPA GENX ELECTRONIC
RESOURCES & SOLUTIONS P. LTD.**

101,103, Vikas Surya Plaza, CU Block
L.S.C.Market, Pitam Pura, New Delhi-110 034
Ph : +91 11 27341616, 27341717, 27341718
E-mail:newindiapublishingagency@gmail.com
www: www.nipabooks.com

For customer assistance, please contact
Phone: + 91-11-27 34 17 17 Fax: + 91-11- 27 34 16 16
E-Mail: feedbacks@nipabooks.com

ISBN: 978-93-91383-19-0

Composed and Designed by NIPA.

Preface

Years ago, the focus was to support life by providing food, fodder, timber, fibre, etc. For this, the farmers were trying to increase the productivity level year after year and they achieved it through the modern high-input agricultural practices. However, during the mid-1980s it was realised that high productivity and economic development at the cost of ecological health and social equity did not lead to long-lasting prosperity. It was also realised that the very existence of life was in danger due to environmental pollution, climate change, disease, loss of biodiversity, conflicts and so on. The world leaders agreed that if the present generation do not change the ways of fulfilling their short-term goals, the future generations would suffer more than the present, and they would be denied their fundamental rights to a healthy productive, life-enhancing environment. Because resources on the earth are limited, it is clear that the world needed to find a way to harmonise ecology with prosperity. So now the focus should be to save the life first and then support it. In this context, the prime focus of farmers, researchers, and policymakers goes to sustainable development through the sustainable agricultural practices.

Keeping this in view, an effort has been made to write a book on 'Sustainable Agriculture' which would be useful to the students as well as the teachers. The people, who watch the agricultural development and sustainability with an interest, would also find it worth reading.

In writing this book, a number of books, periodicals and journals including international publications and university bulletins have been referred. I extend my sincere thanks to the authors and editors of these books and journals. Especially, I do express my gratitude to the authorities of Food and Agriculture Organisation of the United Nations for giving permission to use their literature freely. Every care has been taken to cite the bibliographic references. However, any omissions, misrepresentations, incorrect citations or other mistakes that may have occurred are regretted. Any suggestion to improve the contents of the book will be highly appreciated. I take full responsibility for any errors in this book. Any shortcomings may be intimated so that it will be taken care of in the next edition.

I am grateful to my colleagues in the All India Coordinated Research Project on Integrated Farming Systems, Odisha University of Agriculture & Technology, Bhubaneswar for their help in various ways during the preparation of the manuscript. I thank to the New India Publishing Agency, New Delhi for bringing out the book timely and neatly.

I express my indebtedness and gratitude to my beloved parents who are a constant source of inspiration to me throughout my academic journey. I am also thankful to my wife Jharashree and daughter Prachurya for their constant support and encouragement.

October 13, 2021

Alok Kumar Patra
OUAT, Bhubaneswar

Contents

Glossary

3R approach: The principle of reducing waste, reusing and recycling resources and products is often called the 'Rs'.

Adaptation: The changes in policies, the changes in behaviour and the responses and solutions used to deal with the changing climate.

Additive intercropping: Here the plant population of the main or base crop remains unchanged as a pure crop stand but one or more minor crops are grown simultaneously with it by utilising the inter-row space.

Aggressivity: Aggressivity gives a simple measure of how much the relative yield increase in species 'a' is greater than that for species 'b' in an intercropping system.

Agricultural insurance: Agricultural insurance is a policy which involves the insured farmer paying a little sum (premium), to an insurance company (insurer) to guarantee against loss due to any of the perils (cyclone, flood, drought, etc.) covered for a particular period of time, usually not more than one year with a promise to pay back the value of loss should such occur.

Agricultural waste management system (AWMS): Agricultural waste management system is a planned system in which all necessary components are installed and managed to control and use by-products of agricultural production in a manner that sustains or enhances the quality of air, water, soil, plant, and animal resources.

Agricultural waste: The residues from production and processing of raw agricultural products such as fruits, vegetables, meat, poultry, dairy products and crops are called agricultural wastes.

Agrisilviaquaculture: A form of agroforestry consisting of tree (woody perennial), agricultural crop and freshwater aquatic animal components.

Agrisilvipasture: A form of agroforestry consisting of tree, crop and pasture/animal components.

Agrobiodiversity: The variety and variability of animals, plants and microorganisms that are used directly or indirectly for food and agriculture, including crops, livestock, forestry and fisheries. It comprises the diversity of genetic resources and species used for food, fodder, fibre, fuel and pharmaceuticals.

Agroecology: The study of bringing ecology principles into agriculture, including the diversity of species and genetics, recycling nutrients on the farm, and maintaining healthy soils.

Agroforestry: A collective name for land use systems and technologies where woody perennials (trees, shrubs, palms, bamboos, etc.) are deliberately used on the same land management units as agricultural crops and/or animals, in some form of spatial arrangement or temporal sequence.

Allelopathy: Any direct or indirect harmful effect that one plant has on another or mutually on each other through the production of chemical compounds that escape into the environment.

Alley cropping: A farming system in which arable crops are grown in alleys formed by trees or shrubs, established mainly to hasten soil fertility restoration and enhance soil productivity.

Annidation: Complementary interactions which occur both in space and time are referred to annidation.

Apisilviculture: In this agroforestry system, various honey or nectar producing trees frequently visited by honeybees are planted on the boundary of the agricultural field. Bees benefit the trees and trees in turn provide series of benefits to bees.

Aquaforestry (or aquasilviculture): The aquaforestry system comprises of composite fish culture in farm ponds, and various trees and shrubs, leaves of which are preferred by fish are planted on the boundary and around fishponds. Leaves of these trees are used as feed for fish.

Aquaponics: A method of producing food that combines raising aquatic animals, such as fish or shrimp and plants. The waste from the animals is used as fertiliser for the plants.

Augmenting crop: When sub crops are sown to supplement the yield of the main crop, the sub crops are termed as augmenting crops. Mustard is sown on the borders of wheat fields to harvest mustard leaves for greens during the initial stages.

Bench terrace: Conversion of a steep slope into a series of steps with near-horizontal hedges and near-vertical walls between hedges, using retaining walls (rock) or steep banks (soil) for intensive cropping.

Biochar (or charcoal): Produced by heating organic material (manure or plant leaves, stalks, roots, husks, shells, seeds) in a low oxygen environment. The biochar can be put into soil to increase its fertility and to allow it to hold water.

Biochemical pesticide: These are naturally occurring substances that control pests by non-toxic mechanisms. Biochemical pesticides include substances that interfere with growth or mating, such as growth regulators, or substances that repel or attract pests, such as pheromones.

Bioherbicides: The use of plant pathogens which are expected to kill the targeted weeds are called bioherbicides.

Biological pest control: Biological control is a method of controlling insect pests and diseases using other beneficial organisms which rely on predation, parasitism, or some other natural mechanisms.

Biopesticides: Biopesticides are certain types of pesticides derived from animals, bacteria and trees. These have a narrow target range and a very specific mode of action. Biopesticides have no risk on environment.

Biotech crops: Crops that have been grown from genetically modified seeds.

Biotic community: A group of organisms that interact with each other, depend on each other and inhabit an area.

Blind cultivation: In blind cultivation, the entire field is tilled shallowly, paying little attention to where the rows are. It is the easiest way to destroy the weeds that would be growing within the rows and presenting direct competition to the crop.

Botanical pesticides: Botanical pesticides are raw crushed plant leaves, extracts of plant parts, and chemicals purified from the plants. Pyrethrum, neem, tobacco, garlic, and pongamia formulations are some examples of botanicals.

Carbon footprint: The quantity of greenhouse gas emissions created by a person, family, business or other entity.

Carbon sequestration (or terrestrial carbon sequestration or bio-sequestration): A process where airborne carbon dioxide is removed from the air and stored in plants as leaves, roots, stems, trunks and soils.

Carbon sink: A long-term storage reservoir for carbon, such as soil, wetlands, forests.

Carbon-stock: Carbon stored in vegetation or soil.

Catchment area: An independent unit of treated or untreated land area contributing runoff water to a reservoir.

Climate change: Climate change includes both human-induced global warming and its large-scale impacts on weather patterns.

Climate smart agriculture (CSA): This is a relatively new concept which was launched in 2009 advocating for better integration of adaptation and mitigation actions in agriculture to capture synergies between them and to support sustainable agricultural development for food security under climate change.

Compacted soil: Soils where the air pockets between soil particles have been reduced to such an extent that water can no longer infiltrate the soil, air is not held in the soil and plant roots are unable to be established in the soil.

Companion cropping: In companion cropping the yield of one crop is not affected by other. In other words, the yield of both the crops is equal to their pure crops.

Competitive interaction: In this system the components interact in such a way that the increase in the yield of one component leads to decrease in the yield of other component due to competitive interaction.

Complementary interaction: In a system if the component crops help each other, by creating favourable conditions for their growth in such a way that the system provides a greater yield than the yield of their corresponding sole crops, then the interaction between the components is said to be complementary in nature.

Compost: Organic material, such as leaves, stalks and roots, that has decomposed and is being added to soil as a fertiliser and to rejuvenate soil.

Conservation agriculture: This is an approach to managing agroecosystems for improved and sustained productivity, increased profits and food security while preserving and enhancing the resource base and the environment.

Conservation tillage: Methods for growing annual crops in the previous year's crop residue which reduce soil erosion and retain water and nutrients on the land; the farmers practising conservation tillage leave 30% of the crop residue on the fields.

Continuous grazing: Ggrazing animals on an entire pasture, rather than rotating the animals through subsections of the pasture.

Contour bunding: These are small earthen bunds built horizontally in parallel rows across the hill slope to collect runoff from the catchment to improve soil moisture on the cropped area.

Contour: An imaginary line connecting points of equal elevation on the surface of the soil.

Contour farming: Contour farming is farming with row patterns that run nearly level around the hill.

Contour hedgerow: Dense, narrow hedges of woody perennials, planted along the contour of a slope to prevent soil erosion.

Contour ridge: A series of parallel ridges on the contour of cultivated land which has grass or shrubs planted on them to control soil erosion and improve water management.

Cover crop: Crops grown primarily to cover the soil and to reduce the loss of moisture due to leaching and erosion by wind and water.

Crop diversification: Crop diversification is considered as an attempt to increase the diversity of crops through crop rotation, multiple cropping or intercropping with the aim to improve the productivity, stability and delivery of ecosystem services.

Crop equivalent yield (CEY): The yields from all the crops grown mixed or intercropped or sequentially cropped are converted to the yield of main crop of the system based on price of the produce.

Cropping intensity (CI): Cropping intensity refers to raising of a number of crops from the same field during one agricultural year. It is the ratio of effective crop area harvested to the physical area expressed in percentage.

Cropping intensity index (CII): Cropping intensity index is a measure of farmer's actual land use in area and time relationships for each crop or group of crops compared to the total available land area and time, including the land available for production.

Crop insurance: Crop insurance is a comprehensive yield-based policy meant to compensate farmers' losses arising due to production problems.

Crop rotation: Crop rotation refers to growing of different crops alternatively on the same piece of land in a definite sequence or process of growing different crops in succession on a piece of land in a specific period of time with an objective to get maximum profit from least investment without impairing the soil fertility.

Cropping pattern: The yearly sequence and spatial arrangement of crops or of crops and fallow on a given area.

Cropping system: The cropping patterns taken up for a given piece of land, or order in which the crops are cultivated on a piece of land over a fixed period of time and their interaction with farm resources, other farm enterprises and available technology which determine their make-up.

Cultivated land utilisation index (CLUI): This index is calculated by summing the products of land area planted to each crop, multiplied by the actual duration of that crop and divided by the total cultivated land area available during 365 days.

Cultural pest control: It includes crop production practices that make crop environment less susceptible to pests. Crop rotation, fallowing, manipulation of planting and harvesting dates, manipulation of plant and row spacing, and destruction of old crop debris are a few examples of cultural methods that are used to manage the pests.

Decision support system: A decision support system (DSS) is a computerised programme used to support determinations, judgments, and courses of action in an organisation or a business. In the agricultural sector, it helps farmers to solve complex issues related to crop production.

Deficit irrigation practices: In arid and semiarid regions the irrigation strategies adopted for more effective and rational use of water are called deficit irrigation practices. Such management practices include regulated deficit irrigation, partial root drying and subsurface irrigation.

Deforestation: Deforestation is the purposeful clearing of forested land. Forests have been razed to make space for agriculture and animal grazing, and to obtain wood for fuel, manufacturing, and construction. Deforestation has greatly altered landscapes around the world.

Degraded pasture: Degraded pasturelands are native or planted pastures which have experienced a sharp decrease in carrying capacity, productivity and biomass production.

Degraded soil (or exhausted soil): Soil that is no longer able to support crops or other plant life due to loss of nutrients in the soil.

Double cropping: Cultivation of two crops in succession on a piece of land in a year.

Early season drought: It occurs due to delay in onset of monsoon or due to long dry spells after early sowing.

Ecosystem: All the plants and animals in a given area and their physical environment, including the interactions among them.

Environmental pollution: Environmental pollution is the contamination of the physical and biological components of the earth/atmosphere system to such an extent that normal environmental processes are adversely affected.

Environmental services: Environmental services include the provision of raw materials and energy used to produce goods and services, as well as the removal of waste from human activities, and their role in life support and landscape maintenance.

Environmental sustainability: Environmental sustainability is the responsibility to conserve natural resources and protect global ecosystems to support health and wellbeing, now and in the future.

Erosion: The wearing away of the land surface by running water, wind, ice or other geological agents, including such processes as gravitational creep.

Farmers market: A retail market where farmers sell their produce, meat and eggs directly to the consumers.

Farming system: An appropriate combination of farm enterprises,viz., crops, livestock, forestry, fishery, apiary, poultry, and the means available to the farmer to raise them for profitability. It interacts adequately with the environment without dislocating the ecological and socioeconomic balance.

Feed grains: Grain crops that are used to feed livestock, such as corn, barley, and wheat.

Fertigation: Fertigation is a method of fertiliser application in which fertiliser is incorporated within the irrigation water by the drip system. In this method liquid fertiliser as well as water soluble fertilisers are used.

Filler cropping: Growing of short duration crops in between the newly established perennial crops for few years to fill the space and to utilise the resources. It is an agroforestry system.

Filter strips: A strip of vegetation in a farm field that slows water movement across the field and also traps nutrients to prevent them from running off the land and into water bodies.

Flood forecasting: Flood forecasting is a process of estimating and predicting the magnitude, timing and duration of flooding based on known characteristics of a river basin, with the aim to prevent damages to human life, to properties, and to the environment.

Fodder: Parts of plants which are eaten by domestic animals, these may include leaves, stems, fruit, pods, flowers or pollen.

Foliage: The mass of leaves of plants, usually of trees or bushes.

Food insecurity: A condition where a person does not have enough food to eat on a regular basis.

Food loss: Food is spoiled or discarded during production, processing, storage, and transportation phases.

Food security: A condition where a person has enough food to eat.

Forage: Vegetative material in a fresh, dried, or ensiled state which is fed to livestock (hay, pasture, silage).

Forage crops: Vegetative parts of crops that are used to feed livestock.

Genetic diversity: Genetic diversity is the variety of traits a species has. This creates a unique combination of alleles from the mother and father passed of the offspring, which increase genetic diversity, or the varying traits that different members of a species exibit.

Genetically modified organisms: Seeds, plants or animals whose DNA are modified by insertion of genetic materials from another species. The goal is to create organisms with traits that are unnatural to the species.

Global warming: Global warming is the long-term heating of earth's climate system observed since the pre-industrial period due to human activities, primarily fossil fuel burning, which increases heat-trapping greenhouse gas levels in earth's atmosphere.

Green manure: Green plant material incorporated into the soil to improve its fertility.

Greenhouse gases: A group of gases that hover in the atmosphere and that trap heat near the earth's surface, among them are carbon dioxide, methane, nitrous oxide, hydrofluorcarbons, perfluorcarbons and sulfur hexafluoride.

Groundcover: Living or non-living material which covers the soil surface.

Guard crops: The main crops is grown in the centre, surrounded by hardy or thorny crops such as safflower around pea or wheat, mesta around sugarcane, jowar around maize, etc. with a view to provide protection to the main crop.

Gully: A deep, narrow channel cut into the soil by erosion.

Gully plugging: These are soil and water retaining structures built across gullies in hilly areas.

Hedgerow (or hedge): A closely planted line of shrubs or trees, often forming a boundary or fence.

HEIA: In the high external input agriculture (HEIA), external inputs like chemical fertilisers, high yielding and hybrid seeds, pesticides, irrigation, mechanisation based on fossil fuels are extensively used.

Heirloom seeds: Seeds or cultivars of plants that are no longer grown for large-scale commercial purposes.

High-input modern agriculture: High input modernagriculture uses heavy equipment and large amounts of financial capital, fossil fuel, water, commercial fertilisers, and pesticides to produce single crops, or monocultures.

Homegarden: A land use form on private lands surrounding individual houses with a definite fence, in which several tree species are cultivated together with annual and perennial crops, often with inclusion of livestock.

Indicators of agricultural sustainability: These are a composite set of attributes or measures that embody a particular aspect of agriculture.

Infiltration: The downward movement of water into the soil.

Infiltration rate: A soil characteristic determining or describing the maximum rate at which water can enter the soil under specified conditions, including the presence of an excess of water.

Integrated crop-livestock systems: Farms that include livestock on the landscape as well as crops; they are more beneficial in returning carbon to the soil and maintaining healthy soils.

Integrated pest management (IPM): This is an ecologically based approach to pest control that utilises a multidisciplinary knowledge of crop/pest relationships, establishment of acceptable economic thresholds for pest populations and constant field monitoring for potential problems.

Intensive agriculture: Intensive agriculture, also known as intensive farming, conventional, or industrial agriculture, is a type of agriculture, both of crop plants and of animals, with higher levels of input and output per unit of agricultural land area.

Intercropping: This refers to growing of two or more dissimilar crops simultaneously on the same piece of land.

Invasive species: Invasive species (may be weeds, insect pests, etc.) are a major cause of crop loss and can adversely affect food security.

Irrigation scheduling: Irrigation scheduling is the process used by irrigation system managers to determine the correct frequency and duration of watering.

Land degradation: Land degradation is a process in which the value of the biophysical environment is affected by a combination of human-induced processes acting upon the land. It is viewed as any change or disturbance to the land perceived to be deleterious or undesirable.

Land equivalent ratio (LER): The relative land area under sole crop required to produce the same yield as obtained under a mixed or an intercropping system at the same management level.

Land utilisation index (LUI): Land utilisation index has been defined as the number of days during which the crops occupied the land during a year divided by 365.

Land-use system: The way in which land is used by a particular group of people within a specified area.

Late-season drought: It occurs due to early cessation of rainfall and crop water stress at maturity stage. When the moisture stress strikes during the flowering and milking/ yield formation stages of the crop, it is called the late-season agricultural drought. Monitoring of late-season agricultural drought is a challenge because during this phase, the crop is fully grown and covers the soil.

Leaching (of nutrients): The dissolving and washing away of nutrients down the soil profile by the action of rain water.

LEISA: The low external input sustainable agriculture (LEISA) is a way of farming, where it seeks to optimise the use of locally available resources by maximising the complementary and synergistic effects of different components of the farming systems.

Litter: Layer of decomposing plant material (leaves, branches, etc.) covering the ground, especially under trees.

Live mulching: Living mulch is a cover crop interplanted or undersown with a main crop, and intended to serve the purposes of mulching, such as weed suppression and regulation of soil temperature. Living mulches grow for a long time with the main crops, whereas cover crops are incorporated into the soil or killed with herbicides.

Microbial pesticides: Microbial pesticides are naturally occurring or genetically altered bacteria, fungi, algae, protozoans or viruses. These can be effective and used as alternatives to chemical insecticides.

Microclimate: The specific local climatic conditions near the ground or area around plants up to 2 m height, resulting from the modifications of the general climatic conditions by local differences in relief, exposure and cover, etc.

Microclimate amelioration: Microclimate amelioration is the difference between the local climatic conditions organisms are experiencing and their macroclimate, or the 'free-air' conditions of well-mixed air in nearby open areas.

Mid-season drought: It occurs due to long gaps between two successive rains and stored moisture becoming insufficient during the long dry spell.

Mineralisation: The conversion of an element from an organic form to an inorganic state as a result of microbial decomposition.

Minimum tillage: Minimum tillage is a soil conservation system like strip-till with the goal of minimum soil manipulation necessary for a successful crop production. It is a tillage method that does not turn the soil over, in contrast to intensive tillage, which changes the soil structure using ploughs.

Mitigation: Mitigation involves strategies, processes and technologies that reduce greenhouse gases.

Mixed cropping: Growing of two or more crops simultaneously on the same piece of land, without any definite row arrangement is termed as mixed cropping.

Mixed farming: Cropping systems which involve the raising of crops, animals and/or trees.

Mixed intercropping: It is an intercropping where two or more crops are grown simultaneously intermingled in the same piece of land with no distinct row arrangement.

Mixed row intercropping: It is growing two or more crops simultaneously in the same piece of land intermingled within a distinct row arrangement.

Monoculture: The repetitive growing of the same sole crop on the same land.

Mulch: A layer of loose material on the soil to reduce moisture loss, moderate soil temperature and inhibit weed growth.

Multiple cropping: Growing two or more crops on the same piece of land in one calendar year is known as multiple cropping.

Multiple cropping index (MCI): This measures the sum of area planted to different crops and harvested in single year divided by the total cultivated area times 100.

Multipurpose trees: Trees yielding one or more products and offering environmental benefits as well.

Multistoried cropping (or multitier cropping): Growing plants of different height in the same field at the same time is termed as multistoried cropping. It is mostly practised in orchards and plantation crops for maximum use of solar energy even under high planting density.

Mushroom compost: Mushroom compost is a type of slow-release, organic plant fertiliser. It is the residual compost waste generated or left out after mushroom productions.

It contains lots of salt and organic material along with enzymes and other nutrients that make it suitable habitats for different microbes, synergistically they are found to be beneficial in disease suppression and plant growth promotion.

Natural communities: Natural communities are recurring assemblages of plants and animals found in particular physical environments.

Nitrification inhibitors: Nitrification inhibitors are chemical compounds that slow the nitrification of ammonia, ammonium-containing, or urea-containing fertilisers, which are applied to soil as fertilisers. These inhibitors can help reduce losses of nitrogen in soil.

Nitrogen cycle: The sequence of chemical and biological changes undergone by nitrogen as it moves from the atmosphere into water, soil and living organisms, and upon death of these organisms (plants and animals) is recycled through a part or all of the entire process.

Nitrogen fixation: The biological conversion of elemental nitrogen (N_2) to organic combinations or to forms readily utilised in biological processes, by nitrogen-fixing microorganisms. When brought about by bacteria in the root nodules of leguminous plants, it is referred to as symbiotic; if by free living microorganisms acting independently, it is referred to as non-symbiotic or free fixation.

No-till agriculture (or zero tillage): Keeping crop residues on the land and leaving the earth largely undisturbed during planting.

Nutrient pump: A deep tree root system, that takes up nutrients from deep soil layers and brings them to the surface in the tree and its litter fall.

Nutrient recycling: Nutrient recycling is a cyclic process that encompasses the movement of nutrients from the physical environment to living organisms and back to the environment.

Organic farming: Farming that does not use pesticides and artificial fertilisers but instead uses organic fertilisers and natural pesticide control. Also avoids using antibiotics and hormones for weight gain in animals.

Organic manures: These are organic materials derived from animal, human and plant residues which contain nutrients in complex organic forms.

Overstory (or overstorey): The highest layer of vegetation, often the tree canopy, which grows over lower shrub or plant layers.

Ozone depletion: Ozone depletion, gradual thinning of earth's ozone layer in the upper atmosphere caused by the release of chemical compounds containing gaseous chlorine or bromine from industry and other human activities.

Parallel intercropping: In this practice two crops are selected which have different growth habits and have a zero competition between each other and both of them express their full yield potential.

Partial root drying: Partial root drying is a new irrigation technique, where one half of the root system is allowed to drying conditions by withholding irrigation and the other half is irrigated. Wetted and dried sides of the root system alternate on a 7-14 day cycle. Partial root drying uses biochemical responses of plants to water stress to achieve balance between vegetative and reproductive growth.

Percolation tanks: Percolation tank is an artificially created surface waterbody, submerging in a highly permeable land, so that surface runoff is made to percolate and recharge the ground water storage.

Permaculture: A method of farming that involves caring for the earth and the earth's natural systems, providing for people to use the resources they need and returning waste to the earth's natural systems.

Plant genetic resources: Plant genetic resources are plant genetic materials of actual or potential value. They describe the variability within plants that comes from human and natural selection over millennia.

Pollinator: A bee, moth, butterfly or other insect; bird; bat; or other animal that moves pollen from the male anthers to the female stigma of flowering plants. The wind can also serve as a pollinator.

Precision agriculture: Involves applying pesticides and fertilisers to specific areas within a field based on the particular needs of the soils and plants of that area. It is based on sophisticated tools involving satellites and computer programmes and the concept that a farm field varies in things like soil type, elevation and water retention.

Precision farming: Precision agriculture, satellite farming or site specific crop management is a farming management concept based on observing, measuring and responding to inter and intra-field variability in crops.

Pressure-state-response (PSR) framework: The PSR indicators propose to evaluate the pressures of human activities on environmental states and to provide political and societal responses in order to come back to a desirable state.

Protective plant: Plants grown to protect crops, soil or land from adverse environmental factors.

Protein bank: Protein banks are blocks of forage plants deliberately planted to alleviate fodder shortages in arid, semiarid and mountainous regions, especially during the dry seasons.

Pyrolysis system: Pyrolysis is a process of chemically decomposing organic materials at elevated temperatures in the absence of oxygen. In pyrolysis systems, agricultural waste is heated up to a temperature of 400-600° C in the absence of oxygen to vapourise a portion of the material, leaving a char behind.

Rainwater harvesting: Rainwater harvesting is the collection and storage of rain, rather than allowing it to run off. Rainwater is collected from a roof-like surface and redirected to a tank, cistern, deep pit, aquifer, or a reservoir with percolation, so that it seeps down and restores the groundwater.

Ratoon cropping (or ratooning): It refers to raising a crop with regrowth coming out of roots or stalks after the harvest of the crop.

Real time contingency planning (RTCP): Real time contingency planning is considered as any contingency measure, either technology related (land, soil, water, crop) or institutional and policy based, which is implemented based on real time weather pattern in any crop growing season.

Regenerative agriculture: Using farming techniques that enhance the land, including regenerating topsoil and increasing biodiversity; are resilient to climate change; that provide a livelihood for the farm families and the local community.

Regulated deficit irrigation: Regulated deficit irrigation is an optimising strategy under which crops are allowed to sustain some degree of water deficit and yield reduction.

Relative crowding coefficient (RCC): Relative crowding coefficient is a measure of the relative dominance of one component crop over the other in an intercropping or mixed cropping system. The coefficient (K) is determined separately for each component crop.

Relative economic efficiency (REE): It refers to the capacity of the diversified cropping system for realising net profit in relation to the existing system and expressed in percentage.

Relative productivity efficiency: It refers to the capacity of the diversified system for production in relation to the existing system and expressed in percentage.

Relay intercropping: Growing two or more crops simultaneously during at least a part of the life cycle of each. A second crop is planted after the first crop has reached its reproductive stage but before it is ready for harvest.

Resilience: The ability of nature and mankind to adapt and survive in the face of change.

Resilient agricultural practices: It is an approach that includes sustainbly using existing natural resources through crop and livestock production systems to achieve long-term higher productivity and farm incomes under any environmental change, especially climate variabilities.

Resource use efficiency: Resource use efficiency (fertiliser, water, etc.) is the output of any crop or anything else per unit of the resource applied under a specified set of soil and climatic conditions.

Risk assessment: Risk is an important aspect of the farming. Agricultural production risk derives from the uncertain natural growth processes of crops and livestock. Weather, disease, pests, and other factors affect both the quantity and quality of commodities produced.

Rooftop rainwater harvesting: This is the technique through which rainwater is captured from the roof catchments and stored in reservoirs. Harvested rainwater can be stored in subsurface ground water reservoir by adopting artificial recharge techniques to meet the household needs through storage in tanks.

Rotational grazing: A practice where grazing land is divided into segments called paddocks, with the grazing animals moved from paddock to paddock every few days. The forage grasses, both introduced and native grasses, are allowed to rejuvenate while the paddock is resting between grazing periods resulting in less soil erosion.

Rotational intensity (RI): This is calculated by counting the number of crops in a rotation and multiplied by 100 and then divided by duration of the rotation.

Row intercropping: It is an intercropping where at least one crop is planted in regular rows, and crop or other crops may be grown simultaneously in row or randomly with the first crop.

Runoff: The portion of the precipitation on an area which is not absorbed by the soil but finds its way into the streams after meeting the persisting demands of evapotranspiration and other losses.

Salinisation: It is the accumulation of water-soluble salts of sodium, magnesium and calcium in the soil.

SDG (Sustainable development goal) 2.4.1: Ratio of area under productive and sustainable agriculture and agricultural land area.

Sequential cropping: Sequential cropping refers to growing crops in sequence within a crop year, one crop being sown after the harvest of the other.

Shelterbelts: A wind barrier of living trees and shrubs established and maintained for protection of crop fields.

Shifting cultivation: It is a system of land use in which the site of cultivation is regularly changed, with older sites reverting to forest or bush fallow.

Silvipastoral system: A form of agroforestry systems consisting of the trees (woody perennial) and pasture/animal components.

Social sustainability: Social sustainability addresses the importance of human resilience, success and fulfilment in agriculture. The social sustainability of the sector deals with the social impacts on local communities, workers, and consumers as the main groups of the stakeholders.

Soil amendment: Materials added to the soil to make it more fertile such as compost, biochar, manure and artificial fertiliser.

Soil conservation: A combination of all management and land-use methods that safeguard the soil against depletion or deterioration caused by nature and/or humans.

Soil degradation: It refers to decline in the productive capacity of land due to decline in soil quality caused through processes induced mainly by human activities.

Soil organic matter: The organic fraction of the soil that includes plant and animal residues at various stages of decomposition, cells and tissues of soil organisms, and substances synthesised by the soil population.

Soil rejuvenation: Returning carbon to the soil, especially when it has been significantly reduced.

Soil salinisation: Soil salinity is the salt content in the soil; the process of increasing the salt content is known as salinisation.

Soil sealing: Soil sealing is the covering of the soil surface with an impervious material or the changing of its nature so that the soil becomes impermeable such that *it* reduces the amount of water infiltration into the soil. Soil sealing often affects fertile agricultural land, puts biodiversity at risk, increases the risk of flooding and water scarcity and contributes to global warming.

Soil-water balance: Soil water balance is an account of all quantities of water added, removed or stored in a given volume of soil during a given period of time. The soil water balance equation thus helps in making estimates of parameters, which influence the amount of soil water.

Stover (or crop residue): The residue left on the fields after the grain is harvested, including stalks, leaves and roots.

Strip intercropping: It is growing two or more crops simultaneously in different strips on sloppy lands wide enough to permit independent cultivation, but narrow enough for the crop to interact with each other.

Substitutive intercropping: In the substitutive or replacement series of intercropping, the crops grown together are known as component crops or intercrops. Here, one component crop is introduced by the replacement of the other crop and no crop is sown with its fullest population as seeded in respective sole cropping. In this system, a definite proportion of a crop is sacrificed and the component crop is introduced in that place.

Subsurface drip irrigation: Subsurface drip irrigation is a low-pressure, low-volume irrigation system that uses buried tubes to apply water. The applied water moves out of the tubes by soil matrix suction.

Subsurface dykes: A subsurface dyke a structure that is built in an aquifer with the intention of obstructing the natural flow of ground water, thereby raising the ground water level and increasing the amount of water stored in the aquifer.

Supplementary interaction: If the two components interact in such a way that the yield of one component exceeds the yield corresponding to its sole crop without affecting the yield of other component, the interaction is known to be supplementary in nature.

Supply chain efficiency: Supply chain management implies managing the relationships between the businesses responsible for the efficient production and supply of products from the farm level to the consumers to meet consumers' requirements reliably in terms of quantity, quality and price.

Sustainability criteria: These refer to critical/thresholds values against which the performance of each sub-indicator is assessed to classify the farm in terms of the sustainability level.

Sustainable agriculture: The management and conservation of the natural resource base and the orientation of technological change should be in such a manner as to ensure the attainment of continued satisfaction of human needs for present and future generations. Sustainable agriculture conserves land, water, and plant and animal genetic resources, and is environmentally non-degrading, technically appropriate, economically viable and socially acceptable.

Sustainable land use: Land use that maintains productivity of the land while conserving or enhancing the resources on which future production depends.

Sustainable soil management: Soil management is sustainable if the supporting, provisioning, regulating, and cultural services provided by soil are maintained or enhanced without significantly impairing either the soil functions that enable those services or biodiversity.

Sustainable value index (SVI): This index is used to assess the sustainability of a cropping system on monetary terms. In cropping systems where more than one crop is involved, the economic assessment is ideal one than the biological assessment. To work out the SVI, the monetary values of the economic produce are used instead of yield values. SVI $= \frac{y - Sd}{Y\,max}$, where, Y is the estimated average net profit of a system over years, Sd is the estimated standard deviation and Ymax is the observed maximum profit among all the systems over years.

Sustainable yield index (SYI): This index is used to assess the sustainability of a cropping system or management practice which can be considered for wider adoption. It identifies the practices systems giving the maximum sustainable yield. SYI $= \frac{y - Sd}{Y\,max}$, where, Y is the estimated average yield of different practices or systems over years, Sd is the estimated standard deviation and Ymax is the observed maximum yield among all the systems over years.

Taungya: Method of raising forest crops in combination with an agricultural crop; the agricultural use to which the land is put does not generally continue throughout the rotation of the forest crop but is confined to that period which ends with the closing of the canopy of the forest crop.

Temperature humidity index (THI): It is a single value representing the combination of air temperature and humidity associated with the level of thermal stress THI is calculated from air temperature and relative humidity using the following equation; THI = dry bulb temperature in degree Celsius + (0.36 × dew point temperature in degree Celsius) + 41.2.

Thermoneutral zone: It is defined as the range of ambient temperatures without regulatory changes in metabolic heat production or evaporative heat loss. It is the temperature zone in which the birds are able to keep their body temperature constant with the help of physical heat regulation.

Topsoil: The layer of soil where plant roots are found. It consists of organic material, silt, sand and clay.

Transgenic seeds: Seeds that have been modified using genetic modification techniques.

Trap crops: These crops are grown in the main cropped field in definite rows to trap insect pests.

Triple cropping: Cultivation of three crops in succession on a piece of land in a year

Urease inhibitors: Urease inhibitors are the chemical compounds that block the activity of the enzyme urease.

Water use efficiency: It is a measure of the amount of biomass produced per unit of water used by a plant.

Windbreak: A strip of trees or shrubs or crop plants serving to reduce the force of wind and provide a protective shelter against wind.

1

Introduction to Sustainable Agriculture

The food productivity increased manifolds due to high-input modern agriculture with the adoption of new technologies like high yielding varieties, farm mechanisation, increased use of fertiliser and pesticide, specialised farming practices, water resource development and improved irrigation practices, etc. The Green Revolution started in the early 1960s has led to the attainment of self-sufficiency in food grain production. This has been considered as the greatest agricultural transformation in the history of humankind. The change has brought higher and more stable food production, ensuring food stability and security for a constantly growing world population. The Green Revolution contributed to widespread poverty reduction, averted hunger for millions of people, and avoided the conversion of thousands of hectares of land into agricultural cultivation. At the same time, the Green Revolution also spurred its share of unintended negative consequences, often not because of the technology itself but rather, because of the policies that were used to promote rapid intensification of agricultural systems and increase food supplies. Some areas were left behind, and even where it successfully increased agricultural productivity, the Green Revolution was not always the panacea for solving the myriad of poverty, food security, and nutrition problems facing poor societies (Pingali, 2012). The benefits of Green Revolution have been poorly distributed and thus, hunger still persists in many parts of the world. Some of the adverse effects of modern high-input agriculture are listed below.

1. Overuse of natural resources, causing depletion of groundwater, and loss of forests, wild habitats, and of their capacity to absorb water, causing waterlogging and increased salinity
2. Contamination of the atmosphere by ammonia, nitrous oxide, methane and the products of burning, which play a role in ozone depletion, global warming and atmospheric pollution
3. Contamination of food and fodder by residues of pesticides, nitrates and antibiotics
4. Contamination of water by pesticides, nitrates and agrochemicals causing human and animal health hazards, harm to wildlife, disruption of ecosystems

5. Build-up of resistance to pesticides in pests and diseases including herbicide resistance in weeds
6. Erosion of genetic diversity by focusing on modern varieties, and causing the displacement of traditional varieties and breeds
7. New health hazards for workers in the agrochemical and food processing industries

Added to the above adverse effects, the increasing human as well as livestock population imposed intense pressure on available natural resources. Accordingly, a challenge emerged that required a new vision, holistic approaches for ecosystem management and renewed partnership between science and society. In December 1983, the United Nations General Assembly established the World Commission on Environment and Development under the leadership of former Norwegian prime minister Harlem Brundtland. In 1987, the 'Brundtland Commission' released its final report, *Our Common Future*. It was realised that after decades of effort to raise living standards through industrialisation, many countries were still dealing with extreme poverty. It seemed that economic development at the cost of ecological health and social equity did not lead to long-lasting prosperity. It was also realised that our generation has too often been willing to use the resources of the future to meet our own short-term goals. It is a debt we can never repay. If we fail to change our ways, our future generations would suffer more than we, and they would be denied their fundamental right to a healthy productive, life-enhancing environment. Because resources on the earth are limited, it was clear that the world needed to find a way to harmonise ecology with prosperity. The document famously defines sustainable development as development that meets the needs of the present without compromising the ability of future generations to meet their own needs (WCED, 1987). The Commission successfully unified environmentalism with social and economic concerns on the world's development agenda.

Almost at the same time the realisation of prime importance of staple food production for achieving food security for future generations has brought the concept of 'Sustainable Agriculture' to the forefront and began to take shape in the following three points.

1. The interrelatedness of all the farming systems including the farmer and the family
2. The importance of many biological balances in the system
3. The needs to maximise desired biological relationships in the system and minimise the use of materials and practices that disrupt these relations

In fact, the idea of sustainable agriculture has been around a long time. Since the very first crop was sown and animal was penned, farmers have tried to ensure that their land produces a similar or increasing yield of products year after year. The recent attempts to popularise the concept build on this tradition. Sustainable agriculture has evolved from three perspectives; as a system of production to achieve food self-reliance; as a concept of stewardship; and as a vehicle for sustaining rural communities (MacRae, 1990). A sustainable agriculture is a system of agriculture that will last. It is an agriculture that maintains its productivity over the long run. Sustainable agriculture is both a philosophy and a system of farming. It has its roots in a set of values that reflects an awareness of both ecological and social realities. It involves design and management procedures that work with natural processes to conserve all resources, minimise waste and environmental damage, while maintaining or improving farm profitability. Working with natural soil processes is of particular importance. Sustainable agriculture systems are designed to take maximum advantage of existing soil nutrient and water cycles, energy flows, and soil organisms for food production. Such systems aim to produce food that is nutritious, without being contaminated with products that might harm human health.

PRINCIPLES OF SUSTAINABLE AGRICULTURE

The Food and Agriculture Organisation of the United Nations has established five basic principles for the global agricultural sector to become increasingly productive and sustainable (FAO, 2014). The intention of these five principles is to build a production system that works in favour of the ecosystem, satisfying the human needs. For that, the farmers should work in the economic development of this sector and, so, they can be benefited from a fair, equitable and efficient environment.

Principle 1 : Improving efficiency in the use of resources is crucial to sustainable agriculture

Principle 2 : Sustainability requires direct action to conserve, protect and enhance natural resources

Principle 3 : Agriculture that fails to protect and improve rural livelihoods, equity and social well-being is unsustainable

Principle 4 : Enhanced resilience of people, communities and ecosystems is key to sustainable agriculture

Principle 5 : Sustainable food and agriculture requires responsible and effective governance mechanisms

These five principles are complementary; Principle 1 and Principle 2 directly support the natural system, while Principle 3 directly supports the human system; Principles 4 and Principle 5 underpin both the natural and human systems. For application of the five principles, a range of actions should be taken to enhance sectoral as well as cross-sectoral productivity and sustainability. To be sustainable and productive, agriculture will need to adopt a vision that maximises synergies, mitigates negative externalities and minimises harmful competition between its sectors. The key strategies, policies and technologies under each of the five principles have been prescribed by FAO (table 1.1). The proposed approach, when properly localised, would help identify cross-sectoral synergies, negative externalities and actions that would minimise their impacts in the source sectors or mitigate them in the affected sectors (FAO, 2014).

Table 1.1. Innovative governance and technologies that sustainably increase agricultural production

Principle 1: Improving efficiency in the use of resources is crucial to sustainable agriculture

Sectors	Key Policies and Practices
Crop	1. Genetically diverse portfolio of varieties 2. Conservation agriculture 3. Judicious use of organic and inorganic fertilisers, improved soil moisture management 4. Improved water productivity, precision irrigation 5. Integrated pest management
Livestock	1. Genetically diverse base of breeds 2. Improved resource use efficiency 3. Balanced and precision animal feeding and nutrition 4. Integrated animal health control
Forestry	1. Sustainable management of natural and planted forests 2. Forest area increase and slowing deforestation 3. Improved efficiency of use of wood-based energy 4. Development of innovative renewable forest products 5. Tree improvement to support productivity and resilience
Aquaculture	1. Aquafeed management 2. Integrated multi-trophic aquaculture 3. Robust biosecurity/aquatic animal health 4. Use of best management practices, good aquaculture practices 5. Domestication of aquaculture species 6. Aquaculture certification for animal health and welfare, and food safety 7. Implementing the 'ecosystem approach to aquaculture'
Fisheries	1. Fuel efficiency increase and use of static gears 2. Reduction of fishing costs and capacity 3. Reduction of waste and discards 4. Integration of inland fisheries in water and land planning and management

Principle 2: Sustainability requires direct action to conserve, protect and enhance natural resources

Sectors	Key Policies and Practices
Crop	1. Use better practices for biodiversity, such as *in situ* and *ex situ* conservation of plant genetic resources, integrated pest management, etc. 2. Use better practices for soil: land rehabilitation, appropriate cropping systems, etc. 3. Use better practices for water management, deficit irrigation, preventing water pollution, etc. 4. Set payments for using and for providing environmental services such as pollinators, carbon sequestration, etc. 5. Set policies, laws, incentives, and enforcement to promote the above
Livestock	1. Conserve animal genetics *in situ* and *ex situ* 2. Use grassland for biodiversity, carbon storage and water services 3. Protect water from pollution through waste management 4. Use better practices for reduced emission intensity 5. Set payments for using and for providing environmental services, e.g. grazing fees 6. Set policies, laws, incentives, and enforcement to promote the above
Forestry	1. Conserve biodiversity and forest genetic resources 2. Restore and rehabilitate degraded landscapes 3. Enhance the role of forests in soil protection and conservation 4. Enhance the role of forests in the protection and conservation of water resources 5. Use reduced impact harvesting techniques 6. Certification of forest management
Aquaculture	1. Conserve aquatic genetic resources 2. Promote aquaculture certification for environmental protection 3. Ensure biosecurity: pathogens, use of veterinary drugs, invasive species, biodiversity 4. Use integrated aquaculture-agriculture systems 5. Implementing the 'ecosystem approach to aquaculture'
Fisheries	1. Assess non-target resources 2. Develop and use low-impact fishing gears 3. Build fish passes in dams 4. Rebuild depleted stocks and protect critical habitats 5. Restock inland fisheries 6. Implement the 'ecosystem approach to fisheries': protect vulnerable marine ecosystems, use 'marine protected areas' in fishery management, implement ecolabelling 7. Implement the 'Code of Conduct for Responsible Fisheries' and international action plans 8. Deter illegal (IUU- illegal, unreported and unregulated) fishing

Principle 3: Agriculture that fails to protect and improve rural livelihoods, equity and social well-being is unsustainable

Sectors	Key Policies and Practices
Crop	1. Increase/protect farmers' access to resources, e.g. through equitable land and water tenure systems
	2. Increase farmers' access to markets through capacity-building, credit, infrastructure
	3. Increase rural job opportunities, e.g. in small and medium enterprises sustainability and related activities
	4. Improve rural nutrition: production of more and affordable nutritious and diverse foods, including fruits and vegetables
Livestock	1. Increase/protect farmers' access to resources, such as pasture, water, credit
	2. Increase farmers' access to markets through capacity-building, credit, infrastructure
	3. Increase rural job opportunities, e.g. in small and medium enterprises sustainability and related activities
	4. Improve rural nutrition: production of more and affordable nutritious and diverse foods, including fruits and vegetables
Forestry	1. Improve forest tenure rights and access to forest resources
	2. Promote engagement of local stakeholders including communities and women's groups
	3. Promote the development of small and medium-scale enterprises
	4. Provide forest-based employment including health and safety provisions
	5. Establish payment schemes for environmental services
	6. Integrate forestry in poverty reduction strategies
Aquaculture	1. Aquaculture business-model, especially for small-scale producers
	2. Aquaculture certification including gender considerations
	3. Cooperative marketing
	4. Mainstream aquaculture into rural development processes
	5. Technological, trade, institutional, infrastructure, capacity development, investment and other support services
	6. Aquaculture and nutrition programmes
	7. Implementing the 'ecosystem approach to aquaculture'
Fisheries	1. Improve access and tenure rights
	2. Improve local markets
	3. Promote small/medium enterprises
	4. Enhance gender equity
	5. Integrate fisheries in poverty reduction strategies

Principle 4: Enhanced resilience of people, communities and ecosystems is key to sustainable agriculture

Sectors	Key Policies and Practices
Crop	1. Generalise risk assessment/management and communication
	2. Prepare for/adapt to climate change
	3. Respond to market volatility, e.g. encouraging flexibility in production systems, and savings
	4. Contingency planning for droughts, floods, and pest outbreaks; development; social safety nets
Livestock	1. Generalise risk assessment/management and communication
	2. Prepare for/adapt to climate change
	3. Respond to market volatility, e.g. encouraging flexibility in production systems, and savings
	4. Contingency planning for droughts, floods, and pest outbreaks; development; social safety nets
Forestry	1. Increase resilience of ecosystems to biotic and abiotic hazards including climate change phenomena, pests and diseases, forest fires
	2. Prevent the transmission of pathogens to other countries through international trade
	3. Integrate risk prevention and management into sustainable land-use planning
Aquaculture	1. Assess risks in aquaculture (pathogens, food safety, ecological, environmental including climate change, genetic, social and financial)
	2. Set early warning, preparedness, surveillance systems and contingency plans for aquatic emergencies
	3. Implementing the 'ecosystem approach to aquaculture'
Fisheries	1. Generalise risk assessment/management and communication
	2. Develop multipurpose industries
	3. Assess/value ecosystems and services
	4. Maintain stocks at high level
	5. Adopt precautionary approach
	6. Enhance social safety nets
	7. Prepare for climate change

Principle 5: Sustainable food & agriculture requires responsible & effective governance mechanisms

Sectors	Key Policies and Practices
Crop	1. Increase effective participation 2. Encourage formation of associations 3. Increase frequency and content of consultations among stakeholders 4. Develop decentralised capacity
Livestock	1. Increase effective participation 2. Encourage formation of associations 3. Increase frequency and content of consultations among stakeholders 4. Develop decentralised capacity
Forestry	1. Develop personnel and institutional capacity 2. Support good governance of rural areas 3. Decentralise decision-making and empower local communities to promote participatory forestry 4. Develop financial incentive packages to support private investment and enable equitable distribution of benefits 5. Apply mediation and other conflict resolution mechanisms in resource governance 6. Enhance communication to better articulate the benefits of forests
Aquaculture	1. Compliance with international treaties, standards, agreements on sustainable aquaculture, animal health, food safety 2. Voluntary adoption of good aquaculture practices 3. Contribution and impact assessments 4. Implementing the 'ecosystem approach to aquaculture'
Fisheries	1. Develop local governance capacity 2. Empower local communities 3. Adopt good governance principle 4. Decentralise decision-making

Source: FAO, 2014

DEFINITIONS OF SUSTAINABLE AGRICULTURE

Sustainable agriculture refers to an agricultural production and distribution system that:

1. Achieves the integration of natural biological cycles and controls
2. Protects and renews soil fertility and the natural resource base
3. Reduces the use of nonrenewable resources and purchased (external or off-farm) production inputs
4. Optimises the management and use of on-farm inputs
5. Provides adequate and dependable farm income
6. Promotes opportunity in family farming and farm communities
7. Minimises adverse impacts on health, safety, wildlife, water quality and the environment

There is no generally accepted definition of sustainable agriculture (CGIAR/TAC, 1988). However, as Swindale (1988) explained, sustainability conveys the idea of a balance between human needs and environmental concerns. A common theme among definitions is that sustainable agricultural systems remain productive over time (Senanayake, 1991). The enhancement of the environmental quality and careful use of the resource base on which agriculture depends is viewed as a requisite to sustained agricultural productivity (ASA, 1989). Sivakumar et al., (2000) summarised the definitions of 'sustainable agriculture' given by several authors and organisations (table 1.2) and concluded that 'natural resource use' is a keyword common to almost all of them.

Table 1.2. Selected definitions of sustainability/sustainable agricultural system

Author	Definition of sustainability/sustainable system(s)	Keywords
Conway (1985)	Ability of a system to maintain productivity in spite of major disturbance such as is caused by intense or large perturbation.	Productivity
Rome Forum (1986)	Efforts to achieve sustainable food security should blend the unique features that cater for specific cultural economic and ecological conditions among countries and within different regions in the same country.	Ecological conditions
Davis and Schirmer (1987)	Among the topics considered under sustainable agriculture are resource management issues dealing with soils, land, resources and watersheds; and environmental problems such natural as desertification,soil degradation, etc.	Natural resources; environmental problems
Dover and Talbot (1987)	Those systems whose production can continue indefinitely without undue degradation of other ecosystems	Ecosystem degradation
WCED (1987)	Conserving and enhancing the resource base and merging economics and environment in decision making are among the objectives for sustainable development.	Resource base; environment
BIFAD (1988)	The successful management of resources for agriculture to satisfy changing human needs,while maintaining or enhancing the natural resource base and avoiding environmental degradation.	Management of resources; natural resource base; environmental degradation
BIFAD (1998)	One that should conserve and protect natural resources and allow for long-term economic growth by managing all exploited resources for sustainable yields.	Natural resources; exploited resources
CGIAR/ TAC (1988)	Sustainability refers to successful management of resources for agriculture to satisfy changing human needs while maintaining or enhancing the quality of the environment enhancing and conserving natural resources.	Quality of environment; natural resource conservation
FAO (1988)	The management and conservation of the natural resources. base and the orientation of technological change should be in such a manner as to ensure the attainment of continued satisfaction of human needs for present and future	Natural resource base; technological change

Author	Definition of sustainability/sustainable system(s)	Keywords
	generations. Sustainable agriculture conserves land, water, and plant and animal genetic resources, and is environmentally non-degrading, technically appropriate, economically viable and socially acceptable.	
Knezek et al. (1988)	Resource conserving and uses external and internal resources as efficiently as possible. Environmentally sound, actually enhancing rather than detracting from the natural environment. Economically viable in that it earns a fair return on farm investments.	Resource conserving; environmentally sound
Lynam and Herdt (1988)	The capacity of a system to maintain output at a level approximately equal to or greater than its historical average with the approximation determined by its historical variability.	Resources for maintaining output
Rodale (1988)	The system where the resources used in production are maintained in such a way that they are more or less self-generating and ensure continual improvement beyond conventional expectations.	Self-generating; production resources
American Society of Agronomy (ASA,1989)	That, over the long-term, enhances environmental quality and the resource base on which agriculture depends; provides for basic human food and fibre needs; is economically viable; and enhances the quality of life for farmers and society.	Environmental quality; resource base
Ruttan (1989)	Development of technology and practices that maintain and/or enhance the quality of land and water resources. The improvement in plants and animals and the advances in production practices that will facilitate the substitution of biological technology for chemical technology.	Land and water resources
Lal (1991)	An increasing trend in production over time per unit consumption of the non-renewable or the limiting resource, or per unit degradation of soil and environmental characteristics.	Resource use; environmental characteristics
Okigbo (1991)	A system which maintains an acceptable and increasing level of productivity that satisfies prevailing needs and is continuously adapted to meet the future needs for increasing the carrying capacity of the resource base and other worth-while human needs.	Resource base
Okigbo (1991)	A system in which the farmer continuously increases productivity at levels that are economically viable, ecologically sound, and culturally acceptable, through the efficient management of resources and orchestration of inputs in numbers, quantities, sequences and timing with minimum damage to the environment and danger to human life.	Ecologically sound; management of resources

Source: Sivakumar et al., 2000

These definitions can be supplemented by some fundamental principles of sustainable agriculture.

1. Farm productivity is enhanced over the long term.
2. Adverse impacts on the natural resource base and associated ecosystems are ameliorated, minimised or avoided.
3. Residues resulting from the use of chemicals in agriculture are minimised.
4. Net social benefit (in both monetary and non-monetary terms) from agriculture is maximised.
5. Farming systems are sufficiently flexible to manage risks associated with the vagaries of climate and markets.

OBJECTIVES AND GOALS OF SUSTAINABLE AGRICULTURE

Objectives

1. Make best use of the resources available
2. Minimise use of non-renewable resources
3. Protect the health and safety of farm workers, local communities and society
4. Protect and enhance the environment and natural resources
5. Protect the economic viability of farming operations
6. Provide sufficient financial reward to the farmer to enable continued production and contribute to the well-being of the community
7. Produce sufficient high-quality and safe food
8. Build on available technology, knowledge and skills in ways that suit local conditions and capacity.

Goals

Pretty (1996) identifies the following goals of sustainable agriculture.

1. A more thorough incorporation of natural processes such as nutrient cycling, nitrogen fixation and pest-predator relationships into agricultural production processes
2. A reduction in the use of those off-farm, external and nonrenewable inputs with the greatest potential to damage the environment or harm the health of farmers and consumers, and more targeted use of the remaining inputs used with a view to minimising variable costs
3. A more equitable access to predictive resources and opportunities, and progress towards more socially just forms of agriculture

4. A greater productive use of local knowledge and practices, including innovation in approaches not yet fully understood or widely adopted
5. An increase in self-reliance among farmers and rural people
6. An improvement in the match between cropping patterns and the productive potential and environmental constraints of climate and landscape to ensure long-term sustainability of current production levels
7. Profitable and efficient production with an emphasis on integrated farm management and the conservation of soil, water, energy and biological resources
8. The full participation of farmers and rural people in all processes of problem analysis and technology development, adoption and extension
9. A greater productive use of the biological and genetic potential of plant and animal species

THREE PILLARS OF SUSTAINABILITY

The adjectives biological, ecological, alternative, regenerative, and low-input are commonly used to refer to seemingly similar concepts of agricultural systems. The term sustainable agriculture broadly includes all of these concepts to some degree, and also addresses a specific set of criteria. Thus, three broad areas of concern seem to underlie the concept of sustainable agriculture (Weil, 1990).

1. Economic concerns over economic justice, the survival of owner operated farms, and the long-term profitability of agriculture
2. Environmental concerns over adverse impacts of agriculture on land, water, and wildlife resources
3. Public welfare concerns over food quality and human exposure to toxic chemicals

Basically, sustainable agriculture is a philosophy based on human goals and on understanding the long-term impact of our activities on the environment and other species (Francis, 1990). Broad concepts in sustainable agriculture encompass ecological, economic, and social parameters. Therefore, sustainability is a holistic approach that considers ecological, social and economic dimensions, recognising that all must be considered together to find lasting prosperity (Weil, 1990).

Environmental sustainability

Ecological integrity is maintained, all of earth's environmental systems are kept in balance while natural resources within them are consumed by humans at a rate where they are able to replenish themselves.

Economic sustainability

Human communities across the globe are able to maintain their independence and have access to the resources that they require, financial and other, to meet their needs. Economic systems are intact and activities are available to everyone, such as secure sources of livelihood.

Social sustainability

Universal human rights and basic necessities are attainable by all people, who have access to enough resources in order to keep their families and communities healthy and secure. Healthy communities have just leaders who ensure personal, labour and cultural rights are respected and all people are protected from discrimination.

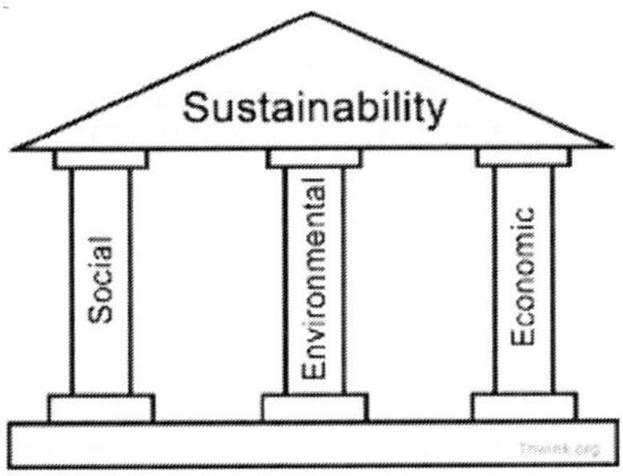

Fig. 1.1. Three pillars of sustainability

Source: https://www.calsense.com/three-pillars-of-sustainability

INPUT–OUTPUT RELATIONSHIPS IN SUSTAINABLE AGRICULTURE

The complex nature of the interrelationships between agricultural production and the natural environment means that 'we are a long way from knowing just what methods and systems in diverse locations will really lead to sustainability' (Youngberg and Harwood, 1989). This seems to be a crucial issue, and leads one to ask, how long should an agrosystem behave 'sustainably' to be considered 'sustainable', and how should sustainability be assessed? With respect to these issues, Monteith (1990) suggests that sustainability can be inferred by examining the trends observed in both farming outputs and inputs (table 1.3). This represents an interesting approach to quantify sustainability, and to observe it as a historical process. There are also weaknesses in the approach. For example, sustainability cannot simply be summarised as a balance between agricultural inputs and outputs. To accept this, would imply an oversimplification of the complex ecological processes upon which farming rests (Rigby and Caceres, 1997).

Table 1.3. Sustainability inference according to trends in farming inputs and outputs

Outputs ↓	**Inputs→**		
	Decreasing	**Constant**	**Increasing**
Decreasing	*indeterminate*	*unsustainable*	*unsustainable*
Constant	*sustainable*	*sustainable*	*unsustainable*
Increasing	*sustainable*	*sustainable*	*indeterminate*

Source: Monteith, 1990

The low input concept has a role in agricultural sustainability; reducing the use of many purchased inputs may enhance sustainability. Excessive use of off-farm purchased inputs may jeopardise sustainability by squeezing farmer profits, disrupting ecosystems with off-target effects, and depleting non-renewable resources. Indeed, in many cases reduction or elimination of certain off-farm inputs may be an important component of strategies to increase sustainability. Low input in the pursuit of sustainability does not necessarily mean an overall reduction in inputs, but rather a switch away from inputs that derive from non-renewable resources (fossil fuels, fertilisers, paleo-groundwater, etc.) or inputs that have negative side effects on the health of ecosystems and people (toxic herbicides, insecticides, etc.) towards inputs that enhance or do not deplete the resource base and have positive rather than negative side effects (Weil, 1990).

ELEMENTS OF SUSTAINABLE AGRICULTURE

Weil (1990) proposed three sets of criteria by which the direction, toward greater or less agricultural sustainability, may be judged. To be judged sustainable, changes to the current system should be an improvement over present practice and policy by all three sets of criteria. An agricultural programme, policy, or practice contributes to agricultural sustainability if it

1. Enhances, or maintains, the number, quality, and long-term economic viability of farming and other agricultural business opportunities in a community or region.
2. Enhances, rather than diminishes, the integrity, diversity, and long-term productivity of both the managed agricultural ecosystem and the surrounding natural ecosystems.
3. Enhances, rather than threatens, the health, safety, and aesthetic satisfaction of agricultural producers and consumers alike.

Using these criteria, certain inputs or practices may be judged 'sustainable' in one situation but not in another. There are several means that are generally, but not always, relevant to achieving greater agricultural sustainability. There are many ways to improve the sustainability of a given farming system, and these vary from region to region, However, there are some common sets of practices among farmers trying to take a more sustainable approach, in part through greater use of on-farm or local resources each contributing in some way to long-term profitability, environmental stewardship and rural quality of life. Weil (1990) has listed the following elements that increase sustainability in agriculture.

Biological diversity

Species diversity is usually a sign of ecological stability. A diversity of crop and animal enterprises on a farm also often lends economic stability. Growing a number of crops on a farm can help reduce risks from extremes in weather, market conditions or crop pests. Increased diversity crops and other plants, such as trees, shrubs and other woody perennials, also can contribute to soil conservation, wildlife habitat and increased populations of beneficial insects. Agroforestry systems such as agrisilviculture, silivipastoral, agrisilvihorticultural, hortisilvipastoral, alley cropping, etc. may be practised to achieve both ecological and economic stability.

Crop rotations

Along with increasing diversity, crop rotations are known to reduce soil erosion, ameliorate pest problems, improve soil fertility when legumes are included in the rotation, and also produce an additional, unaccounted for 10 to 15% increase in crop yields.

Animal integration

Plant and animal production are so complimentary that integration of animals into farming systems often seems essential to developing a sustainable system. Concentration of livestock production separate from crop production makes it difficult to justify the kinds of rotations that enhance sustainability. It also makes manure a serious waste disposal problem instead of a valuable on-farm resource. New management-intensive grazing systems may be introduced to take animals out barn into the pasture to provide high-quality forage and reduced feed cost. Rotational grazing improves the sustainability in forage production in the pasture.

Cover crops

Growing plants such as sunhemp, horsegram, cowpea in the off-season after harvesting a grain or vegetable crop can provide several benefits, including weed suppression, erosion control, and improved soil nutrients and soil quality.

Soil conservation and soil health

Many soil conservation practices, including contour cultivation, contour bunding, graded bunding, vegetative barriers, strip cropping, cover cropping, reduced tillage, etc. help prevent loss of soil due to wind and water erosion. The soil management practices should be designed to enhance the viability and diversity of soil organisms (microbes, arthropods, earthworms, etc.) and biological functions (organic matter and nutrient cycling, etc.).

Nutrient management

Proper management of nitrogen and other plant nutrients can improve the soil and protect environment. Increased use of farm nutrient sources such as manure and leguminous cover crops also reduces purchased fertiliser costs.

Integrated pest management

Integrated pest management is a sustainable approach to managing pests by combining biological, cultural, physical and chemical tools in way that minimises economic, health and environmental risks.

Knowledge-based farming

Sustainable agriculture will probably depend increasingly on sophisticated information on weather prediction, pest life cycles, soil processes, markets, and the like, melded into a holistic understanding of the agroecosystem and the agricultural community. An important aspect of this arena is the ability of the individual farmer to assimilate and, in many cases, to generate the information and insights needed to allow him or her to adapt to changing circumstances. Sustainability implies a dynamic, self-adjusting system.

Human scale farm size

There is no simple answer to how big is too big for sustainability, but the trend of the past few decades toward ever larger farms, operated by ever fewer farmers who are ever more remote from their fields is one that does not stand up well to the criteria of sustainability, both in terms of substituting management for non-renewable resources, and in terms of sustaining the numbers of economic opportunities in farming.

Minimal dependence on non-renewable resources

The basic resources of agriculture are sunlight, air (carbon and oxygen), water, soil (including nutrient elements therein), germplasm (of animals, plants and microbes), and farmers (with their skills, knowledge and labour). These resources are renewable and can sustain agriculture in perpetuity, if managed well. Agriculture cannot be sustained in the long run if it depends on the one-way, dead-end street of non-renewable resources use; e.g., burning fossil fuels, dispersing pesticides made from fossil fuels, or mining nutrients and dispersing them without recycling. Soil can only be considered renewable if not allowed to erode away faster than it forms.

Marketing

Farmers realise that improved marketing is required for their produce to enhance profitability. Without a favourable market the economic aspect of agriculture may not be sustainable. Direct marketing of agricultural product from farmers to consumers should be promoted.

BENEFITS OF SUSTAINABLE AGRICULTURE

Contributes to environmental conservation

Sustainable agriculture helps to replenish the land as well as other natural resources such as water and air. By adopting sustainable practices, farmers will reduce their reliance on nonrenewable energy, reduce chemical use and save scarce resources. This replenishment ensures that these natural resources will be able to sustain life for future generations considering the rising population and demand for food.

Saves energy for future

Modern agriculture is heavily dependent on nonrenewable energy sources, especially petroleum. Sustainable agricultural systems have reduced the need for fossil fuels or nonrenewable energy sources and a substitution of renewable sources or labour to the extent that is economically feasible.

Diversifies crops and farm products

Diversity of crops and farm products promotes recycling of products and by-products and thus, reduces the need for external high cost inputs. Diversity buffers the farm against sudden fall in prices of one or a few farm products and increase in costs of production inputs. It also buffers the farm against unpredictable weather. With sustainable agriculture, it is possible to produce more than one product in small areas at a time with high efficiency.

Increases self-reliance

Sustainable agriculture uses locally available renewable resources, appropriate and affordable technologies and minimises the use of external and purchased inputs, thereby increasing self-sufficiency and self-reliance, and insuring a source of stable income for the farmers. This encourages people to stay on farming.

Public health safety

Sustainable agriculture avoids hazardous pesticides and fertilisers. As a result, farmers are able to produce fruits, vegetables and other crops that are safer for consumers, workers, and surrounding communities. Through careful and proper management of livestock waste, sustainable farmers can protect humans from exposure to pathogens, toxins, and other hazardous pollutants.

Prevents pollution

Sustainable agriculture means that any waste a farm produces remains inside the farm's ecosystem. Organic farm wastes are converted to useful manures through the process of composting and vermicomposting. In this way, the waste cannot cause pollution.

Prevents air pollution

Agricultural activities affect air quality by smoke from agricultural burning; dust from tillage, traffic and harvest; pesticide drift from spraying; and nitrous oxide emissions from the use of nitrogen fertiliser. In sustainable agriculture, there are options to improve air quality by incorporating crop residue into the soil, using appropriate levels of tillage, and planting windbreaks, cover crops or strips of native perennial grasses to reduce dust.

CONCERNS OF AGRICULTURAL SUSTAINABILITY

Land degradation

One of the most critical threats to sustainable agriculture is land degradation. Vast areas of cropland, grassland, woodland and forest in Asia and Pacific have already been lost, and many more are threatened. The land degradation has both on-site and off-site effects. On-site effects include lowering of productive capacity causing either reduced outputs or need for increased inputs. Off-site effects of water erosion occur because of the deterioration in water quality, river sedimentation, biodiversity loss, and natural calamities. The irrigated agriculture, especially through canal systems, has resulted in land degradation at many places due to the twin problems of waterlogging and salinisation. A decline in soil productivity, particularly of organic carbon and nitrogen, deterioration in soil physical characteristics, and decreasing water availability are identified as the causes of this slowdown in productivity.

Much of the land degradation results from unstable use of land, over intensive cultivation and inappropriate management practices. In order to meet their basic food needs, smallholders and the rural poor have been pushed into using ecologically fragile areas, forced to crop intensively on steep slopes that are vulnerable to erosion. Land degradation has also resulted from excessive use of mineral fertilisers and over-intensive livestock-keeping. Pesticides and heavy metals diminish soil respiration, microbial activity, soil chemical action, restrain significant soil procedures like ammonification and nitrification, decrease earthworm dominance, and smother algal activity. These possibly unsafe substances may accumulate in soil and cause long-term consequences for yield and quality, and may harm soil microflora.

Lal (1991) estimated that the per capita arable land will progressively decline from about 0.3 ha in 1990 to 0.15 ha by 2050 and 0.14 ha by 2100. The availability of per capita net sown area in India has reduced from 0.33 ha in 1951 to 0.14 ha in 2001 and is expected to decline further to 0.09 ha by 2050. A minimum economic holding size of 2 ha of unirrigated land and 1 ha of irrigated land has been suggested in India for sustaining a family of 5 or 6 persons (Ghosh et al., 2020).

Water and its availability

Rice requires two or three times more water for cultivation than other cereals. Rice being the staple food across Asia and the Pacific region, agriculture is the principal user of water, consuming around 70% of total withdrawals. It is estimated that by 2050, Asia and Pacific would need an additional 2.4 billion cubic meters of fresh water/day. World population is expected to increase to over 10 billion by 2050, and this population will need food and fibre to meet its basic needs. Combined with the increased consumption of calories and more complex foods, which accompanies income growth in the developing world, it is estimated that agricultural production will need to expand by approximately 70% by 2050 (World Bank, 2020). While 2 litres of water are often sufficient for daily drinking, it takes about 3,000 litres to produce the daily food needs of one person.

In India, the utilisable surface-water resource is estimated to reduce by 7% due to deforestation and soil erosion, while the loss due to water pollution is put at 20% (IEG, 2000). Some of the overuse of water in agriculture reflects government policy. Many governments have subsidised the construction of inefficient irrigation systems, along with free electricity supply. This encourages farmers to withdraw too much water from rivers, over-pump groundwater and generally waste freshwater resources. In addition, intensive agriculture and industrial effluents are creating high levels of pollution. Hence, there is an urgent need to preserve and maintain the water quality of surface and groundwater resources for production and other purposes. It is estimated that by 2050 about 22% of the geographical area and 17% of the population would be under absolute water scarcity. Paradigm changes would be required in water resources development and management to avert such worst scenario. The problem of the falling groundwater table in central Punjab, where rice is a predominant crop, is because of the overdraft of water. A holistic strategy is required to overcome the water crisis, which includes crop diversification, delayed transplanting of rice, and adoption of water-saving agronomic practices etc. (Ghosh et al., 2020)

Deforestation

Forests provide critical ecosystem services to the agricultural sector, including pollination and watershed protection, and support to fisheries. Millions of poor people and small-scale enterprises depend on forests for food, fibre, fodder and other materials, but are finding this increasingly hard as the natural forests are shrinking. Human activity has always modified the natural ecosystems in such a way that environment becomes more favourable for non-agricultural activities. Apart from diversion of forest area to non-forest activities like industries, infrastructure development and human habitation, unrestricted exploitation of timber as well as other wood products for commercial purposes, and slash and burn method of cultivation are major causes of forest degradation (Ghosh et al., 2020). Due to deforestation, watersheds that once provided drinking water and irrigation water have now become unreliable due to extreme fluctuations in water flow (FSI, 2005). Another serious consequence of deforestation is the loss of biodiversity, i.e., the extinction of thousands of species and varieties of plants and animals. Global warming is another consequence of deforestation.

The Asia-Pacific region has around half the world's total area of mangroves. These are under severe strain as a result of the extraction of timber and coastal development, including for the production of environmentally damaging, export-oriented cultivation of shrimps. According to the India State of Forest Report, 2017 the total forest and tree cover is spread across 24.39% of the geographical area of the country as against the target of 33%.

Depletion of soil organic carbon

The carbon stores in arid and semiarid lands show a high temporal and spatial variability, some parts acting as C sources and others as C sinks. A decline in soil organic carbon (SOC) content is a common phenomenon when land use changes from natural vegetation to cropping, reasons being reduction in C inputs, increased rate of decomposition due to mechanical disturbance of the soil, higher soil temperatures due to exposure of the soil surface, more frequent wetting and drying cycles and increased loss of surface soil rich in organic matter through erosion (Ghosh et al., 2020).

Soil organic carbon content in most Indian soils is low but it is also dynamic in nature. Farming practices affect both quantity and quality of organic matter. Lands deteriorated due to depletion of SOC are widely distributed across the country, ranging from cultivated areas of subtropical belt to the areas under shifting cultivation. This problem is acute in high-intensive cultivated areas of rice-wheat cropping system in Indo-Gangetic plains. In the major rice-wheat regions of north-western India, SOC has decreased from 0.5% in 1960s to 0.2% in 2010s (Bhattacharyya et al., 2015). The removal or *in situ* burning of crop residues, no or least addition of organic manures, and intensive cultivation are major reasons for depletion of SOC.

Loss of biodiversity

Agrobiodiversity is the backbone of a nation's food security and the basis of economic development as a whole. However, the rapid and large-scale global extinction of species also caused loss of biodiversity. In the 20th century, it happened thousand times higher than the average rate during the preceding 65 million years. Overexploitation, habitat destruction, pollution and species extinction are major causes of biodiversity loss. Other factors include fires, which adversely affect regeneration in some cases. Agricultural biodiversity is also at risk from climate change (IPCC, 2007). It is reported that genetically modified crops threaten human health and the environment and will allow large corporations to tighten their grip over agricultural production by narrowing down the wide range of varieties now available just to a few.

Over the years biodiversity in India is under pressure due to the massive commercialisation of agriculture leading to almost extinction of traditional farming systems. Indian farmers grew more than 30,000 different varieties of rice before green revolution. Unfortunately, this enormous diversity has reduced during post-green revolution period. There has been a loss of several thousand rice varieties. Species and communities at particular and possibly critical risk include those with limited climatic ranges, limited dispersal ability and those with specialised habitat requirements (Upadhyay et al., 2008).

Climate change

Climate change is the most dominant environmental concerns to the agricultural planning. Food security is also being threatened by climate change which will have many complex effects, bringing advantages in some places, but disadvantages in many others. For example, higher concentrations of carbon dioxide in atmosphere may increase photosynthesis in several crops, thereby increasing yields. However, this potential gain could easily be outstripped by the effects of higher temperatures and more variable rainfall. The carbon dioxide has the least global warming potential among major greenhouse gases but due to its much higher concentration in the atmosphere, it is the major contributor towards global warming and climate change. Moreover, this may create significant uncertainties for agricultural production due to longer dry or rainy season, intense heat during summer, more or scanty rains during rainy months, frequent floods, droughts, storms, etc.

The changes in temperature, precipitation, CO_2 concentration, changes in frequency of infestation by pests and diseases caused agriculture vulnerable (Mendelssohn, 2014). In the case of an annual crop, the duration between sowing and harvesting will shorten and the crop may experience terminal heat stress.

The shortening of crop duration may have an adverse effect on productivity. In India, impact of 1-2° C increase in mean air temperature is expected to decrease rice yield by about 0.75 t/ha in efficient zones and 0.06 t/ha in coastal regions; and impact of 0.5° C increase in winter temperature is projected to reduce wheat yields by 0.45 t/ha. Productivity loss of 4–6% in rice, 6% in wheat, 18% in maize, 2.5% in sorghum, 2% in Indian mustard and 2.5% in potato are projected due to climate change (Naresh Kumar et al., 2012). The diversity and dominance of weeds and pests are likely to increase with climate change. Climate change alters the seed dormancy and germination of horticultural crops (Aggarwal et al., 2004). Higher temperature restricts the seed germination and also detrimental to the fruit setting in trees (Ghosh et al., 2020).

REFERENCES

Aggarwal, P. K., Joshi, P. K., Ingram, J. S. I. and Gupta, R. K. 2004. Adapting food systems of the Indo-Gangetic plains to global environmental change: Key information needs to improve policy formulation. *Environmental Science Policy* **7**: 487-498.

ASA. 1989. Decision reached on sustainable agriculture. *Agronomy News* January 1989. American Society of Agronomy, Madison, Wisconsin, USA. p. 15.

Bhattacharyya, R., Ghosh, B. N., Mishra, P. K., Mandal, B., Rao, C. S., Sarkar, D., Das, K., Anil, K. S., Lalitha, M., Hati, K. M. and Franzluebbers, A. J. 2015. Soil degradation in India: Challenges and potential solutions. *Sustainability* **7**(4): 3528-3570.

BIFAD. 1988. Environment and Natural Resources: Strategies for Sustainable Agriculture. Board for International Food and Agricultural Development Task Force, U.S. Agency for International Development, Washington DC.

CGIAR/TAC. 1988. Sustainable Agricultural Production: Implications for International Agricultural Research. Consultative Group for International Agricultural Research/ Technical Advisory Committee, TAC Secretariat, FAO, Rome.

Conway, G. R. 1985. Agroecosystem analysis. *Agricultural Administration* **20**: 1-25.

Davis, T. J. and Schirmer, I. A. (Eds.). 1987. Sustainability issues in agricultural development. Proceedings of the Seventh Agriculture Sector Symposium. The World Bank, Washington DC.

Dover, M. and Talbot, L. M. 1987. To Feed the Earth: Agroecology for Sustainable Development. World Resources Institute, Washington DC.

FAO. 1988. Report of the FAO Council, 94th Session, 1988. Food and Agriculture Organisation of the United Nations, Rome.

FAO. 2014. Building a common vision for sustainable food and agriculture: principles and approaches. Food and Agriculture Organisation of the United Nations Rome. pp. 18-31.

Francis, C. A. 1990. Sustainable agriculture: myths and realities. *Journal of Sustainable Agriculture* **1**(1): 97.

FSI. 2005. State of Forest Report. Forest Survey of India, Ministry of Environment and Forests, Dehradun, Uttarakhand. pp. 171.

Ghosh, P. K., Nath, C. P., Hazra, K. K., Kumar, P., Das, A. and Mandal, K. G. 2020. Sustainability concern in Indian agriculture: needs science-led innovation and structural reforms. *Indian Journal of Agronomy* **65**(2): 131-143.

http://www.fao.org/assets/infographics/FAO-Infographic-water-thirsty-en.pdf

https://liverur.eu/5-principles-of-sustainable-agriculture-according-to-fao/

https://www.conserve-energy-future.com/methods-and-benefits-of-sustainable-agriculture.php

https://www.mcgill.ca/sustainability/files/sustainability/what-is-sustainability.pdf

IEG. 2000. Integrated Water Resource Development and Management for Sustainable Frame Work (Effect of land degradation and pollution), Institute of Economic Growth, New Delhi.

IPCC. 2007. In: Solomon, S., Qin, D., Manning, M., Chen, Z., Marquis, M., Averyt, K. B., Tignor, M., Miller, H. L. (eds.). Climate Change 2007: The Physical Science Basis. Contribution of I to the Fourth Assessment Report of the Intergovernmental Panel on Climate Change. Cambridge University Press, Cambridge, U. K. pp. 996.

Knezek, B. D., Hesterman, O. B. and Wink, L. 1988. Exploring a new vision of agriculture. *National Forum* **68**: 10-13.

Lal, R. 1991. Soil structure and sustainability. *Journal of Sustainable Agriculture* **1**(4): 67-92.

Lynam, J. K. and Herdt, R. W. 1988. Sense and sustainability: sustainability as an objective in international agricultural research. CIP-Rockefeller Foundation Conference on Farmers and Food Systems, Lima, Peru, 26–30 September 1988.

MacRae, R. 1990. A History of Sustainable Agriculture. https://www.eap.mcgill.ca/AASA_1.htm

Mendelsohn, R. 2014. The impact of climate change on agriculture in Asia. *Journal of Integrative Agriculture* **13**(4): 660-665.

Monteith, J. L. 1990. Can sustainability be quantified? *Indian Journal of Dryland Agriculture Research Development* **5**(1-2): 1-5.

Naresh Kumar, M., Murthy, C. S., Sesha Sai, M. V. R. and Roy, P. S. 2012. Spatiotemporal analysis of meteorological drought variability in the Indian region using standardised precipitation index. *Meteorological Applications* **19**(2): 256-264.

Okigbo, B. N. 1991. Development of sustainable agricultural production systems in Africa. International Institute for Agricultural Research, Ibadan, Nigeria. 66 pp.

Pingali, P. L. 2012. Green Revolution: Impacts, limits, and the path ahead. *Proceedings of the National Academy of Sciences of the United States of America* **109**(31): 12302-12308.

Pretty, J. 1996. Sustainable Agriculture: Impacts on Food Production and Challenges for Food Security. IIED Gatekeeper Series No. SA60

Rigby, D. and Caceres, D. 1997. The sustainability of agricultural systems. Rural resources/ rural livelihoods working paper series 10. Institute for Development Policy and Management, University of Manchester, Precinct Centre, Manchester, UK.

Rodale, R. 1988. Agricultural systems: the importance of sustainability. *National Forum* **68**(3): 2-6.

Rome Forum. 1986. Sustainable food and nutrition security. Report and Recommendations. Association for research studies on agriculture, food and forestry systems in emerging regions. FAO, Rome.

Ruttan, V. W. 1989. Sustainability is not enough. Better crops with plant food. Potash and Phosphate Institute, Atlanta, Georgia, USA. pp. 6-9.

Senanayake, R. 1991. Sustainable agriculture: definitions and parameters of measurement. *Journal of Sustainable Agriculture* **1**(4): 7-28.

Sivakumar, M. V. K., Gommes, R. and Baier, W. 2000. Agrometeorology and sustainable agriculture. *Agricultural and Forest Meteorology* **103**: 11-26.

Swindale, L. D. 1988. Agricultural development and the environment: a point of view. Sustainable Agricultural Production: Implications for International Agricultural Research. TAC Secretariat, FAO, Rome. pp. 7-14.

Upadhyay, R. C., Singh, S. V. and Ashutosh. 2008. Impact of climate change on livestock. 36th Dairy Industry Conference, Varanasi, Uttar Pradesh, India.

WCED. 1987. Our Common Future. World Commission on Environment and Development. Oxford University Press, Oxford. http://www.un-documents.net/ocf-02.htm

Weil, R. R. 1990. Defining and Using the Concept of Sustainable Agriculture. *Journal of Agronomic Education* **19**(2):126-130.

World Bank. 2020. Water in agriculture. https://www.worldbank.org/en/topic/water-in-agriculture

Youngberg, G. and Harwood, R. 1989. Sustainable farming systems: needs and opportunities. *American Journal of Alternative Agriculture* **4**(3&4): 100.

2

Indicators of Agricultural Sustainability

Indicators of agricultural sustainability are a composite set of attributes or measures that embody a particular aspect of agriculture. These are quantified information, which help to explain how things are changing over time. Smyth and Dumanski (1993) define indicators as environmental attributes that measure or reflect environmental status or condition of change. According to Glen and Pannell (1998) an indicator is a quantitative measure against which some aspect or aspects of policy performance or management strategy can be assessed. Sustainability indicators look at economic, social and environmental information in an integrated manner. Tunstall (1992) and Gallopín (1997) identified the following functions of indicators.

1. To assess conditions and changes
2. To compare across place and situations
3. To assess conditions and trends in relation to goals and targets
4. To provide early warning information
5. To anticipate future conditions and trends

CRITERIA FOR SELECTING INDICATORS

At least the following three criteria should guide the development of sustainability indicators.

Policy relevance: Indicators should address the issues of primary concern to a country or district and receive the highest priority. In some cases policy makers may already share concern about an aspect of sustainability and be ready to use indicator information for addressing the issue.

Predictability: Indicators should allow a forward-looking perspective that can promote planning and decisions on issues before they become too severe. Anticipatory decision-making is at least as important to sustainable agriculture as is recognition of existing problems.

Measurability: Indicators should allow planners and analysts the means to assess how these were derived, either qualitatively or quantitatively, and decide how these can best be applied in the planning and decision-making process.

Selection of effective indicators is the key to the overall success of any monitoring programme. An indicator must be selected carefully so that it can measure and describe explicitly the condition of sustainability.

The ecological indicators should have the following characteristics (Dale and Beyeler, 2001). The indicators should

1. Be easily measurable,
2. Be sensitive to stresses on the system,
3. Respond to stress in a predictable manner,
4. Be anticipatory, meaning that they signify an impending change in the ecological system,
5. Predict changes that can be averted by management actions,
6. Be integrative, meaning that the full suite of indicators provides a measure of coverage of the key gradients across the ecological systems (such as soils, vegetation types and temperature),
7. Have a known response to natural disturbances, anthropogenic stresses, and changes over time, and
8. Have low variability in response.

Operational indicators for measuring agricultural sustainability may be categorised under three components of sustainability.

Economic

- Crop productivity
- Net farm income
- Benefit-cost ratio of production
- Per capita food grain production

Social

- Food self sufficiency
- Equality in income and food distribution
- Access to resources and support services
- Farmers, knowledge and awareness of resource conservation

Ecological

- Amount of fertilisers / pesticides used per unit of cropped land
- Amount of irrigation water used per unit of cropped land

- Soil nutrient content
- Depth of groundwater table
- Quality of groundwater for irrigation
- Water use efficiency
- Nitrate content of groundwater and crops

Howlett (1996) identified the following purposes of the use of indicators at the farm level.

1. To develop capacity and commitment of farmers towards more sustainable land use, and to allow farmers to evaluate their own practices.
2. For the simple diagnosis of problems and improvements to farming practices, and development of appropriate research and extension activities.
3. To improve the relationship between the researcher, farmer and extension agent, and through this to encourage farmer participation, the incorporation of indigenous knowledge, and ultimately to an increase in the adoption of improved technologies.
4. To assess and monitor the spatial and temporal sustainability of different farming systems, and to use this for the evaluation, prediction, planning and management of these systems by farmer, researcher, extension agent and planners.

Measuring sustainability is most challenging and complex. RIEDC (1997) indicated general broad measurable components under each hierarchical level of components of sustainability. Hoang (2013) had analysed productive performance of crop production systems in an integrated analytical framework considering economic, institutional, physical, social and technological factors and indicated that in a dynamic analysis to make efficiency framework to be forward looking, climate change innovations in crop science to be incorporated.

Table 2.1. Sustainability indicators

Hierarchical level	**Sustainability indicators (Economic, social and environmental)**
Cropping system/ farming system	Non-negative trends in 1. Farm productivity 2. Net farm income 3. Total factor productivity 4. Nutrient balance 5. Soil quality 6. Residues in soil, plant, products 7. Farm water use efficiency 8. Farmer skills and education 9. Debt service ratio 10. Health 11. Time spent on other social cultural activities
Agroecosystem (watershed, agroecozone, etc.)	Non-negative trends in 1. Regional production 2. Regional income 3. Regional total factor productivity 4. Regional nutrient balance 5. Income distribution 6. Species diversity 7. Soil loss 8. Surface water quality 9. Groundwater quality 10. Regional social and economic development indicators
Global, national, regional systems	Indefinitely meet the demands at acceptable social, economic and environmental costs.

Source: RIEDC, 1997

Table 2.2. Indicators of key natural resources in rainfed cropping systems

Indicators	**Key Management Aspects**
Nutrient balance	Organic matter - rate of change Nitrogen cycling - especially when using rain legumes in rotation with cereals Monitoring status of phosphorus, sulphur and potassium Micronutrients
Erosion	Vegetation cover - includes trees as well as stubble Soil surface cover - stubble retained (30% sufficient to prevent wind and water erosion) Stream bank Sheet and gully erosion
Productivity, yield and quality	Water use efficiency - i.e., actual versus potential (in some areas the potential is much less than the actual) (biomass/grain yield/net return), recharge (dryland salinity and nutrient leaching) Pasture composition - legume and perennial Matched animal versus pasture production - appropriate enterprise selection/capability Maintenance of genetic base/improvement
Soil structure	Infiltration Permeability/water storage Stability Waterlogging Compaction
pH	Change Toxicity - deficiency Indicator plants
Energy efficiency	Energy input vis-a-vis energy output of the whole agricultural system
Biological factors	Soil macro/micro flora and fauna Animal health Plant health (root growth and other) Pests (animals and plants)
Farm management skills	Understanding - a good indicator would be the understanding of the farmers of their own technical system
Rainfall / precipitation	Performance of rainfall in a year as % of normal and its coefficient of variation Distribution of area based on rainfall amount Categorisation of the amount of rainfall Number of districts having mean annual rainfall of 750-1250 mm and moisture availability period for at least 150 days Area affected due to drought

Source: Subba Rao and Mandal, 2007

INDICATOR FRAMEWORKS

Several sets of methodological frameworks or guidelines have been identified for the measurement of sustainability indicators at the farm or community levels. These have all tended to come from an approach focussed on sustainable agriculture and/or sustainable land management. The United Nations, World Bank, Organisation for Economic Cooperation and Development (OECD), European Environment Agency (EEA) and many other organisations and national governments are currently producing indicators or proposed indicators of sustainable development and sustainable agriculture. In addition to the various frameworks used, there are differing dimensions, aspects or properties of sustainable agricultural systems that are proposed as criteria for sustainability assessment. The most widely accepted framework for agricultural sustainability is referred to as Pressure-State-Response (PSR framework). This framework is used by OECD, SCOPE (Scientific Committee on Problems of the Environment) and some other organisations working in the field.

Pressure-state-response (PSR) framework

The Pressure-State-Response framework can facilitate better understanding of actions and activities that are affecting the state of the system, and appropriate response for addressing them both for the agency and stakeholders. The Pressure-State-Response (PSR) framework was conceived by Statistics Canada (Friend and Rapport, 1979), then further developed and adopted internationally in many countries (Waheed et al., 2009). A typical example of a PSR framework is shown in Fig. 2.1.

Pressure refers to the driving forces that create environmental impacts; state refers to the condition(s) that prevail when a pressure exists and response refers to the mitigation action(s) and levers that could be applied to reduce or eliminate the impacts. Thus, a PSR framework states that human activities exert pressure (such as pollution emissions or land use changes) on the environment, which can induce changes in the state of the quality and quantity of the environment (such as changes in ambient pollutant levels, habitat diversity, water flows). Society then responds to the changes in the pressures or the state with environmental and economic policies/ programmes intended to prevent, reduce or mitigate pressures and/or environmental damage (Gallopín, 1997).

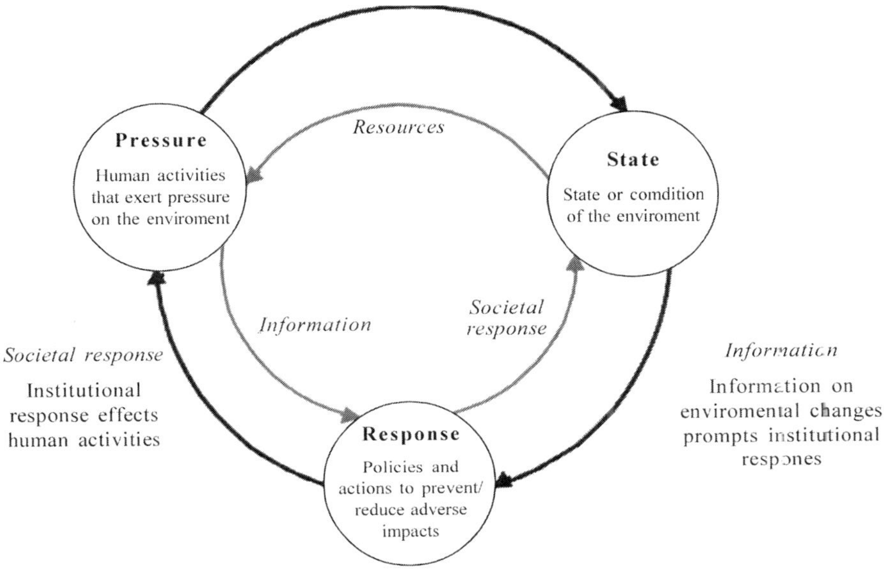

Fig. 2.1. Pressure-state-response (PSR) framework
Source: OECD, 1999; Woodhouse et al., 2000

Table 2.3. Examples of PSR framework

Issue	Pressure (driving force)	State (resulting condition)	Response (mitigating action)
Soil erosion	Hillside farming	Declining yield	Terracing, perennial cropping
Water quality	Agro-industrial processing	Fish die-off	Water treatment, technology adjustment
Condition of grassland	Livestock grazing	Soil erosion	Stock rotation, de-stocking, revegetation
Environmental pollution	CO_2 emissions	Rising global temperatures	Introduction of energy taxes

Source: Subba Rao and Mandal, 2007

SDG INDICATOR 2.4.1

Agriculture plays an essential role in ensuring a better future for all. As a fundamental connection between people and the planet, it can help achieve multiple sustainable development goals (SDGs). Yet, to ensure enough food is produced for a population of nearly 10 billion by 2050 without critically degrading natural resources, a transition to sustainable agricultural systems is needed. FAO's Sustainable Food and Agriculture (SFA) approach, articulated in the following five key principles, places people at the centre, focusing on efficient use of economic resources and environmental protection.

1. Increase productivity, employment and value addition in food systems
2. Protect and enhance natural resources
3. Improve livelihoods and foster inclusive economic growth
4. Enhance the resilience of people, communities and ecosystems
5. Adapt governance to new challenges

In September 2015, the United Nations General Assembly adopted the 2030 Development Agenda and an associated 17 Sustainable Development Goals (SDGs). The resultant SDGs are accompanied by 169 targets under the various goals and a set of 232 indicators to monitor progress toward the SDGs. Responsibility for the development of indicators is given to the United Nations Statistical Commission (UNSC) which established an Inter-Agency Expert Group for SDG indicators (IAEG-SDG). Under the auspices of the IAEG-SDG, various agencies were given custodianship for the finalisation of the appropriate indicators for the different SDG targets. FAO was given custodianship of 21 indicators across six SDGs (FAO, 2019).

SDG 2.4.1 is a set of tools used to measure concrete progress towards the achievement of SDG Target 2.4, one of eight targets under SDG 2, i.e., 'end hunger; achieve food security and improved nutrition and promote sustainable agriculture'. Specifically, Target 2.4 aims at 'by 2030, ensure sustainable food production systems and implement resilient agricultural practices that increase productivity and production, that help maintain ecosystems, that strengthen capacity for adaptation to climate change, extreme weather, drought, flooding and other disasters and that progressively improve land and soil quality' (FAO, 2019).

Methodology for constructing the indicator

Indicator 2.4.1 reflects the multiple dimensions of sustainability, economic, environmental and social. Specific areas within a dimension are called themes (e.g. land productivity, biodiversity, decent employment, etc.). Variable used to measure performance of the farm in relation with a given theme is termed as sub-indicator. Sustainability criteria refer to critical/thresholds values against which the performance of each sub-indicator is assessed to classify the farm in terms of the sustainability level.

Deriving an indicator for sustainable agriculture involves several steps. Although these steps are presented in a linear fashion, in practice, a degree of iteration was required through processes of discussion and investigation. The description of the relevant approach for assessing sustainability performance depends on the sub-indicator, but at the same time, the choice of sub-indicator will likely be closely informed by the data collection instruments (FAO, 2019).

1. Determining the scope of the indicator: The choice made for indicator 2.4.1 is to focus on crops and livestock production, thus excluding forestry, fisheries and aquaculture.
2. Determining the dimensions to be covered: The choice made for indicator 2.4.1 is to include environmental, economic and social dimensions in the sustainability assessment.
3. Choosing the scale for the sustainability assessment: The choice made for indicator 2.4.1 is farm level with aggregation to higher levels.
4. Selecting the data collection instrument(s).
5. Selecting the themes within each dimension, and choosing a sub-indicator for each theme. The sub-indicators should satisfy a number of criteria.
6. Assessing sustainability performance at farm level for each sub-indicator: Specific sustainability criteria are applied in order to assess the sustainability level of the farm for each theme according to the respective sub-indicators.
7. Deciding the periodicity of monitoring the indicator.
8. Modality of reporting the indicator: The set of sub-indicators are presented in the form of a dashboard. The dashboard offers a response in terms of measuring sustainability at farm level and aggregating it at national level.

Characteristics of Indicator 2.4.1

By tracking the proportion of agricultural land area by its suitability status, SDG Indicator 2.4.1 provides an assessment of progress towards sustainable agriculture. In doing so, it supplies decision-makers with strategic information for evidence-based policies and action. The indicator is expressed by the following formula.

$$\text{SDG } 2.4.1 = \frac{\text{Area under productive and sustainable agriculture}}{\text{Agricultural land area}}$$

This implies the need to measure both the extent of land under productive and sustainable agriculture (the numerator), as well as the extent of land area under agriculture (the denominator). The denominator is a function of the scope of the indicator. It is the agricultural land area managed by agricultural holdings, defined as the sum of agricultural area utilised by agricultural holdings that are owned (excluding rented-out), rented-in, leased, sharecropped or borrowed.

The construction of the indicator must respect the following three conditions.

1. The indicator must reflect the priorities as they are expressed in the SDG target 2.4 and therefore consider issues related to resilience, productivity, ecosystem maintenance, adaptation to climate change and extreme events, and soils.
2. The preferred data source is the farm survey.
3. The need to define productive and sustainable agriculture implies the use of criteria to distinguish between sustainable and unsustainable areas.

Measurement scope of the Indicator 2.4.1

Indicator 2.4.1 focuses on agricultural land, and therefore primarily on land that is used to grow crops and raise livestock. This choice of scope is fully consistent with the intended use of a country's agricultural area as the denominator of the aggregate indicator.

The following aspects are included within the scope of SDG Indicator 2.4.1.

- Both intensive and extensive production systems
- Subsistence agriculture
- State and common land when used exclusively and managed by the holding
- Food and non-food crops and livestock products
- Crops grown for fodder or for energy purposes
- Agroforestry (trees on the farm)
- Aquaculture, to the extent that it takes place within the agricultural area

From an environmental perspective, the scope of the indicator focuses on the environmental impacts of farming, i.e., the direct impacts that farming practices, farmer choices and farming methods have on the environment. This implies that all possible impacts that are beyond this scope are not considered. From a social perspective, the approach also focuses on farming as a source of livelihood. Thus, the social impact of farming activities in terms of household livelihood and food security is included.

The following aspects are excluded from the scope of SDG Indicator 2.4.1.

- State and common land not used exclusively by the agriculture holding
- Nomadic pastoralism
- Production from gardens and backyards, production from hobby farms
- Holdings focusing exclusively on aquaculture
- Forest and other wooded lands outside of an agricultural holding
- Food harvested from the wild

In terms of agricultural value chain, the scope of the assessment stops at the farm gate. It does not extend to the sustainability of the transportation, storage, processing, distribution and marketing of agricultural products. Likewise, the approach does not take into consideration the sustainability of supply chains that provide inputs to agricultural production. The impacts of agricultural production systems on the health of end-consumers and their dietary outcomes (except for the farm household itself) are also outside the scope of the indicator.

Defining themes and sub-indicators

Hayati (2017) identified a large number of potential sustainability themes across the three dimensions of sustainability and, for each theme, usually a large number of possible sub-indicators. The key considerations in the selection of themes are relevance and measurability. In terms of relevance, the relationship between the associated sub-indicator and sustainable agriculture outcomes at farm level should be strong. Following this approach, only sub-indicators that are responsive to farm level policies aimed at improving sustainable agriculture are considered. In terms of measurability, only a 'core' set of themes and sub-indicators for which measurement and reporting is expected in the majority of countries are selected.

There are many relevant subthemes and sub-indicators. However, it is not possible to consider all of them to measure progress towards sustainable agriculture. The subject is too complex, and the factors influencing sustainable agriculture are too diverse across countries, for reaching a consensus on an indicator that covers all sustainability issues while remaining manageable and universally valid. FAO therefore proposes to measure Indicator 2.4.1 through a core set of 11 themes for global reporting purposes.

Selecting the most appropriate sub-indicator for each theme is a distinct step in the process. For any given theme, indeed, there may be multiple sub-indicators that are relevant and/or measurable. Consequently, in selecting the sub-indicators for Indicator 2.4.1 the following six key criteria have been considered (FAO, 2019).

1. **Policy relevance:** The indicator must be easily understood (reasons why it is selected) and the results easily interpreted by policy makers (is agricultural sustainability decreased and why? Which policies need to be implemented to address the issue?).
2. **Universality:** The indicator must be relevant for all countries in the world, both developing and developed.
3. **International comparability:** The way indicators are computed must ensure comparability across countries in order to ensure global reporting.

Comparability, however, does not necessarily mean the use of absolute standards. For instance, agricultural wages may be compared with the national minimum wage rate, even if these wage rates vary from one country to another. Similarly, compliance with national environmental standards or nationally recognised certification systems can be considered in computing environmental sub-indicators, even if national criteria vary from one country to another.

4. **Measurability:** Many themes are important sustainability issues but their measurement is difficult, complex or would involve costs that cannot be sustained in the framework of a regular monitoring exercise. To the extent possible, alternative measures have been proposed to maintain indicators that are considered relevant while offering feasible measurement solutions.
5. **Cost effectiveness:** Cost effectiveness is related to measurability. The cost associated with indicator measurement has systematically been considered in relation with the accuracy and reliability of the results obtained through different measurement options.
6. **Minimum cross-correlation between sub-indicators:** In selecting a limited set of themes and sub-indicators, efforts were made to reduce cross-correlation between different sub-indicators. High cross-correlation between sub-indicators would imply that two or more sub-indicators capture the same sustainability issue. In this case, the inclusion of one single sub-indicator, instead of several, would be sufficient to adequately measure agricultural sustainability performances.

Sub-indicators may be of five broad types:

1. Impact/outcome indicators record what the state or change in state of environmental, economic and social factors and associated flows of benefits or costs.
2. Awareness indicators record the level of awareness and knowledge of interviewed persons in relation with a given sustainability issue. Awareness is considered a prerequisite step towards addressing sustainability issues.
3. Behaviour indicators that capture peoples' attitudes in relation to a given sustainability issue. While behaviour is influenced by awareness, the two can also be disconnected.
4. Practice indicators that measure specific and codified agricultural methods applied on a farm.
5. Perception indicators record peoples' views about a specific issue.

The list of selected themes and sub-indicators is provided in table 2.4.

Table 2.4. List of themes and sub-indicators

Dimensions	No.	Theme	Sub-indicators
Economic	1	Land productivity	Farm output value per hectare
	2	Profitability	Net farm income
	3	Resilience	Risk mitigation mechanisms
Environmental	4	Soil health	Prevalence of soil degradation
	5	Water use	Variation in water availability
	6	Fertiliser pollution risk	Management of fertilisers
	7	Pesticide risk	Management of pesticides
	8	Biodiversity	Use of biodiversity-friendly practices
Social	9	Decent employment	Wage rate in agriculture
	10	Food security	Food insecurity experience scale
	11	Land tenure	Secure tenure rights to land

Source: FAO, 2019

For each sub-indicator, criteria to assess sustainability levels are developed. The concept of sustainability implies an idea of continuous progress and improvement towards better performances across all themes and such performances can therefore be more or less sustainable. In order to capture the concept of continuous progress towards sustainability, three sustainability levels such as desirable, acceptable and unsustainable are considered for each sub-indicator.

SDG Indicator 2.4.1 measures progress towards more sustainable and productive agriculture. For many sub-indicators, it is likely that changes will be relatively limited from a year to another. It is therefore recommended that the survey be conducted every three years. The survey's sampling design must respond to the need to capture the structure and the different typologies of agricultural holdings.

Alternative data sources to construct the indicator

It has also been reported that alternative data sources like remote sensing and GIS can be more cost-effective and sometimes provide more reliable results than farm surveys. The possible instruments/sources of information for each sub-indicator are furnished in table 2.5.

Table 2.5. Possible data collection instruments for each sub-indicator

S.No.	Sub-indicators	Possible Data Collection Instruments
1.	Farm output value per hectare	Agricultural surveys, household surveys linked with administrative records and market surveys, remote sensing, agricultural and livestock census
2.	Net farm income	Agricultural surveys, household surveys linked with administrative records and market surveys, agricultural and livestock census
3.	Risk mitigation mechanisms	Household surveys with agricultural information, community surveys, administrative records
4.	Prevalence of soil degradation	Environmental monitoring systems, soil sampling, remote sensing calibrated with ground observations, GIS data/maps/models calibrated with ground observations and samplings
5.	Variation in water availability	River flows records, water level records, abstraction records, remote sensing, GIS information/ maps/ hydrogeological models, administrative sources, householdsurveys
6.	Management of fertilisers	Environmental monitoring systems (soil, water quality), agricultural surveys, GIS data/maps and models based on sale data, agricultural surveys and administrative sources
7.	Management of pesticides	Environmental monitoring systems (soil, water quality), agricultural surveys, models based on active substance sale data, agricultural surveys and administrative sources
8.	Use of biodiversity-friendly practices	Environmental monitoring systems including remote sensing (land use/ land cover), GIS data/ maps
9.	Wage rate in agriculture	Labour force survey, household survey with agricultural module, administrative data
10.	Food insecurity experience scale	Household surveys, health data
11.	Secure tenure rights to land	Household surveys with agricultural module, administrative/ legal sources

Source: FAO, 2019

The use of such instruments can be considered, but several aspects need to be carefully taken into account prior to using alternative data sources. It should be confirmed that the alternative source gives results of at least same quality as the surveys. In order to produce consistent and reliable data as per recommended periodicity, it is advised that the use of alternative data sources may be considered when the available datasets fulfil the following criteria (FAO, 2019).

1. Can be reflected in or attributed to agricultural land area in the country, considering different farm typologies and agricultural regions;

2. Can be associated with the country's agricultural productions systems, particularly crops, livestock and the combinations in between;
3. Capture the same aspect/phenomenon as the proposed farm survey with at least a documented same quality, considering scientific standards;
4. Are representative of the situation at the national level (with respect to agricultural land area) taking into account main agricultural region types;
5. Are compliant with international/national standards and classifications systems in order to ensure the indicator to be internationally comparable;
6. Data are available at the same level of territorial disaggregation as the farm survey.
7. The ways and means to adjust for under-coverage and non-coverage (when needed) should be clearly devised and described;
8. Data collection year and periodicity are homogenous across the sub-indicators.

Alternative data sources may also be used to complement and/or validate farm survey data. This combined approach has the potential to improve the validity and soundness of results. The information from other sources may be used and leveraged in different ways depending on quality and regularity of its collation.

REFERENCES

Dale, V. H. and Beyeler, S. C. 2001. Challenges in the development and use of ecological indicators. *Ecological Indicators* **1**: 3-10.

FAO. 2019. SDG Indicator 2.4.1: Proportion of agricultural area under productive and sustainable agriculture. Food and Agriculture Organisation, Rome. http://www.fao.org/sustainable-development-goals/indicators/241/en/

Friend, A., Rapport, D. 1979. Towards a comprehensive framework for environmental statistics: a stress-response approach. Statistics Canada, 1979.

Gallopín, G. 1997. Indicators and their use: information for decision making. In: Moldan, B. and Billharz, S. (eds.) Sustainability Indicators. Report on the project on indicators of sustainable development. John Wiley & Sons, Ltd., Chichester, West Sussex, UK.

Glenn, N. A. and Pannell, D. J. 1998. The Economics and Application of Sustainability Indicators in Agriculture, Paper presented at the 42nd Annual Conference of the Australian Agricultural and Resource Economics Society, University of New England, Armidale, January 19-21, 1998.

Hayati, D. 2017. A Literature Review on Frameworks and Methods for Measuring and Monitoring Sustainable Agriculture. Technical Report No. 22. Global Strategy Technical Report. FAO, Rome.

Hoang, Viet-Ngu. 2013. Analysis of productive performance of crop production systems: An integrated analytical framework. *Agricultural Systems* **116**: 16-24.

Howlett, D. J. B. 1996. Development of social, economic and biophysical indicators for sustainable land management in the South Pacific. In: Sustainable Land Management in the South Pacific. Howlett, D.J.B. (ed.) Network Document No. 19. International Board for Soil Research and Management, Bangkok, Thailand.

OECD, 1999. Towards Sustainable Development: Indicators to Measure Progress. Organisation for Economic Co-operation and Development Conference, Rome, Italy, December 15-17, 1999.

RIRDC, 1997. Sustainability indicators for agriculture, Rural Industries Research and Development Corporation, Australia. 54 pp.

Smyth, A. J. and Dumanski, J. 1993. FESLM: An international framework for evaluating sustainable land management. World Soil Resources Report 73. Food and Agriculture Organisation of the United Nations, Rome, Italy.

Subba Rao, A. and Mandal, K. G. 2007. Modern concepts of agriculture: Sustainable agriculture. Indian Institute of Soil Science, Bhopal. https://www.scribd.com/document/88818986/Revised-Sustainable-Agriculture.

Tunstall, D. 1992. Developing environmental indicators: definitions framework and issues. Background materials for the World Resources Institute Workshop on Global Environmental Indicators, Washington DC, December 1992.

Waheed, B., Khan, F. and Veitch, B. 2009. Linkage-Based Frameworks for Sustainability Assessment: Making a Case for Driving Force-Pressure-State-Exposure-Effect-Action (DPSEEA) Frameworks. *Sustainability* **1**: 441-463.

Woodhouse, P., Howlett, D. and Rigby, D. 2000. Sustainability indicators for natural resource management & policy. Department for International Development Research Project No. R7076CA, Institute for Development Policy and Management, University of Manchester, UK.

3

Low External Input and Sustainable Agriculture

Low external input agriculture developed as a concept in the 1980s, gaining ground with the establishment of the Information Centre for Low External Input and Sustainable Agriculture in 1984 (Kessler and Moolhuijzen, 1994). Low external input agriculture developed as a response to the criticisms surrounding Green Revolution technology being neither sustainable nor feasible for many small-scale farmers around the world. The concept aimed to redesign the agricultural system by optimising the use of biological resources, keeping changes to the natural ecosystem as well as the use of external inputs at a minimum (Pimentel et al., 1989).

Today, low external input agriculture (LEIA) is described as an approach referring to a set of agronomic practices that aim to reduce the use of inputs from outside the production system. These inputs may include water, energy, seeds, chemicals, and the like. According to the FAO, LEIA does not mean the elimination so much as a reduction of external inputs, such as pesticides and fertilisers, placing a greater emphasis on improved agronomic practices, integrated pest management, labour and overall farm management, towards maintaining yields (Corsi and Muminjanov, 2019). LEIA is an approach that can be applied to various production systems in various geographical contexts. The other terms in use with the same concept as LEIA include low external input sustainable agriculture (LEISA) (Kessler and Moolhuijzen, 1994) and low input farming systems (LIFS).

One of the most promising paradigms that has emerged for the benefit of small scale resource-poor farmers is low external input and sustainable agriculture, which can enable such farmers to achieve higher income and attain sustainability by following two considerations (Gold, 2007).

1. Optimising the use of locally available resources, thereby achieving a synergetic effect among the various components of the farming system (soil, water, animals, plants, etc.) so that they complement each other in the production of output.

2. Minimising the use of external inputs, except where there is a serious deficiency and where the effect on the system will be to increase recycling of nutrients.

Thus, the low external input sustainable agriculture (LEISA) is a way of farming, where it seeks to optimise the use of locally available resources by maximising the complementary and synergistic effects of different components of the farming systems. External inputs are used in a complementary way. Here the aim is not to maximise short-term production, but to attain an adequate and sustainable level over the longer term. To achieve these goals, LEISA must tap the most viable indigenous knowledge and practices and ecologically friendly technologies in a given ecological and sociocultural setting, since the experience in one agroecological setting may not be appropriate in other areas (FAO, 1996).

The practice of LEISA is guided by the following four basic principles.

1. Securing favourable soil conditions for plant growth particularly managing organic matter and enhancing soil life.
2. Optimising the nutrient availability and balancing the nutrient flow, particularly by means of nitrogen fixation, nutrient acquisition and complementary use of external fertilisers.
3. Minimising the losses due to plant and animal pests by means of prevention and safety treatment.
4. Minimising losses due to flows of solar radiation, air, water by way of microclimate management, water management and erosion control.

HIGH EXTERNAL INPUT AGRICULTURE

In the high external input agriculture (HEIA), external inputs like chemical fertilisers, high yielding and hybrid seeds, pesticides, irrigation, mechanisation based on fossil fuels are extensively used. However, here the problem is related to excessive and unbalanced use of agrochemicals, salinisation from irrigation, and loss of plant diversity which can in the long run harm humans, animals, and the soil. Increasingly, HEIA areas are experiencing a decrease in the effectiveness of these external inputs because of biological factors and also because of a loss in profitability as the soil productivity is reduced over time. But HEIA could rapidly increase agricultural production to meet the demand for food for the increasing population. The other benefits of HEIA include

- Many problems related to diseases caused by mal-nutrition and deficiencies can be effectively managed.
- New improved varieties give yields within a short period of time.
- Mechanisation solves the problem of labour shortage.

- Income and profit margins of the products may be increased.
- Productivity of land is improved.
- A market facility for production is increased.

With all the above benefits, the high external input agriculture on long-run developed serious threats mostly on ecological and environmental issues.

- The environmental balance is collapsed due to lack of biodiversity by planting a few cash crops.
- Soil erosion is increased due to constant furrowing by machinery.
- Dependence on imported machinery, chemical fertiliser, pesticides, hybrid seeds and other inputs is increased.
- Extensive use of pesticides disturbs the natural mechanism of controlling pest and diseases as the artificial pesticides kill both pests and their natural enemies.
- Use of agrochemicals adversely affects the soil *p*H, cation exchange capacity, soil structure, soil texture and soil organisms that consequently lead to reduced microbial activities in the soil.
- The large farmers get benefited with high capital investment while small-scale farmers who were short of capital run into debt.
- Traditional varieties of seeds and their genetic resources face extinction due to introduction of hybrids.
- Conventional agricultural knowledge and techniques are neglected and extinguished.

Table 3.1. The characteristics of HEIA and LEISA

	Characteristics of HEIA	Characteristics of LEISA
1.	The farming pattern depends heavily on external and chemical inputs. Although yields have increased substantially, contributing to raising total production, farmers and the environment have had to pay the price for keeping up with this development.	LEISA relies on the optimal use of natural processes.
2.	The focus of agricultural development and research has mainly been on maximising yields coupled with increasing specialisation of production.	The focus is on the sustainability of farming system.
3.	There is a great damage to the environment.	This is environmentally sound and has the potential to contribute to the long-term sustainability of agriculture.

Characteristics of HEIA	Characteristics of LEISA
4. The continuing drop in prices of farm produce and the rising costs of agricultural inputs make farming increasingly unprofitable.	LEISA puts greater emphasis is on the long-term sustenance and promotes a balance between the profit and livelihood.
5. HEIA depends on the higher production and profit, without consideration of the local needs and local market.	Sustainable ecological practices depend largely on local agroecological conditions and on local socioeconomic circumstances, as well as on farmers' individual needs and aspirations.
6. Primarily one or two commodity driven development, lack of diversity in the farming practices; as a result, there is greater risk of failure and price fluctuation. The number of products and commodities are very less.	One way of LEISA is to diversification of farms; with a range of crops and/or animals, farmers will suffer less from price fluctuations or drops in yield of single crops. Maintaining diversity will also provide a farm family with a range of products to eat or sell throughout a large part of the year.
7. Under HEIA system, soil quality deteriorates, and there is resurgence of pests, lack of resilience in the soil-plant system.	LEISA maintains a healthy soil, recycling nutrients on the farm, and utilising approaches such as integrated pest management.
8. In HEIA, there is lack of use of indigenous technologies.	Best technologies, for example, soil and water conservation (terraces, ditches, and vegetation strips on sloping land), better timing of operations, improved crop spacing and densities, manure or compost and water application based on local conditions are used to increase the productivity.

CRITERIA FOR LEISA

Important characteristics of LEISA systems are that they are based on a preventive approach whereby the problem is tackled at its roots, as opposed to the more symptom-curing nature of modern high input agriculture. Ecological and biological principles are the basis of the farm system. Nature works for the farmer, the farmer does not have to work against it. They are often based on local knowledge and production systems adapted to modern requirements and much less on external expertise. They are generally more labour-intensive compared to the mechanisation- and chemical-intensive character of the modern agriculture.

Ecological criteria

1. Balanced use of nutrients and organic matter
2. Efficient use of water resources

3. Diversity of genetic resources
4. Efficient of genetic resources
5. Efficient use of energy sources
6. Minimal negative environmental effects
7. Minimal use of external inputs

Economic criteria

1. Sustained farmer livelihood systems
2. Competitiveness
3. Efficient use of production factors
4. Low relative value of external inputs

Social criteria

1. Wide-spread and equitable adoption potential, especially among small farmers
2. Reduced dependency on external institutions
3. Enhanced food security at the family and national level
4. Respecting and building on indigenous knowledge, beliefs and value systems
5. Contribution to employment generation

ECOLOGICAL PRINCIPLES OF LEISA

The principal objective of LEISA is to maintain the agricultural production at an optimum level using less external inputs in an eco-friendly environment. To achieve this objective, the LEISA practices are concentrated heavily on maintaining a living soil, creating biodiversity, recycling of resources and natural pest management. Basic ecological principles of low external input sustainable agriculture are discussed below.

A living soil

Soil can be regarded as a non-renewable resource, as soil formation is a very slow process. The soil provides a medium to anchor plant roots, but is also a very complex ecosystem. The climate, animals, plants and human beings influence the physical, chemical and biological characteristics of the soil. Adequate amount of water, air and nutrients in the soil is essential to maintain crop production at a sustainable level. Favourable soil structure is essential to retain water, nutrients and the growth of root systems of the plants. It is important that soil should be free from poisonous substances.

A productive agricultural soil is full of life, with millions of microorganisms which all interact chemically and physically with their soil environment. These processes regulate the release of nutrients from minerals and organic matter to feed the plants. A living soil has a better structure and can absorb and retain more water and air than a sterile soil. Sound ecological production therefore begins with improving the soil.

The following practices can help improving the condition of the soil.

1. Growing legumes to fix atmospheric nitrogen and provide it to the crop
2. Feeding the soil with as much organic matter as possible through green manure, compost, cover crops, crop residues and agroforestry practices
3. Keeping the soil covered at all times with mulch or cover crops
4. No or reduced tillage which enhances water availability and soil conservation
5. Cultivating a range of crops and animals to reduce risks of disease and pest outbreaks, maintain a balanced nutrient supply and provide resilience
6. Planting trees on contours or making terraces to prevent soil erosion by wind or water

Biological diversity

The diversity of different species of plants and animals, and the genetic variation within each species, provides the vital resource of biological diversity, which enables life on earth. Healthy ecosystems are relatively stable and the diversity they contain enables them to adapt to changing circumstances.

For many small farmers the available agrobiodiversity is the basis of survival. A mix of different locally adapted crops and animals and different varieties of the same increases on-farm diversity, increasing the chances of producing something even under adverse conditions. These principles of traditional farming can be further developed and used systematically in ecological farming (LEISA, 2006).

1. Intercropping in time and space: Planting different crops together in different combinations or formations, or in sequence, can optimise the use of available resources and reduce the pressure of pests.
2. Different plant species can also be used to support the ecological functioning of the whole farming system: Trees or bushes for windbreaks, flowering plants which provide food and habitat for beneficial insects that help control pests, shade trees for light-sensitive plants, trees to provide green manure and fuel wood.
3. Integration of different crops or weeds with animals for better utilisation of resources: Fish in rice fields, integrated crop-chicken-fish systems and other combinations of crops and animals.

Water

Growing population, rapid urbanisation and increasing industrial and agricultural production are all increasing competition for and pressure on water. As agriculture is one of the major users of water, and one of the major polluters of water resources, it is imperative that water use in agriculture is as efficient as possible and that leaching of surplus nutrients and in small scale farming it is important to make the best possible use of the limited amounts of available water. Infiltration can be improved by keeping the soil covered, through minimum disturbance of the soil, adding organic matter from cover crops and mulching.

When introduced, water-harvesting systems are generally multipurpose. Farm ponds, earth dams and subsurface tanks will often serve as a source of drinking water and water for livestock during periods of water scarcity. Water harvesting can open up new livelihood options. Subsistence farmers, who invest in water harvesting systems with a storage component, often diversify their farming system to include cash crop production, for the local market during off-season when prices are high. This diversification increases the resilience of farm households, as they are better equipped to cope with periods of climatic hazards such as droughts and floods.

Energy

Solar energy is captured by plants that are able to transform it into biomass. This is the basis for all higher life forms, animals as well as humans, and is a process that is unique for green plants. Biomass contains stored energy as well as nutrients, and agriculture should focus on maximising the amount of solar energy, which is captured and transformed into plant growth and thereby food and fodder resources. But additional energy is required for cooking and heating and is useful for irrigation, threshing and processing. There are many opportunities to make use of renewable energy, most frequently fuel wood, straw, crop residues and even manure are used. There are also other possibilities to make use of renewable energy; small scale bio-digesters which use manure, solar energy devices, small scale hydropower generators, wind-power and woodlots for fuel wood.

Exploiting animal-plant interaction

In nature, nothing functions in isolation; everything depends on the other factors present. In animal production, to optimise the performance of cattle, it is very important that management practices should enhance the ecological functioning of the web of living organisms within the production system - climate, soil and soil life, vegetation and cattle - by influencing their interactions.

For cattle production, it is important that the breed is selected first, then the pasture suited to that breed and finally the soil is corrected with proper fertiliser or amendment to make the pasture grow. This order has to be reversed. The pasture has to be adapted to the soil and the cattle to the pasture, and all of it has to fit the climate. In addition, the forage crops are to be grown. Of course, in dry areas, forage yields depend strongly on the availability of water. In a well-structured soil, roots are able to explore a larger soil volume for more water and nutrients. Integrating deep-rooting crops and trees into the pasture system will further increase the production of biomass and the overall performance of the system.

In native grassland, cattle always first eat the plants it likes most. The plants that are not eaten get old, hard and are not tasty. The eaten plants sprout again and are grazed on another time. This goes on until these palatable plants disappear. But the less appreciated plants continue to grow and multiply and with time the entire pasture gets hard, rough and has little nutritive value. Then the ranchers set fire to the pasture. Many plants die, and only those that can protect their growing points against fire survive. Thus the pasture becomes worse and the forage volume smaller. Thus, all organic matter that nourishes soil microorganisms is burned out, resulting in their death. The soil compacts, water runs off and the vegetation gets scantier. Thus, the division of pastures into smaller subunits for grazing rotation is fundamental for sustainability of pasturelands.

However, to prevent 'global climate change' by 'greenhouse gases', it is important to reduce methane emission by cattle. This obliges farmers to speed up animal production per unit area and to reduce the slaughter age to get a lower ratio of kg methane/kg animal protein. The use of grains for animal feed has to be reduced as well, giving priority to human consumption. This then increases the dependence on forage. But, as grass cellulose is the main source of methane emission, management practices that contribute to an increase of forage yield per unit area and maintain stocking rate without weight losses are needed.

Towards local resources-based integrated crop-livestock systems

The present livestock production systems in most industrialised countries are in direct competition with human needs. Livestock presently consume almost 50% of world cereal grain supplies. In the intensive large-scale production systems, increasingly promoted by corporate agriculture, livestock wastes contaminate soil and water resources, create less than favourable working conditions for the personnel involved in feeding and cleaning, and decrease employment opportunities. To meet food needs in 2050, it is necessary to develop livestock production systems, which do not depend on cereal grain.

In developing countries like India, instead of grain-based livestock systems, alternative production systems must be developed which make optimal use of locally available resources. Close integration of livestock in the farming system, with recycling of all excreta, will be the basis of agriculture, which can be highly productive and also sustainable. In tropical countries, especially in the humid zone, there are many crops and farming systems that considerably exceed the productive capacity of grain cereals. Key plants in this scenario are sugarcane, cassava, and the palm family, especially the oil and sugar palms.

PROMISING LEISA TECHNIQUES AND PRACTICES

Farming techniques encompass all human activities on a farm that aim at enhancing agricultural production. These activities can involve skilful management of farm resources, assets, inputs and/or outputs. They combine human knowledge, insights and skills with technical means and are mainly oriented towards managing the physical and biological components and processes of the farm (Reijntjes et al., 1992). A technology system involves the whole complex of techniques used in a farm system. A LEISA technology system is a combination of deliberately chosen techniques oriented towards sustainability. LEISA techniques have productive, protective, reproductive and/or social functions, and complement each other. Under the complex and diverse conditions of LEIA farmers, techniques are farm-specific and depend on the local availability of skills, assets and inputs. In view of the often high variability of climatic conditions, an important characteristic of LEISA systems is flexibility in the choice of techniques. Promising LEISA techniques and practices are those which farmers in specific areas have found to be effective in making their farm systems more sustainable (Reijntjes et al., 1992).

Use of mineral fertiliser

Nutrient management is managing the amount, source, placement, form, and timing of the application of nutrients and soil amendments to ensure adequate soil fertility for plant production and to minimise the potential for environmental degradation, particularly water quality impairment. Soil fertility traditionally dealt with supplying and managing nutrients to meet crop production requirements, focusing on optimisation of agronomic production and economic returns to crop production. Contemporary nutrient management deals with these same production concerns, but recognises that ways of farming must now balance the limits of soil and crop nutrient use with the demands of intensive animal production.

Many tropical soils are very poor in plant nutrients or have specific nutrient deficits and crop growth is generally poor in these soils. With the harvest of crop plants nutrients are also removed permanently from the soil. Thus, to maintain or enhance the yield level, external inputs of nutrients are required to replenish the removed nutrients by harvesting of previous crops. Like plants, soil microbes and soil animals also need mineral nutrients. Mineral fertiliser normally increases the availability of biomass which in turn, increases the deposition of organic manure in the soil. This may enhance soil life. On the other hand, enhanced soil life increases the efficiency of mineral fertilisers. Application of mineral fertilisers in low to moderate dose with organic manures and micronutrients enhance soil balance, nutrient availability and, hence, the level and sustainability of crop production. The efficiency and recycling of fertiliser can be further increased by controlling weeds, pests, diseases, erosion and leaching; by rotating shallow- and deep-rooting crops; by split doses of nitrogen and by applying it below the soil surface and near the root zone.

Continuous use of easily soluble concentrated mineral fertiliser can have harmful effects on soil. This may disturb soil life and lead to acidification, micronutrient depletion, soil degradation, poor crop health and lower crop yields (Reijntjes et al., 1992). Ammonium sulphate is a very strong biocide which hinders nitrogen fixation and kills nematodes and earthworms. Among mineral fertilisers, calcium ammonium nitrate is less harmful than urea, which is still to be preferred over ammonium sulphate. This is because urea and ammonium sulphate acidify quickly, the later being quicker, leading to high levels of toxic soluble aluminium in the soil (Smaling, 1990). Mineralisation of rock phosphate can be accelerated in acid conditions. Rock phosphate can be composted together with manure and plant residues for better result. Rock phosphate may also be applied to legumes to enhance nitrogen fixation. The choice of mineral fertilisers will depend on soil type and other variables. It is always better to go for soil testing which will help farmers find the right kind and amounts of mineral and organic fertilisers. The availability and price of mineral fertilisers may limit their use. Appropriate and balanced levels of application can improve the economics of using mineral fertilisers.

Use of organic manures

Organic manures are organic materials derived from animal, human and plant residues which contain nutrients in complex organic forms. They are the sources of plant nutrients. They release nutrients after their decomposition and provide organic acids that help to dissolve soil nutrients and make them available for the plants. Organic manures can be grouped into bulky organic manures and concentrated organic manures based on the concentration of the nutrients. The organic manures provide several benefits.

1. Organic manures supply plant nutrients including micronutrients needed for optimum plant growth.

2. Continued use of manures builds organic matter in soils and improves soil structure. This modification of soil structure helps improve water holding capacity, aeration, friability and drainage.
3. Organic manures improve soil condition for better penetration of roots into deeper layers.
4. Organic manures increase the availability of nutrients through improvement in cation exchange capacity.
5. Carbon dioxide released during decomposition acts as a CO_2 fertiliser.
6. Organic manures increase soil microbial biomass carbon which improves rhizosphere environment.
7. Plant parasitic nematodes and fungi are controlled to some extent by altering the balance of microorganisms in the soil.

But in many developing countries, locally available valuable organic resources are not used to their fullest advantage for agriculture. This is because of inefficient handling and use of organic farm by-products for the purposes other than manures. Improved collection, composting, storage and transport of dung and urine can reduce nutrient losses. The nutrient content of manure is directly related to the species, sex and age of the animals and the quality and quantity of the feed and bedding material. The quality and quantity of manure can be improved by choosing appropriate animal species and adjusting their numbers to the available feed resources; and by balancing the protein/energy content in feed and providing good bedding material and housing for livestock. Nutrient cycling within a farm system can also be improved through the use of manure from poultry. By making simple night-housing for backyard poultry that run freely by day, smallholders can collect manure, which can then be applied on small plots.

Farmers in the hill agriculture follow a system of close integration of crop, livestock and forestry/grassland management. Farmers maintain traditional practices such as terracing, manure management, legumes intercropping, and mulching where appropriate. Farm yard manure quality can be improved by better decomposition and the N-content can be increased through proper management of urine and manure. Liquid manure can be prepared from urine and various plant extracts rich in minerals or secondary plant compounds. These 'manure teas' have been found to be effective liquid fertilisers on crops such as vegetables and also for organic pest and disease management. The use of urea fertiliser is expected to decline in the areas where this liquid manure is used.

Organic manures are not free from some limitations. Nutrients from organic manures are not immediately available to the plants. They are released slowly and over a longer period of time than from most commercial fertilisers. So if there is an immediate need for nutrients, organic manures cannot readily supply nutrient to plants. Many organic manures have low nutrient content and therefore need to be applied in larger quantities to obtain sufficient quantities of nutrients for crops. If there are large distances between the places where dung, feed and bedding material are collected, the places where manure is stored and the fields where it is applied, manure handling will require much time and energy. Handling and transportation cost becomes more. This may also be labour-intensive. Some organic manures need composting before its application to the field. Improperly processed organic manures may contain harmful pathogens and may also contaminate produce. The use of fresh manure may introduce new weeds into fields since certain weed seeds remain alive even after passage through animals and composting process. Cultural taboos can make handling of manure unacceptable to some people.

Crop residue management and conservation tillage under LEISA system

Conservation tillage is a term that covers a broad range of soil management systems that leave residue cover on the soil surface, substantially reducing the effects of soil erosion from wind and water. These practices minimise nutrient loss, decreased water storage capacity, crop damage, and improve soil quality. The soil is left undisturbed from harvest to planting except for nutrient amendment. Weed control is accomplished primarily with herbicides, limited cultivation, and in more sustainable systems, with cover crops. At least 30% of crop residues must be left in the field to qualify for conservation tillage system. Some specific types of conservation tillage are minimum tillage, zone tillage, no-till, ridge-till, mulch-till, reduced-till, strip-till and rotational tillage.

Converting farm wastes into useful organic manure under LEISA system

There is nothing such as organic waste in a farming system. The so called farm wastes such as green leaves, dry leaves, weeds, cow dung and urine, fodder wastes from cattle sheds, cow dung gas slurry, etc. can be converted to valuable organic manures. These items are put in alternate layers (cow dung in between layers) in a heap and left for 45 days. The heap is turned once or twice. In 45 days, the items get semi-decomposed. This partially decomposed material can be further used in two ways. By releasing earthworms to this material, vermicompost can be prepared in another 45 days. The other way is to add coir pith, tank silt and quarry dust to semi-decomposed material, which turns into good quality manure in 45 days.

Practice of green manuring and cover crops

Trees, shrubs, cover crops, grain legumes, grasses, weeds, ferns and algae provide green manure, an inexpensive source of organic fertiliser to build up or maintain soil organic matter and fertility. The cumulative effects of continued use of green manures are important, not only in terms of nitrogen supply but also with regard to soil organic matter and phosphate and micronutrients which are mobilised, concentrated in the topsoil and made available for plant growth. Green manuring adds organic matter to the soil and increases the availability of certain plant nutrients like P, K, Ca, K, Mg and Fe. Leguminous crops add nitrogen to soil through atmospheric nitrogen fixation. Green manuring improves the soil structure and other soil physical properties. Deep-rooted green manure crops in a rotation can help recover nutrients leached to the subsoil. Green manure crops hold plant nutrients that would otherwise be lost by leaching. The practice of green manuring also facilitates infiltration of water thus decreasing runoff and soil erosion. Rapid establishment of a cover crop is crucial to its potential to suppress weeds. If a cover crop is being introduced into a crop rotation, good potential to suppress weeds is a key to farmer acceptance, as beneficial effects from weed control can be observed immediately, while effects of improved nutrient supply may occur only in the longer run.

Green manure crops can be planted in different combinations and configurations in time and space.

1. **Improved fallow:** Natural fallow vegetation is replaced with green manure crops to speed up regeneration of soil fertility and permit permanent cultivation. These green manures may be left to grow for one or several years, or only during the dry season.
2. **Alley cropping:** It is a form of simultaneous fallow in which fast-growing trees or shrubs, usually legumes are planted in rows and are regularly cut back. The prunings are used as mulch or worked into the soil in the alleys between the rows.
3. **Integration of trees into cropland:** Tree legumes are grown among the crops and are regularly cut for mulch material to maintain soil fertility and conserve soil moisture. For example, *Erythrina poeppigiana* is grown in coffee plots.
4. **Live mulching:** In this practice, the rows of food crops are sown into a low but dense cover crop of grasses or legumes, strips of the cover crop are removed by hand or killed by herbicides when the food crops are to be sown, thus reducing soil tillage operations to zero.

Green manure and cover crop species should fit the agroecological condition. In general, these crops should have the characteristics such as easy establishment, vigorous growth under local conditions, ability to cover weeds quickly, ability to either fix atmospheric nitrogen or concentrate plenty of phosphorus, should also have multiple uses.

Various nitrogen-fixing leguminous and non-leguminous species, particularly trees, creepers and bushes can be used as green manures. Some guidelines for selecting green manure species for improving fallow land are given in table 3.3. Using grain legumes for green manuring brings quick economic benefit but, as they tend to accumulate nutrients in the grain, which is then harvested, their positive effect on subsequent crop yields is usually low. The green manure and cover crops can be grown intercropped with another food crop. Mixtures of green manure crops are often more successful than sole crops, as they are less susceptible to pest attacks and combine different characteristics needed for improved fallow, such as quick soil cover and deep rooting. As legume growth depends on the presence of suitable *Rhizobium* strains, inoculation may be necessary. Applying organic manures improve quick establishment of legume. Tropical grasses such as *Pennisetum purpureum* or *Panicum maximum* can produce much biomass and accumulate phosphorus and potassium more quickly than many legumes.

Table 3.2. Fallow plants for improving soil fertility

Calliandra calothyrsus	*Calopogonium mucunoides*	*Canavalia ensiformis*
Canavalia gladiata	*Centrosema pubescens*	*Crotalaria juncea*
Crotalaria lanceolata	*Crotalaria ochroleuca*	*Crotalaria retusa*
Desmodium ascendens	*Desmodium hetrophyllum*	*Desmodium intortum*
Dolichos lablab	*Gliricidia sepium*	*Grevillea robusta*
Indigofera sp.	*Leucaena leucocephala*	*Mimosa invisa*
Mucuna pruriens	*Mucuna utilis*	*Phaseolus aureus*
Phaseolus lunatus	*Phaseolus mungo*	*Psophocarpus tetragonolobus*
Pueraria phaseoloides	*Sesbania bispinosa*	*Sesbania cannabina*
Sesbania grandiflora	*Sesbania rostrata*	*Sesbania sesban*
Stylosanthes guianensis	*Stylosanthes hamata*	*Tephrosia bracteolata*
Tephrosia candida	*Tephrosia vogelii*	*Vigna aconitifolia*
Vigna radiata	*Vigna umbellata*	*Vigna unguiculata*

Source: Copijn, 1985

Table 3.3. Criteria for selection of plants for improved fallow

Criteria	Effects
High biomass production	Mobilisation of nutrients from soil into vegetation; suppression of weeds
Deep rooting system	Pumping up of weathered and/or leached nutrients from soil layers not occupied by roots of main crop
Fast initial growth	Quick soil cover for effective soil protection; suppression of weeds
More leaf than wood (low C/N ratio)	Easy decomposition of organic matter leading to enhanced availability of nutrients for succeeding crops; easy to handle during cutting and/or incorporation into the soil
Nitrogen fixing	Increase of nitrogen availability
Good affinity with mycorrhiza	Mobilisation of phosphorus leading to improved availability for crops
Efficient water use	Possibility to grow after main cropping season on residual soil moisture or with less rainfall
Non-host for crop-related pests and diseases	Decrease in pest and disease populations
No rhizomes	Controllable growth
Easy and abundant seed formation	Propagation in farmers' fields
Useful 'by-products' (e.g. fodder, wood)	Integration of animal husbandry and forestry

Source: Copijn, 1985

The main problem in adoption of green manuring is that the land on which green manure crop is raised could have been profitably utilised for growing a crop of economic importance. Another problem is adequate soil moisture either through rainfall or irrigation is essential for *in situ* decomposition of green manure crop. Under rainfed condition, if sufficient rainfall is not received proper decomposition may not take place, and the germination of the succeeding crop is hampered. There is a possibility of incidence of diseases and pests, even nematodes.

Mulching

Mulch can be defined as a shallow layer at the soil/air interface with properties that differ from the original soil surface layer (Stigter, 1984). Any material, such as straw, plant residues, leaves, loose soil or plastic film which is placed on the soil surface to reduce evaporation, runoff, erosion or to protect plant roots from extremely low or high temperature, to suppress weed growth is called as mulch; and the practice of applying mulches is called mulching.

Mulching has the following advantages.

1. Mulching favourably influences the soil moisture regime by controlling evaporation from soil surface, improving infiltration and soil moisture retention and facilitating condensation of water at night due to temperature reversals.
2. Mulching suppresses weed growth.
3. Mulching invariably decreases soil erosion and often reduces runoff rate. Mulch cover protects the soil from raindrop impact and surface sealing, increases the infiltration rate and decreases runoff velocity through physical resistance to water flow.
4. Mulch has a moderating influence on the soil thermal regime.
5. Crop residue mulch improves soil aeration by promoting free exchange of gases between the soil and the atmosphere.
6. Mulching improves soil structural properties directly by preventing the raindrop impact and indirectly by promoting the biological activity.
7. Organic mulches add organic matter and plant nutrients to soil upon decomposition. Thus, they improve carbon sequestration. Cation exchange capacity is substantially influenced by organic matter content in soils containing predominantly low activity clays.
8. Soil biological activity is either directly influenced by supply of food substrates by organic mulches or indirectly influenced by both organic and inorganic mulches through alteration of soil hydrothermal regime.
9. Mulching reduces the need for tillage.

The mulch materials include organic residues (grass clippings, leaves, hay, straw, shredded bark, sawdust, wood chips, shredded newspaper, cardboard, wool, etc.), compost, rubber mulch, plastic mulch, organic sheet mulch (various products such as biodegradable alternative to plastic mulch), rock and gravel mulch, living mulch etc. The effects of mulch depend on its composition and colour, the amount applied, the timing of application and the rate at which the mulch decomposes. This rate depends, in turn, on the form and timing of application and the meteorological conditions in the soil and air. Mulch is usually applied towards the beginning of the growing season, and may be reapplied as necessary. It serves initially to warm the soil by helping it retain heat. This allows early seeding and transplanting of certain crops in cold regions, and encourages faster growth. As the season progresses, the mulch stabilises temperature and moisture, and prevents sunlight from germinating weed seed.

Possible constraints to mulching are insufficient availability of mulch material, unsuitable soils, soil compaction and pest problems (rodents, insects, fungi, persistent weeds). The infestation of termite is aggravated due to mulching especially in the humid areas. There are also some problems of mulching with respect to its adoption by the farmers. Mulch hinders the operation of sowing (if it is applied before sowing), fertiliser applications, irrigation, etc. Sometimes thick mulching in areas of high rainfall may lead to waterlogging conditions and provide an environment for disease infestation. Plastic mulches usually require pick up and disposal at the end of the season and their manufacture and disposal entail significant environmental cost, as these are not biodegradable. Plastic mulch can interfere with the recharge of soil profile by rain or overhead irrigation.

Practices for preventing land degradation

A good plant cover is very important for preventing soil degradation and achieving soil rehabilitation. A vegetative cover has a number of beneficial effects on soil aeration, soil moisture and organic matter content, physical characteristics and biological activity in the soil. In addition, a plant cover protects against soil erosion. Cover crops can be quite aggressive creepers and may compete with the main crop. On the other hand if leguminous plants are used as cover crops they add nutrients because they fix atmospheric nitrogen and make it available for the crop. Selection of crops is very important while dealing with degraded soils.

Integrated pest management

Integrated pest management (IPM) is an ecologically based approach to pest control that utilises a multidisciplinary knowledge of crop/pest relationships, establishment of acceptable economic thresholds for pest populations and constant field monitoring for potential problems. The following LEISA management practices may be included in IPM (LEISA, 2006).

1. Use of resistant varieties
2. Crop rotation
3. Cultural practices
4. Optimal use of biological control organisms
5. Use of certified seed
6. Protective seed treatments
7. Disease-free transplants or rootstock
8. Timeliness of crop cultivation
9. Improved timing of pesticide applications
10. Removal of infested plant material

The term biointensive IPM emphasises a range of preventive tactics and biological controls to keep pest population within acceptable limits. Reduced risk pesticides are used if other tactics have not been adequately effective, as a last resort and with care to minimise risks. Biological control is, generally, man's use of a specially chosen living organism to control a particular pest. This chosen organism might be a predator, parasite, or disease, which will attack the harmful insect. A complete biological control programme may range from choosing a pesticide which will be least harmful to beneficial insects, to raising and releasing one insect to have it attack another, almost like a 'living insecticide'.

Windbreaks

Windbreaks are barriers used to reduce and redirect wind. They usually consist of trees and shrubs, but may also be perennial or annual crops and grasses, fences, or other materials. The reduction in wind speed behind a windbreak modifies the environmental conditions or microclimate in the sheltered zone. The direction from which wind is blowing is called windward side and direction to which wind is blowing is called leeward side. Besides influencing the microclimate, hedges can be useful in keeping animals out of fields and/or producing fruits, herbs, fodder, mulch, thatching material or fuel. They also play a role in balancing pest populations. The improvement in microclimate as a result of windbreaks can lead to an increase in crop yield and may compensate for the land lost to planting trees. The total biomass production of both crops and hedges may be raised considerably.

Height, density, number of rows, species composition, length, orientation, and continuity determines the effectiveness of a windbreak in reducing wind speed and altering the microclimate. Windbreaks are most effective when its orientation is such that it is at right angles to prevailing winds. The purpose and design of each windbreak is unique, thus the orientation of individual windbreaks depends on the design objectives. To control soil erosion, windbreaks should be planted to block the prevailing winds during the times of greatest soil exposure. Constraints to planting windbreaks may be excessive competition for light and root space with the adjacent crop, and pest transmission from plants in the hedge to certain crops. Establishment may be difficult because of long dry periods and free-roaming animals. Allowing natural regrowth of trees on the borders of fields or contour lines can be a viable alternative to deliberately planting trees.

Water harvesting

In rainfed agriculture, efficient water management is of great importance. Rainfall may be too low or too high or too irregular, creating high risks of yield losses and/or unfavourable growth conditions or damage by erosion. Where there is not enough rain to grow a crop or where rainfall is very irregular, water-harvesting techniques can be used to concentrate rain or flood water in such a way that it can be used for crop growth. Thus, water harvesting, defined in its broadest sense as the collection of runoff for its productive use. More specifically, it is the collection and storing of water on the surface of the soil for subsequent use. It is relevant to areas where the rainfall is reasonably distributed in time, but inadequate to balance potential evapotranspiration of crops. Water harvesting supports a flourishing agriculture in many dry areas. The sustainability of the various water harvesting techniques is found to depend largely upon the timing and the amount of rainfall. However, the techniques should be site-specific. Water harvesting is mainly practised in arid and semiarid regions, where surface runoff often has an intermittent character. Water harvesting is based on the utilisation of runoff and requires a runoff producing area and a runoff receiving area. Because of the intermittent nature of runoff events, storage is an integral part of the water harvesting system.

Water harvesting not only secures and increases crop production in regions where rainfall is normally insufficient; it can also serve to control soil erosion and to recharge aquifers tapped for irrigation. An additional benefit is the improvement in soil fertility. Silt, manure and other organic matter are harvested together with the water. The soil profile stays moist longer, stimulating soil life and improving humus formation, nutrient availability and the soil's capacity to hold water.

Water may be stored directly in the soil profile or in small reservoirs, tanks, and aquifers. Water harvesting system should therefore have the following four components.

1. Runoff producing catchments
2. Runoff collection scheme
3. Runoff storage facility
4. Cultivated or cropped area

The water harvesting structures which are commonly built for surface storage and/or groundwater recharge are check dams, percolation tanks, subsurface dykes, contour bunding, gully plugging, etc.

Water ponds/tanks: This is the most commonly used method to collect and store rainwater in ponds or tanks. Most ponds have their own catchments, which provide the requisite amount of water during the rainy season. Where the catchments are too small to provide enough water, water from nearby streams is diverted through open channels to fill the ponds. In some places water from irrigation canals is also used to fill ponds. Ponds are excavated in different shapes and sizes depending upon the nature of the terrain, availability of land, water requirements of the village community, etc.

Check dams: These are concrete or masonry structures built across small streams for surface storage and incidental benefit of groundwater recharge. The design of these structures is done considering the volume of water that can be stored in the upstream channel.

Percolation tanks: Percolation tank is an artificial reservoir which is constructed across stream, submerging a land area with adequate permeability to facilitate sufficient percolation to collect surface water runoff and allow it to percolate within the permeable land. This is one of the effective methods of refilling groundwater table. Moderate to high porosity of soil and/ or underlying rocky strata is the main criteria for the choice of percolation tanks. Percolation tanks should normally be constructed in a terrain with highly fractured and weathered rock for speedy recharge. Too high permeability may result in the percolated water escaping in the downstream as regenerated surface flow which will defeat the purpose of water conservation. A long-term evaluation of the pattern of rainfall for the benefitted area must be studied so that the percolation tank gets filled up fully during monsoon. The amount of water a percolation tank would be able to seep into the ground for compensating depleted groundwater, depends upon the amount and nature of precipitation of the area, thickness of topsoil and weathered zone, type of vegetation, evaporation from the surface of wet soil, profile of underlying hard rock, the topographical features of the sub-basin, and the status of soil.

Subsurface dykes: Subsurface dyke or underground dam is a subsurface barrier across stream which retards the base flow and stores water upstream below ground surface. By doing so, the water levels in upstream part of groundwater dam rises saturating otherwise dry part of aquifer. The site where subsurface dyke is proposed should have shallow impervious layer with wide valley and narrow outlet. After selection of suitable site, a trench of 1-2 m wide is dug across the breadth of stream down to impermeable bed. The trench may be filled with clay or brick/ concrete wall up to 0.5 m below the ground level. Since the water is stored within the aquifer, submergence of land can be avoided and land above the reservoir can be utilised even after the construction of the dam. No evaporation loss from the reservoir and no siltation in the reservoir take place. The potential disaster like collapse of the dams can also be avoided.

Contour bunding: These are small earthen bunds built horizontally in parallel rows across the hill slope to collect runoff from the catchment to improve soil moisture on the cropped area. These help in augmenting soil moisture and prevent erosion of topsoil. Contour bunds are effective methods to conserve soil moisture in watershed for long duration. Flowing water is intercepted before it attains the erosive velocity by keeping suitable spacing between bunds. Spacing between two contour bunds depends on the slope, the area and the permeability of the soil. Lesser the permeability of soil, the close should be spacing of bunds. This is also known as contour furrows and desert strip farming.

Gully plugging: These are soil and water retaining structures built across gullies in hilly areas. These are built with locally available materials like stone boulders, earth, brushwood, etc. The sites for gully plugs may be chosen whenever there is a local break in slope to permit accumulation of adequate water behind the bunds.

In situ water conservation through various land configurations

Several methods like broad bed and furrow system, ridging and tied ridging, strip cropping, etc. may be practised for *in situ* conservation of water.

Broad bed and furrow system: The broad bed and furrow system (BBF) system has been mainly developed at the International Crops Research Institute for the Semiarid Tropics (ICRISAT) in India. It is a modern version of the very old concept of encouraging controlled surface drainage by forming the soil surface into beds. The recommended ICRISAT system consists of broad beds about 100 cm wide separated by sunken furrows about 50 cm wide. The preferred slope along the furrow is between 0.4 and 0.8% on vertisols. Two, three, or four rows of crop can be grown on the broad bed, and the bed width and crop geometry can be varied to suit the cultivation and planting equipment. The major objectives are (i) to encourage moisture storage in the soil profile, (ii) to dispose safely of surplus surface runoff without causing erosion, (iii) to provide a better drained and more easily cultivated soil in the beds, and (iv) to reuse the runoff water stored in small tanks. The intercropping or sequential cropping is also possible on the broad-beds. Small amounts of life-saving irrigation applications can be very effective in dry spells during the rains, particularly on soils with lower storage capacity than the deep vertisols.

Ridging and tied ridging: This method is also known as furrow blocking, furrow damming and basin listing. The principle is to increase surface storage by first making ridges and furrows, then damming the furrows with small mounds or ties. Tied-ridging is system of land configuration where semi-permanent ridges are prepared with crossties along the furrows to trap runoff of water. The ridges are generally laid across the main slope at a grade of 0.4-1%. Normally once constructed the ridges are not destroyed for a period of six seasons depending on the crop rotations practised by the farmer. Planting is done on top of the ridges. In subsequent seasons land preparation simply involves planting on top of the ridges without any tillage operation. For emergence and good establishment of the crop, planting is recommended only when the ridges are optimally moist. In drier areas planting may also be carried out in the furrows where most of the runoff water collects.

Strip cropping: In strip cropping, the crop is sown in narrow, tilled rows along contours on a hillside. The strips of land between the rows, which are left untilled in natural grasses, slow the flow of rainwater down the slope and prevent it from washing away the topsoil. More water penetrates into the soil and provides moisture for the crop. The grass strips also provide a natural habitat for insects, many of which will prefer this to the crop. Organic matter and fertiliser are concentrated in furrows where the seed is planted. By tilling only the strips to be planted, the farmer saves labour in comparison with cultivation of the entire field. In subsequent seasons, the rows are replanted and the strips of grass cut back periodically, but they are never dug up.

CONSTRAINTS FOR ADOPTION OF LEISA

The following major constraints for adoption of LEISA techniques by the farmers have been identified by Kessler and Moolhuijzen (1994).

Reduced production and rehabilitation period

In most cases, introduction of a LEISA technique decreases agricultural production for a certain period. For example, when shifting from chemical fertilisers to organic manures it takes several years to build up the organic matter content and fauna of the soil for organic inputs to become effective. So in initial years the expected returns are not realised and this may strongly discourage the farmer to adopt the technology. Similarly, in case of alley cropping although the shrubs/trees occupy a good portion of agricultural land the return from the alley cropping is much delayed. The duration of the transition period varies depending upon the ecology of the site like soil, climate vegetation factors, etc.

Labour requirement

Most of the LEISA techniques are labour intensive. Labour shortage for small families and the migrant absentees is acute; particularly to collect huge quantity of organic manure and other related activities. Farmers' decisions to invest extra labour on LEISA techniques are greatly influenced by the expected benefits from it in relation to the availability of alternate source of income, may be from less risky off-farm activities.

Land ownership

The adoption of LEISA techniques is more where the farmers own their land and usufruct rights. Tenant farmers are discouraged and even prohibited going for the LEISA techniques.

Gender biasness

Women appear more rapid in adopting LEISA techniques. However, government organisations and non-government organisations are mainly staffed by men and their activities are oriented towards men. The constraints of LEISA, as for the agricultural development in general, affect more to women than men; women generally have less land ownership than men, and women have less control over the profits of their labour investments in LEISA techniques.

Partnership Technology Development Approach

An important instrument to develop LEISA techniques is the Partnership Technology Development (PTD) approach by which site-specific and socioeconomically adapted techniques are developed and tested on-farm, based on the farmers' indigenous knowledge, skills and priorities, with support from the capacities of the external agents and agricultural science. However, in most cases, the LEISA techniques are introduced with the top-down way for getting quick results. PTD is time consuming, both for extensionists and farmers, and it requires a relatively high level training of the extensionists. But the extension agencies commonly lack sufficient dedicated and well-trained staff. On-farm demonstrations of LEISA techniques are rarely designed properly, nor cover for a sufficient long period to draw a firm conclusion. The potential of PTD to motivate the farmers is strongly related to the possibilities of obtaining short-term economic benefits.

Government policies

Often the policy-makers lack knowledge and appreciation of LEISA approaches. Government agencies provide subsidy and favourable credit system to the external inputs only for farm activities. There is no clear-cut policy to offer incentives to the farmers for adopting LEISA techniques or compensate the loss incurred during initial years of adoption.

Information dissemination

The availability of technical information and examples of successful implementation of LEISA techniques are important elements of PTD approach to motivate the farmers to adopt the LEISA techniques. However, this information is often unavailable.

Low External Input Sustainable Agriculture (LEISA) is receiving increased attention, both as a sustainable alternative to Green Revolution-like strategies making intensive use of external inputs, and as a strategy of sustainable agriculture in resource-poor environments where no or very few external inputs are used. To achieve these goals, LEISA must tap the most viable indigenous knowledge and practices and ecologically friendly technologies in a given ecological and sociocultural setting, since the experience in one agroecological setting may not be appropriate in other areas. There are many cases of farmers using LEISA under different agroecological zones in different countries. These experiences should be documented to learn more of the principles, constraints, and potentials in order to provide policy-makers, development workers and farmers with alternative and viable strategies to develop sustainable farming systems.

REFERENCES

Copijn, A. N. 1985. Bio-ecological soil regeneration methods. AME Programme, ETC Foundation. Leusden, the Netherlands.

Corsi, S. and Muminjanov, H. 2019. Conservation Agriculture: Training guide for extension agents and farmers in Eastern Europe and Central Asia. FAO, Rome. http://www.fao.org/3/i7154en/i7154en.pdf

FAO. 1996. Low External Input Sustainable Agriculture (LEISA) in Selected Countries of Asia: Regional Workshop Report and Case Studies. Food and Agriculture Organisation of the United Nations, Rome.

Gold, M. V. 2007. Sustainable Agriculture: Definitions and Terms. Related Terms - Low input agriculture. Special Reference Briefs Series no. SRB 99-02. Updates SRB 94-05. Alternative Farming Systems Information Center. US Department of Agriculture. https://www.nal.usda.gov/afsic/sustainable-agriculturedefinitions-and-terms-related-terms#term19

Kessler, J. J. and Moolhuijzen, M. 1994. Low external input sustainable agriculture: expectations and realities. *Wageningen Journal of Life Sciences* **42**(3): 181-194.

LEISA. 2006. Farming with biodiversity: the ecological basis for sustainable agriculture. *LEISA magazine*, Special issue 2006. http://www.leisa.info

Pimentel, D., Hunter, M. S., LaGro, J. A., Efronymson , R. A., Landers, J. C., Mervis, F. T., McCarthy, C. A. and Boyd, A. E. 1989. Benefits and risks of genetic engineering in agriculture. *BioScience* **39**: 606-614.

Reijntjes, C., Haverkort, B. and Waters-Bayer, A. 1992. Farming for the future: an introduction to low-external-input and sustainable agriculture. MacMillan Education Ltd., London. pp. 164-209.

Smaling, E. M. A. 1990. Two scenarios for the sub-Sahara: one leads to disaster. *Ceres* **126**: 20-24.

Stigter, C. J. 1984. Mulching as a traditional method of microclimate management. *Archives for Meteorology, Geophysics, and Bioclimatology, Series B* **35**: 147-154.

4

Conservation Agriculture for Sustainability

Conservation agriculture is an approach to managing agroecosystems for improved and sustained productivity, increased profits and food security while preserving and enhancing the resource base and the environment. Food and Agriculture Organisation of the United Nations (FAO) defines conservation agriculture as a farming system that promotes maintenance of a permanent soil cover, minimum soil disturbance (i.e., no tillage), and diversification of plant species. It enhances biodiversity and natural biological processes above and below the ground surface, which contribute to increased water and nutrient use efficiency and to improved and sustained crop production. Conservation agriculture is largely the product of the collective efforts of a number of previous agricultural movements, including no-till agriculture, agroforestry, green manures/cover crops (GMCC), direct planting/seeding, integrated pest management, and conservation tillage among many others. Yet conservation agriculture (CA) is distinct from each of these so-called agricultural packages, even as it draws upon many of their core principles.

Table 4.1. Some distinguishing features of conventional and conservation agriculture systems

Conventional agriculture	Conservation agriculture
Cultivating land, using science and technology to dominate nature	Least interference with natural processes
Excessive mechanical tillage and soil erosion	No-till or drastically reduced tillage (biological tillage)
High wind and soil erosion	Low wind and soil erosion
Residue burning or removal (bare surface)	Surface retention of residues (permanently covered)
Water infiltration is low	Infiltration rate of water is high
Use of *ex situ* FYM/composts	Use of *in situ* organics/composts
Green manuring (incorporated)	Brown manuring/cover crops (surface retention)
Kills established weeds but also stimulates more weed seeds to germinate	Weeds are a problem in the early stages of adoption but decrease with time
Free-wheeling of farm machinery, increased soil compaction	Controlled traffic, compaction in tramline, no compaction in crop area
Monocropping/monoculture, less efficient rotations	Diversified and more efficient rotations

Conventional agriculture	Conservation agriculture
Heavy reliance on manual labour, uncertainty of operations	Mechanised operations, ensure timeliness of operations
Poor adaptation to stresses, yield losses greater under stress conditions	More resilience to stresses, yield losses are less under stress conditions
Productivity gains in long-run are in declining order	Productivity gains in long-run are in incremental order

Source: Sharma et al., 2012

HISTORY OF CONSERVATION AGRICULTURE

Tillage, particularly in fragile ecosystems, was questioned for the first time in the 1930s, when the dustbowls devastated wide areas of the mid-west United States. Concepts for reducing tillage and keeping soil covered were introduced and the term conservation tillage was introduced to reflect such practices aimed at soil protection. Seeding machinery developments allowed then, in the 1940s, to seed directly without any soil tillage. At the same time theoretical concepts resembling today's conservation agriculture concept were developed by Edward Faulkner who published a book called 'Ploughman's Folly' (Faulkner, 1945) and Masanobu Fukuoka who published a book called 'One Straw Revolution' (Fukuoka, 1975). But it was not until the 1960s for no-tillage to enter into farming practice in the USA. In the early 1970s no-till reached Brazil, where farmers together with extension scientists transformed the technology into the production system which today is called conservation agriculture. Yet it took another 20 years before CA reached significant adoption levels. During this time farm equipment and agronomic practices in no-till systems were improved and developed to optimise the performance of crops, machinery and field operations.

While tillage-based agriculture has been researched for several centuries, CA is only half-a-century old and the functioning of CA systems can only be understood as the agroecosystems evolve under the new production management. From the early 1990s, the uptake of CA started growing exponentially, leading to a revolution in the agriculture of Brazil, Argentina, Paraguay and Uruguay. During the 1990s, this development increasingly attracted attention from farmers and researchers in Europe, Asia, Africa and Australia, and from development and international research organisations such as FAO, World Bank, International Fund for Agricultural Development (IFAD), German Agency for International Cooperation (GIZ), Norwegian Agency for Development Cooperation (NORAD), French Agricultural Research Centre for International Development (CIRAD), Australian Centre for International Agricultural Research (ACIAR) and the Consultative Group on International Agricultural Research (CGIAR) system. The improvement of conservation and no-tillage practices within an integrated farming concept such as CA also led to a cropping system diversification and increased adoption of CA in industrialised countries after 2000. Conservation agriculture crop production systems are now becoming popular and experiencing increased interest worldwide. The total area under CA was 117 million hectares in 2010 (Kassam et al., 2010) and 180.44 million hectares in 2015 (Kassam et al., 2019), registering an increase of 54.2% in five years.

Table 4.2. Cropland area under conservation agriculture (M ha) by region in 2015-16

Region	CA cropland area of the region	Percent of global CA cropland area	Percent of cropland area in the region
South America	69.90	38.7	63.2
North America	63.18	35.0	28.1
Australia & New Zealand	22.67	12.6	45.5
Asia	13.93	7.7	4.1
Russia & Ukraine	5.70	3.2	3.6
Europe	3.56	2.0	5.0
Africa	1.51	0.8	1.11
Global total	180.44	100	12.5

Source: Kassam et al., 2019

OBJECTIVES OF CONSERVATION AGRICULTURE

The conservation agriculture offers a range of productivity, socioeconomic and environmental benefits to producers and to society at large on a sustainable basis. These are based on five overall objectives.

1. Simultaneous achievement of increased agricultural productivity and enhanced ecosystem services
2. Enhanced input-use efficiency, including water, nutrients, pesticides, energy, land and labour
3. Judicious use of external inputs derived from fossil fuels (such as mineral fertilisers and pesticides) and preference for alternatives (such as recycled organic matter, biological nitrogen fixation and integrated pest management)
4. Protection of soil, water and biodiversity through use of 'minimum soil disturbance' and maintaining organic matter cover on the soil surface to protect the soil and enhance soil organic matter and soil biodiversity
5. Use of managed and natural biodiversity of species to build systems' resilience to abiotic, biotic and economic stresses, with an underlying emphasis on improving soils' content of organic matter as a substrate essential for the activity of the soil biota

The farming practices required to implement these objectives will differ according to local conditions and needs, but will have the following management practices.

1. Minimising soil disturbance by mechanical tillage and, whenever possible, seeding or planting directly into untilled soil, in order to maintain soil organic matter, soil structure and overall soil health
2. Enhancing and maintaining organic matter cover on the soil surface, using crops, cover crops or crop residues
3. Diversification of species, both annuals and perennials - in associations, sequences and rotations that can include trees, shrubs, pastures and crops, all contributing to enhanced crop nutrition and improved system resilience
4. Use of well adapted, high yielding varieties and good quality seeds
5. Enhanced crop nutrition, based on healthy soils
6. Integrated management of pests, diseases and weeds
7. Efficient water management

Sustainable crop production intensification is the combination of all seven of these improved practices applied in a timely and efficient manner. Such sustainable production systems are knowledge and management-intensive and relatively complex to learn and implement. They offer farmers many possible combinations of practices to choose from and adapt, according to their local production conditions and constraints (Pretty, 2008; Pretty et al., 2011; Kassam et al., 2018).

PRINCIPLES OF CONSERVATION AGRICULTURE

Conservation agriculture practices perused in many parts of the world are built on ecological principles making land use more sustainable (Wassmann, 2009; Behera et al., 2010; Lal, 2013). Adoption of CA for enhancing resource use efficiency and crop productivity is the need of the hour as a powerful tool for management of natural resources and to achieve sustainability in agriculture (Bhan and Behera, 2014). The FAO has determined that CA has three key principles that farmers can proceed through in the process of conservation agriculture. These principles are linked and must be considered together for appropriate design, planning and implementation processes.

These three principles are

1. Minimum mechanical soil disturbance: (i.e., no tillage) through direct seed and/or fertiliser placement
2. Permanent soil organic cover: (at least 30%) with crop residues and/or cover crops

3. Species diversification: through varied crop sequences and associations involving at least three different crops

Conservation agriculture principles are universally applicable to all agricultural landscapes and land uses with locally adapted practices. CA enhances biodiversity and natural biological processes above and below the ground surface. Soil interventions such as mechanical tillage are reduced to an absolute minimum or avoided, and external inputs such as agrochemicals and plant nutrients of mineral or organic origin are applied optimally and in ways and quantities that do not interfere with, or disrupt, the biological processes. CA facilitates good agronomy, such as timely operations, and improves overall land husbandry for rainfed and irrigated production. Complemented by other known good practices, including the use of quality seeds, and integrated pest, nutrient, weed and water management, etc., CA is a base for sustainable agricultural production intensification. It opens increased options for integration of production sectors, such as crop-livestock integration and the integration of trees and pastures into agricultural landscapes (FAO, 2010).

Conservation agriculture can contribute to sustainable agriculture and rural development by improving input use efficiency, increasing farm income, sustaining or increasing agricultural land, and protecting the natural resource base (FAO, 2007).

Minimum tillage and soil disturbance

The first key principle in CA is practising minimum soil disturbance which is essential for maintaining minerals within the soil, stopping erosion, and preventing water loss from the soil. In the conventional agriculture, soil tillage is considered as a main process in the introduction of new crops to an area. It is believed that tilling the soil would increase fertility within the soil through mineralisation that takes place in the soil. However, in conservation agriculture, tillage is considered as a harmful practice that destroys organic matter that is found within the soil cover and causes severe erosion and crusting which leads to a decrease in soil fertility.

The soil biological activity produces very stable soil aggregates as well as various sizes of pores, allowing air and water infiltration. This process can be called 'biological tillage' and it is not compatible with mechanical tillage. With mechanical soil disturbance, the biological soil structuring processes will disappear. Minimum soil disturbance provides/maintains optimum proportions of respiration gases in the rooting-zone, moderate organic matter oxidation, porosity for water movement, retention and release and limits the re-exposure of weed seeds and their germination (Kassam and Friedrich, 2009).

Direct seeding involves growing crops without mechanical seedbed preparation and with minimal soil disturbance since the harvest of the previous crop. Planting refers to the precise placing of large seeds (for example, maize and beans); whereas seeding usually refers to a continuous flow of seed as in the case of small cereals (for example, wheat and barley). The equipment penetrates the soil cover, opens a seeding slot and places the seed into that slot. The size of the seed slot and the associated movement of soil are to be kept at the absolute minimum possible. Ideally the seed slot is completely covered by mulch again after seeding and no loose soil should be visible on the surface.

No-till farming has caught on as a process that can save soil organic levels for a longer period and still allow the soil to be productive for longer periods (FAO, 2007). Additionally, the process of tilling can increase time and labour for producing that crop. Minimum soil disturbance also reduce destruction of soil micro- and macro-organism habitats that is common in conventional ploughing practices.

Land preparation for seeding or planting under no-tillage involves slashing or rolling the weeds, previous crop residues or cover crops; or spraying herbicides for weed control, and seeding directly through the mulch. Crop residues are retained either completely or to a suitable amount to guarantee the complete soil cover, and fertiliser and amendments are either broadcast on the soil surface or applied during seeding.

Permanent soil organic cover with crop residues and live mulches

Keeping the soil covered is a fundamental principle of CA. A permanent soil cover is important to protect the soil against the deleterious effects of exposure to rain and sun; to provide the micro- and macro- organisms in the soil with a constant supply of 'food'; and alter the microclimate in the soil for optimal growth and development of soil organisms, including plant roots. In turn it improves soil aggregation, soil biological activity and soil biodiversity and carbon sequestration (Ghosh et al., 2010). Crop residues are left on the soil surface, but cover crops may be required if the gap is too long between harvesting one crop and establishing the next. Cover crops improve the stability of the CA system, not only on the improvement of soil properties but also for their capacity to promote an increased biodiversity in the agroecosystem. The principle of managing the top soil to create a permanent organic soil cover can allow for growth of organisms within the soil structure. This growth will break down the mulch that is left on the soil surface. The breaking down of this mulch will produce a high organic matter level which will act as a fertiliser for the soil surface.

If CA is in practice for many years and enough organic matter was being built up at the surface, then a layer of mulch would start to form. This layer helps prevent soil erosion from taking place and ruining the soil's profile. The presence of mulching also reduces the velocity of runoff and the impact of raindrops thus reducing soil erosion and runoff. The layer of mulch that is built up over time will become like a buffer zone between soil and mulch and this will help reduce wind and water erosion (Hobbs et al., 2008). This type of ground cover also helps keep the temperature and moisture levels of the soil at a higher level rather than if it was tilled every year (FAO, 2007).

The presence of a mulch layer (of dead vegetation) in conservation agriculture inhibits the evaporation of soil moisture, yet leads to greater water infiltration into the soil profile. The percentage of rainwater that infiltrates the soil depends on the amount of soil cover provided. Vegetative cover is important in CA for the protection of the soil against the impacts of raindrops; to keep the soil shaded; and maintain the highest possible moisture content. Straw residues function as a cushion that reduces the pressure on the soil under wheels and hooves and so they play an important role in reducing soil compaction.

While commercial crops have a market value, cover crops are mainly grown for their effect on soil fertility or as livestock fodder. Cover crops are grown during fallow periods, between harvest and planting of commercial crops, utilising the residual soil moisture. Their growth is interrupted either before the next crop is sown, or after sowing the next crop, but before competition between the two crops starts. In regions where smaller amounts of biomass are produced, such as semiarid regions or areas of eroded and degraded soils, cover crops are very useful.

The cover crops

- Protect the soil during fallow periods
- Recycle nutrients (especially P and K) and mobilise them in the soil profile in order to make them more readily available to the following crops
- Improve the soil structure and break compacted layers and hard pans
- Permit a rotation in a monoculture
- Can be used to control weeds and pests
- Provide an additional source of organic matter to improve soil structure
- Provide 'biological tillage' of the soil; the roots of some crops, especially cruciferous crops are able to penetrate compacted or very dense layers, increasing water percolation capacity of the soil
- Utilise easily leached nutrients (especially N)

Species diversification

The third principle is the practising diverse crop rotations or crop interactions. The rotation of crops is not only necessary to offer a diverse 'diet' to the soil microorganisms, but as they root at different soil depths, they are capable of exploring different soil layers for nutrients. Nutrients that have been leached to deeper layers and that are no longer available for the commercial crop can be 'recycled' by the crops in rotation. This way the rotation crops function as biological pumps. Furthermore, a diversity of crops in rotation leads to a diverse soil flora and fauna, as the roots excrete different organic substances that attract different types of bacteria and fungi, which in turn, play an important role in the transformation of these substances into plant available nutrients. Cropping sequence and rotations involving legumes helps in minimal rates of build-up of population of pest species, through life cycle disruption, biological nitrogen fixation, control of off-site pollution and enhancing biodiversity (Kassam and Friedrich, 2009).

Crop rotation also has an important phytosanitary function as it prevents the carryover of crop-specific pests and diseases from one crop to the next via crop residues. Rotational crops will act as a natural insecticide and herbicide against specific crops. Not allowing insects or weeds to establish a pattern will help to eliminate problems with yield reduction and infestations within fields (FAO, 2007). Crop rotation can also help build-up soil infrastructure. Establishing crops in a rotation allows for an extensive build-up of rooting zones which will allow for better water infiltration (Hobbs et al., 2007). Organic molecules in the soil break down into phosphates, nitrates and other beneficial elements which are thus better absorbed by plants. Ploughing increases the amount of oxygen in the soil and increases the aerobic processes, hastening the breakdown of organic material. Thus, more nutrients are available for the next crop but, at the same time, the soil is depleted more quickly of its nutrient reserves.

The effects of crop rotation can be summarised as follows.

- Higher diversity in plant production and thus in human and livestock nutrition
- Reduction and reduced risk of pest and weed infestations
- Greater distribution of channels or biopores created by diverse roots (various forms, sizes and depths)
- Better distribution of water and nutrients through the soil profile
- Exploration for nutrients and water of diverse strata of the soil profile by roots of many different plant species resulting in a greater use of the available nutrients and water

- Increased nitrogen fixation through certain plant-soil biota symbionts and improved balance of N/P/K from both organic and mineral sources
- Increased humus formation

Thus, conservation agriculture is a set of soil management practices that minimise the disruption of the soil's structure, composition and natural biodiversity. CA is often confused with other terms including no-till/zero-till, conservation tillage/ minimum tillage/reduced tillage, direct planting/direct drilling, direct seeding, organic farming. Though these terms are frequently used in CA practices, there is distinct difference among these terms.

No-till/ Zero till: No-till (NT) and zero till (ZT) are technical components used in conservation agriculture that simply involve the absence of tillage/ploughing operations on the soil. Crops are planted directly into a seedbed not tilled after harvesting the previous crop. It is not necessary that everyone utilising no-till technologies adopts other important components of CA. Another major difference is that NT or ZT do not necessarily leave residue mulch, which is an important practice in CA. Without the residue mulch many of the benefits of CA are lost or decreased.

Conservation tillage/ minimum tillage/ reduced tillage: These are tillage operations that leave at least 30% of the soil surface covered by plant residues in order to increase water infiltration and cut down on soil erosion and runoff. Conservation tillage is an intermediate form of CA since it keeps some soil cover as residue from the previous crop. But some tillage is usually done.

Direct planting/ direct drilling: In direct planting or direct drilling special equipment (e.g. NT drill) are used to plant seeds directly into crop residues left on the soil surface without preparing a seedbed beforehand.

Direct seeding: This term is usually associated with growing rice or any other cereal crop without producing seedlings that are then transplanted into the main field. However, the term direct seeding is used in CA systems as synonymous with no-till farming, zero tillage, no-tillage, direct drilling, etc.

Organic farming: Organic agriculture does not permit the use of synthetic chemicals to produce plant and animal products, relying instead on the management of soil organic matter and biological processes. But organic farming uses the principles of CA to some extent and one objective similar to CA is to maintain and improve soil health. Unlike organic farming, CA does allow farmers to apply synthetic chemical fertilisers, fungicides, pesticides and herbicides. These are used to control pests and weeds, particularly during the early transition years. As soil physical, chemical and biological health improves over time; the use of agrochemicals may be significantly reduced or, in some cases, phased out entirely.

ADVANTAGES OF CONSERVATION AGRICULTURE

Conservation agriculture requires a new way of thinking about agricultural production in order to understand how one could possibly attain higher yields with less labour, less water and fewer chemical inputs. In spite of these challenges, conservation agriculture is spreading to farmers throughout the world as its benefits become more widely recognised by farmers, researchers, scientists and extensionists alike.

Economic considerations

Farmers using conservation agriculture technologies typically report higher yields with less water, fertiliser and labour inputs, thereby resulting in higher overall farm profits. CA increases the productivity of land, labour, water, nutrients and soil biota.

Land: Conservation agriculture improves soil structure and protects the soil against erosion and nutrient losses by maintaining a permanent soil cover and minimising soil disturbance. Furthermore, CA practices enhance soil organic matter levels and nutrient availability by utilising the previous crop residues or growing green manure/ cover crops and keeping these residues as surface mulch rather than burning. Thus, arable land under CA is more productive for much longer periods of time.

Labour: Because land under no-till is not cleared before planting and involves less weeding and pest problems following the establishment of permanent soil cover/crop rotations. Much of the reduced labour comes from the absence of tillage operations under CA, which use up valuable labour days during the planting season. The labour and time saving under CA enables farm families to invest their time in more rewarding activities such as processing of raw produce or other value adding activities, enhancing their farm income than the crop production would do.

Water: Conservation agriculture requires significantly less water use due to increased infiltration and enhanced water holding capacity from crop residues left on the soil surface. Mulches also protect the soil surface from extreme temperatures and greatly reduce surface evaporation, which is particularly important in tropical and subtropical climates. CA reduces the vulnerability of crops against unexpected extreme climatic events. It reduces crop water requirements by 30%, making better use of soil water and facilitating deeper rooting of crops, so that drought periods can be tolerated better.

Nutrients: Soil nutrient supplies and cycling are enhanced by the biochemical decomposition of organic crop residues at the soil surface that are also vital for feeding the soil microbes. While much of the nitrogen needs of primary food crops can be achieved by planting nitrogen-fixing legume species, other plant essential nutrients often must be supplemented by additional chemical and/or organic fertiliser inputs. In general, soil fertility is built up over time under conservation agriculture, and fewer fertiliser amendments are required to achieve optimal yields over time. Legumes in CA rotations provide increased *in situ* availability of nitrogen, thus diminishing the need for large amounts of applied nitrogenous fertilisers.

Soil biota: Insect pests and other disease causing organisms are held in check by an abundant and diverse community of beneficial soil organisms, including predatory wasps, spiders, nematodes, springtails, mites and beneficial bacteria and fungi, among other species. Furthermore, the burrowing activity of earthworms and other fauna create tiny channels or pores in the soil that facilitate the exchange of water and gases and loosen the soil for enhanced root penetration.

Energy savings: The benefits of lower costs for fuel and/or draft animals are straightforward. Farmers who use tractors to plough are able to reduce their fuel use for ploughing by two-thirds. Enhanced soil structure makes soil more easily trafficked by farm machinery for the remaining operations, further reducing fuel costs, as well as equipment wear and tear.

Time savings: Time savings allow farmers to plant earlier. Depending on the agricultural environment, this can improve the odds that the crop will mature and be ready to harvest before the onset of late-season drought. It may allow farmers to squeeze in a third cropping per year in what had been a system producing only two crops per year, adding perhaps a crop of high-value vegetables or another cash crop.

Environmental considerations

Intensive agriculture is often considered to be a damaging factor to ecosystems in view of the negative impact it has on rural landscape, soil resources, water quality and biodiversity. The impact of agriculture on ecosystems through erosion, pollution of waterbodies and dust and smoke emissions is also felt outside the actual agricultural area. However, conservation agriculture represents an environment-friendly set of technologies. Because it uses resources more efficiently than conventional agriculture, these resources become available for other uses, including conserving them for future generations. The significant reduction in fossil fuel use under no-till agriculture results in fewer greenhouse gases being emitted into the atmosphere and cleaner air in general. A reduced application of agrochemicals under CA also significantly lessens pollution levels in air, soil and water. Society gains from CA, regardless of farm size by diminished erosion and runoff, less downstream sedimentation and flood damage to infrastructure, better recharge of groundwater, more regular stream flow throughout the year, resulting in a more assured community water supply and quality.

While the economic advantages of conservation agriculture are often realised with the first cropping, the environmental advantages emerge more slowly. Most environmental advantages require majority adoption in a community, subsequently maintained year after year, to avoid being negated by environmental pollution flowing from conventional farms.

Natural resource base: CA reverses soil degradation processes and builds up soil fertility and productive capacity. It facilitates a better infiltration of rainwater, enabling the recharge of groundwater resources while at the same time reducing the pollution of waterbodies through reduced erosion and leaching. CA conserves and enhances natural resources while maintaining and sustainably increasing production levels.

Air: A rising problem across regions is rural air pollution caused by farmers burning crop stubble and other residues. If conservation agriculture becomes prevalent enough in a farm area, leaving most residues in the field for soil protection, fires and the haze they cause can become a thing of the past. Undisturbed soil is less prone to rise as dust, which is another rural health hazard when ploughed soil is left bare between crops.

Surface water: Improved water infiltration into healthy soil and reduced water erosion lessens the nuisance and threats to public and environmental health that occur when agricultural runoff causes excessive silting of rivers, reservoirs, lakes and micro catchments. As conservation agriculture eliminates soil erosion and greatly reduces runoff, widespread adoption in a farm area can be expected to enhance the cleanliness and biodiversity of surface waterbodies, protecting their value as fisheries, sources of drinking water and venues for recreation.

Landscape: The avoidance of ploughing in CA facilitates the introduction of trees and hedgerows into the agricultural landscape in a closer vicinity of field crops than under tillage based agriculture. The greater diversity in the crop rotations also contributes to a more diverse and pest-free landscape. It also increases biodiversity in the agricultural production systems.

Climate change: CA can contribute to reduced greenhouse gas emissions from agricultural crop production through reduced fuel use, better aeration of soils that reduces nitrous oxide emissions and, in no-till non-flooded rice, methane emissions. In addition it binds atmospheric carbon in the soil in the form of soil organic matter. With this, CA helps to mitigate climate change.

Under CA systems, it is possible to harness many of the above ecosystem services mainly because the ecosystem functions that generate these services are enhanced and protected, so that agriculture is not in competition with nature but works in harmony with it. However, the environmental advantages of conservation agriculture emerge more slowly than the economic advantages which are often realised with the first cropping. Most environmental advantages require majority adoption in a community, subsequently maintained year after year, to avoid being negated by environmental pollution flowing from conventional farms.

Sociocultural considerations

The rural community ultimately benefits from better nutrition, overall health and less pressure on curative health services.

Equity considerations: Conservation agriculture has the benefit of being accessible to many small-scale farmers who need to obtain the highest possible yields with limited land area and inputs. Perhaps the biggest obstacle for the technology spreading to more small-scale farmers worldwide has been limited access in certain areas to certain specialised equipment and machinery, such as no-till planters. This problem can be remedied by available service providers renting equipment or undertaking conservation agriculture operations for farmers who would not otherwise have access to the needed equipment. Formulating policies that promote adoption of CA are also needed. As more and more small-farmers gain access to CA technologies, the system becomes much more 'scale neutral'.

Active role for farmers: As with any new agricultural technology, CA methods are most effective when used with skilful management and careful consideration of the many agroecolgical factors affecting production on any given farm or field. Rather than being a fixed technology to be adopted in blueprint-like fashion, CA should be seen as a set of sound agricultural principles and practices that can be applied either individually or together, based on resource availability and other factors. For this reason, farmers are encouraged to experiment with the methods and to evaluate the results for themselves- not just to 'adopt' CA technologies. Selecting among different cover crop species, for example, needs to be determined in relation to particular agroecological conditions of the farm, including soil type, climate, topography as well as seed availability and what the primary function of the GMCC will be. Similarly, planting distances, irrigation requirements and the use of agrochemicals to control weeds and pests among other considerations, must be decided based on what the farmer needs as well as the availability of these and other resources.

CONSTRAINTS FOR ADOPTION OF CONSERVATION AGRICULTURE

Socioeconomic and psychological factors are significant in the decision making process to adopt the CA. Farmers who have a strong conservation ethic, for example, may be willing to accept reduced profits in return for feeling that they have contributed to welfare of future generations. Environmentally concerned farmers may also be willing to invest in practices that will enhance the environmental quality of their lands and water resources, as well as enhance the economic value of land when it is passed on to next generation.

The most important factor in the adoption of CA is overcoming the bias or mindset about tillage. Convincing the farmers that successful cultivation is possible even with reduced tillage or without tillage is a major hurdle in promoting CA on a large scale (Hobbs and Govaerts, 2010). CA is now, considered a route to sustainable agriculture. The constraints which hampered the adoption of CA have been identified by various researchers (Singh and Meena, 2012; Meena and Singh, 2013) which can be listed under four broad categories; technical, extension, financial and production. These prevailing constraints strictly control the adoption behaviour at farmers' level.

Technical constraints

- Lack of appropriate seeders especially for small and marginal farmers
- Non-availability of quality drill
- Lack of regular monitoring of machines
- Lack of training/ capacity building
- Non-availability of spare parts in local market
- Lack of local manufacturers of machines

Extension constraints

- Lack of knowledge about the potential of CA to agriculture leaders and farmers
- Lack of extension support from state extension agencies
- Lack of extension literature
- Lack of attention by mass media
- Lack of knowledge of extension agencies
- Inadequate extension facility at disposal of input agencies
- Lack of cooperation from fellow farmers

Financial constraints

- Lack of credit facilities
- Lack of money to buy new machines and inputs
- No subsidy on machines
- High cost of drill

Production constraints

- Wide spread use of crop residues for livestock feed and fuel
- Scarcity of crop residues due to less biomass production of crops under rainfed situations
- Burning of crop residues by the farmers for timely sowing of the next crop
- Non-availability of skilled and scientific manpower
- Free grazing of animals that compact the soil and remove all the scil cover, leaving it open to erosion
- Difficulty in getting seeds of cover crops and herbicides

STRATEGY FOR IMPLEMENTATION OF CONSERVATION AGRICULTURE INNOVATIONS

After evaluating determinants and constraints in adoption of CA, it is the prime function of extension workers to diffuse new ideas and practices among farmers. It is their task to expedite the process of getting ideas from their sources of origin to those who can adopt or use them. The following strategies have been suggested by Meena and Singh (2012).

Implementing situations and prevailing constraints

Factors limiting the agricultural production should be rectified before the full benefits from implementation of CA can be realised. This might refer to technical factors, such as soil compaction, insufficient drainage, soil chemical properties, as well as socioeconomic factors such as availability of adequate technology, investment capital, land use rights, livestock pressure, customary practices or access to markets. These will have to be addressed in order to establish CA in a sustainable manner.

Transforming the agricultural system

The transition phase usually takes about two years; however, the full benefits of this system often become visible only after five years. In CA, mechanical tillage is replaced by biological tillage and soil fertility is essentially managed through soil cover management, crop rotations and weed management. Fertilisers, water harvesting technologies and irrigation can complement CA, and minimum tillage might be necessary in some cases particularly during the transition.

Changing the attitude

Changing the attitude or mentality of the farming community is a difficult task but it paves the way to success for task like implementing the conservation agriculture. Proper knowledge about the concepts of conservation agriculture is also inevitable like soil is a habitat for roots and soil organisms, any damage to this habitat endangers soil fertility and leads to land degradation, soil fauna creates a stable soil structure, etc.

Encouragement, support and capacity building

Promotion of CA should be done simultaneously through policies, education, research and extension institutions in the field. Adoption by farmers is supported most effectively through farmers' groups, study tours, networks and NGOs. Research and extension institutions and the private sector have a major role in providing farmers with appropriate and affordable technologies.

Policies and incentives

Conservation agriculture implies a radical change from traditional agriculture. There is need for policy analysis to understand how CA technologies integrate with other technologies, and how policy instruments and institutional arrangements promote or deter CA (Raina et al., 2005). Thus, policies should focus on access to market, credit and input supplies, and rural infrastructures. Policies should support the development of farmers' groups. Incentives should encourage diversification and CA practices, especially during the transition phase. Inadequate policies and subsidies that support conventional practices might constrain CA adoption. Land use and customary rights must also be taken into account and eventually adapted to favour CA adoption by farmers and rural communities.

Support from international organisations

Food and Agricultural Organisation is promoting the adoption of the CA concept at policy level as well as stimulating farmer-based movements and collaboration between the research sector and farmer groups. Due to its positive effects on food security, biodiversity, land and water resources, carbon sequestration and sustainable development, CA is a major opportunity to implement the international conventions on combating desertification, on biodiversity and on climate change.

CHALLENGES IN CONSERVATION AGRICULTURE

Understanding the system

Conservation agriculture systems are much more complex than conventional systems. Site specific knowledge has been the main limitation to the spread of CA system (Derpsch et al., 2010). Managing these systems efficiently will be highly demanding in terms of understanding of basic processes and component interactions, which determine the whole system performance. Thus, the need is to recognise conservation agriculture as a system and develop management strategies.

Building a system perspective

A system perspective is built working in partnership with farmers. A core group of scientists, farmers, extension workers and other stakeholders working in partnership mode will therefore be critical in developing and promoting new technologies. This is somewhat different than in conventional agricultural R&D, the system is to set research priorities and allocate resources within a framework, and little attention is given to build relationships and seek linkages with partners working in complementary fields (Bhan and Behera, 2014).

Technological challenges

While the basic principles which form the foundation of conservation agriculture practices, that is, no tillage and surface managed crop residues are well understood, adoption of these practices under varying farming situations is the key challenge. These challenges relate to development, standardisation and adoption of farm machinery for seeding with minimum soil disturbance, developing crop harvesting and management systems (Bhan, 2011).

Site specificity

Adapting strategies for conservation agriculture systems will be highly site specific, yet learning across the sites will be a powerful way in understanding why certain technologies or practices are effective in a set of situations and not effective in another set. This learning process will accelerate building a knowledge base for sustainable resource management (Bhan and Behera, 2014).

Long-term research perspective

Conservation agriculture practices, e.g. no-tillage and surface maintained crop residues result in resource improvement only gradually, and benefits come about only with time. In conservation agriculture, benefits in terms of yield increase may not come in the early years of practices. Understanding the dynamics of changes and interactions among physical, chemical and biological processes is basic to developing improved soil-water and nutrient management strategies (Abrol and Sangar, 2006). Therefore, research in conservation agriculture must have longer term perspectives and should focus on several innovative features to address the challenge.

OPERATIONAL FACTORS FOR SUCCESS OF CONSERVATION AGRICULTURE

Maintaining permanent or semi-permanent soil cover

Plants are left growing or killed and their residues left to decompose *in situ*. The primary function of this is to protect organic matter-enriched topsoil against chemical and physical weathering. Plant residues intercept energy from falling raindrops, provide a barrier from strong winds, and moderate temperatures, improving water infiltration and decreasing surface evaporation from sunlight. Surface cover also favours enhanced levels of biological activity by providing food for soil microbes, especially in tropical and subtropical areas.

Minimum soil disturbance

No-till does not involve any loosening of the soil except for a very small area immediately surrounding where the seed is planted. This lack of soil disturbance serves to maintain overall soil structure, including aggregate stability and porosity, both of which promote the exchange of water and gases and provide habitat to an abundant and diverse population of soil biota.

Regular crop rotations

Well-balanced crop rotations can neutralise many of the pest and disease problems associated with not tilling the soil, including the proliferation of insect pests and other harmful bacteria, viruses and fungi, by increasing the diversity and abundance of beneficial soil biota that can help keep pest and disease problems in check. Rotating crops also interrupts the life cycle of many weeds, thereby leading to a reduction in overall weed growth.

Utilisation of green manures/cover crops

Cover crops are grown specifically to help maintain soil fertility and productivity by decreasing erosion and/or by adding fresh plant residues to the soil. Leguminous cover crops offer the added advantage of being able to fix nitrogen from the atmosphere and add it to the soil, thereby increasing overall nitrogen availability for other crops. Cover crops are usually mowed, sprayed with chemical herbicides or otherwise killed before or during soil preparation for the next economic crop. It is generally recommended to leave a week or two between the killing of the cover crop and the planting of a primary crop in order to allow for some decomposition to occur as well as to lessen the effects of nitrogen immobilisation and allelopathic effects.

No burning of crop residues

Since crop residues are the principal element of permanent soil cover, they must never be burned or otherwise removed from the soil surface. Rather, plant residues are left on the soil surface in order to protect organic matter enriched topsoil from erosion while also adding fresh organic matter upon decomposition. Burning not only creates significant air pollution but also dramatically increases mineralisation rates, leading to the rapid depletion of soil organic matter and nutrients from the soil. However, in some situations farmers need to think of the tradeoff between removing residues to feed their animals and leaving them to feed the soil. A win-win situation would do both and as yields and biomass increase over time, both become more feasible.

Integrated disease and pest management

Conservation agriculture depends heavily on enhanced biological activity to help control insect pests and other disease causing soil organisms. Integrated pest management entails the judicious use of crop rotations and other beneficial plant associations as well as chemical pesticides, herbicides and fungicides to control insect pest and disease problems. Over time, the enhanced biological activity and abundance brought on by no-till and other CA technologies results in decreased applications of agrochemicals.

Reduction in fossil fuel use and greenhouse gas emissions

Compared to conventional tillage, which often requires 4-8 tractor passes in a typical growing season, no-till significantly reduces the use of tractors and other heavy farm machinery and thus diesel. Furthermore, the increased levels of soil organic matter and plant-available nitrogen typically found in CA soils greatly reduces the need for applying large amounts of chemical fertilisers, many of which require significant fossil fuel energy to process. Thus, overall fossil fuel use and greenhouse gas emissions are greatly reduced.

Controlled / limited human and mechanical traffic over agricultural soils

The number of tractor passes over a given field is significantly reduced under CA, as compared to conventional tillage systems. However, increased bulk densities have been reported under CA, though this can be corrected by limiting the use of heavy farm machinery when soils are wet and most prone to compaction and/or by converting to a permanent raised bed system.

Timely availability of inputs and equipment

The farmers should have access to the right equipment. Timely availability of right seeds for the cover crops should be ascertained. Other inputs like fertiliser and/or manure, herbicides, etc. should be available timely. Conservation agriculture generally saves work. However, it may require more work in the first year.

Timely field operations

A key principle of conservation agriculture is the need for timely implementation. The land should be prepared well before the rain starts and planting soon after an effective rainfall event. Weeding should be at appropriate times and intervals. The pests and diseases should be effectively controlled before their spread too widely.

Precise operations

Precise measurements of row and plant spacing, evenness of depth and placement of soil amendments and covering of seed are also important. Planting should be done on the same lines each season. A key benefit is that compaction of the soil by feet, hooves and wheels will then only occur in the inter-row spaces and not over the crop lines. Residual fertility builds up in the rows and the crop roots of each consecutive crop provide organic matter.

Efficient use of inputs

Precise application of soil amendments leads to very little wastage because only the crops get benefitted, and not the surrounding soil and weeds. This gives higher yields and huge savings on costly inputs. Inputs like time, energy, draught power, etc. are used more effectively under conservation agriculture. Since land preparation can start soon after harvesting up until the rains start, labour inputs can be spread out more evenly over the year.

POLICY AND INSTITUTIONAL SUPPORT FOR CONSERVATION AGRICULTURE

In India and elsewhere much research work on conservation agriculture has been conducted. However, its percolation to farmers is very limited. So there is a need to think about the problems faced at the implementing level and devise a strategy involving all who are concerned. One of the reasons for poor percolation of the technology to the farmers is the mindset about tillage by the majority of farmers (Hobbs and Govaerts, 2010). Even the success rate is reported to be at low level where CA is practised. FAO (2001) has reported that this is partly because policy environments are not favourable. Under such situations, farmers' participatory on-farm research to evaluate/refine the technology in initial years followed by large scale demonstration in subsequent years is needed. In India, efforts are being initiated through a network research project for on-farm evaluation and demonstration of CA technology for its promotion. The following are some of the important policy considerations identified by Bhan and Behera (2014) for promotion of conservation agriculture.

Shift in focus from food security to livelihood security

The 'food security' policy based on cereal production must now replace a well-articulated policy goal for livelihood security. This will help the diversification of dominant rice-wheat cropping systems in the Indo-Gangetic Plains, the cultivation of which in conventional tillage practice has overexploited the natural resources in the region. The nature of cropping patterns and the extent of crop diversification are influenced by policy interventions. The government policies that directly or indirectly affect crop diversification include pricing policy, tax and tariff policies, trade policies and policies on public expenditure and agrarian reforms (Behera et al., 2007).

Development of CA equipments for diversified cropping systems

The success of conservation agriculture depends upon the development and standardisation of equipments for seeding, fertiliser placement and harvesting with minimum soil disturbance in residue management for different edaphic conditions. Thus, ensuring quality and availability of equipment through appropriate incentives will be important. In these situations, the subsidy support from national or local government to firms for developing low cost machines will help in the promotion of CA technologies.

Changes in the plant growing micro-environment due to CA technologies

These include changes in moisture regimes, root environment, emergence of new pathogens and shift in insect-pest scenario, etc. The requirement of plant types suited to the new environment, and to meet specific mechanisation-needs could be different. Therefore, there is a need to develop complementary crop improvement programmes, aimed at developing cultivars which are better suitable to new systems. Farmers' participatory research would appear promising for identifying and developing crop varieties suiting to a particular environment or locations.

Policy support for capacity building by organising training on CA

Availability of trained human resources at ground level is one of the major limiting factors in adoption of CA. Efforts to adequately train all new and existing agricultural extension personnel on CA should be made in relevant departments. In the long-term, CA should be included in curricula from primary school to university levels, including agricultural colleges. Inclusion of conservation and sustainability concepts in the course curricula with a suitable blend of biophysical and social sciences would be important for sustainable resource management.

Institutionalisation of CA

Institutionalising CA into relevant government ministries and departments and regional institutions is required for sustainability of the technology. Local, national and regional policy and decision makers could spearhead and support the formulation and development of strategies and mechanisms for scaling up the technology. CA could be integrated into interventions such as seed, fertiliser and tillage and draft power support programmes as a way of further enhancing productivity.

Support for the adaptation and validation of CA technologies in local environments

Adaptive research is required to tailor CA principles and practices to local conditions. This should be done in collaboration with local communities and other stakeholders. Issues that should be addressed include crop species, selection and management of crop and cover crop and rotations, maintenance of soil cover and CA equipment. The resource poor and small holder farmers do not have economic access to new seeds, herbicides and seeding machineries, etc. (Sharma et al., 2012). This calls for policy frame work to make easily available critical inputs.

Generation of resource database and linkages

There is a need for generating a good resource database on conservation agriculture practices and technologies. The agencies involved in conservation agriculture R&D should complement each others' work. Much attention should be given to build relationships and seek linkages with partners working in the complementary fields. Besides resources, systematic monitoring of the socioeconomic, environmental and institutional changes should become an integral part of the major projects on CA.

Rewards for environmental services and fines for faulty practices

Adopters of CA improve the environment through carbon sequestration, prevention of soil erosion or the encouragement of groundwater recharge. It provides ecosystem services, thus, farmers could be rewarded for such services, which have a great impact on the quality of life for all. Similarly, there should be a strong policy intervention for prohibiting unscientific and environment polluting practice like residue burning by imposing a fine.

Building partnership

CA systems are very complex and their efficient management needs understanding of basic processes and component interactions which determine the system performance. A system perspective is the best to build working in partnership with farmers, who are at the core of farming systems. Scientists, farmers, extension agents, policy makers and other stakeholders in the private sector working in partnership mode will be important in developing and promoting new technologies. Instead of using a top-down approach where the extension agent places CA demonstrations in farmer fields and expects the farmer to adopt, a more participatory system is required where the farmers are enabled through provision of equipment and training to experiment with the technology and find out for themselves whether it works and what fine-tuning is needed to make it successful on their land.

Credit and subsidy

There is a need to provide credit to farmers to buy the equipment, machinery, and inputs through banks and credit agencies at reasonable interest rates. The government should also provide a subsidy for the purchase of such equipment by farmers.

REFERENCES

Abrol, I. P. and Sangar, S. 2006. Sustaining Indian agriculture - conservation agriculture the way forward. *Current Science* **91**(8): 1020-2015.

Behera, U. K., Amgain, L. P. and Sharma, A. R. 2010. Conservation agriculture: principles, practices and environmental benefits. In: Behera, U. K., Das, T. K. and Sharma, A. R. (eds.). Conservation Agriculture. Division of Agronomy, Indian Agricultural Research Institute, New Delhi. pp. 28-41.

Behera, U. K., Sharma, A. R. and Mahapatra, I. C. 2007. Crop diversification for efficient resource management in India: Problems, prospects and policy. *Journal of Sustainable Agriculture* **30**(3): 97-217.

Bhan, S. 2011. Conservation agriculture - Problems and prospects in Indian context. In: Sharma, A. R. and Behera, U. K. (eds.). Resource Conserving Techniques in Crop Production. Scientific Publishers, Jodhpur, India. pp. 16-23.

Bhan, S. and Behera, U. K. 2014. Conservation agriculture in India – problems, prospects and policy issues. *International Soil and Water Conservation Research* **2**(4): 1-12.

Derpsch, R., Friedrich, T., Kassam, A. and Li, H. W. 2010. Current status of adoption of notill farming in the world and some of its main benefits. *International Journal of Agricultural and Biological Engineering* **3**: 1-25.

FAO. 2001. Conservation agriculture - case studies in Latin America and Africa. FAO Soils Bulletin No. 78. FAO, Rome.

FAO. 2007. FAO-SARD Initiative. http://www.fao.org/sard/initiative

FAO. 2010. What is Conservation Agriculture? http://www.fao.org/ag/ca/1a.html

Faulkner, E. H. 1945. Ploughman's Folly. Michael Joseph, London. 142 pp.

Fukuoka, M. 1975. One Straw Revolution, Rodale Press, English translation of shizen noho wara ippeon no kakumei, Hakujusha Co., Tokyo. 138 pp.

Ghosh, P. K., Das, A., Saha, R., Kharkrang, E., Tripathy, A. K., Munda, G. C., and Ngachan, S.V. 2010. Conservation agriculture towards achieving food security in north east India. *Current Science* **99**(7): 915-921.

Hobbs, P. R., and Govaerts, B. 2010. How conservation agriculture can contribute to buffering climate change. In: Reynolds, M. P. (ed.). Climate Change and Crop Production. CAB International. pp. 177-199.

Hobbs, P. R., Sayre, K. and Gupta, R. 2008. The role of conservation agriculture in sustainable agriculture. *Philosophical transactions of the Royal Society of London Series B Biological Sciences* **363**(1491): 543-455.

Kassam, A., and Friedrich, T. 2009. Perspectives on nutrient management in conservation agriculture. Invited paper, IV World Congress on Conservation Agriculture, 4-7 February 2009, New Delhi, India.

Kassam, A., Friedrich, T. and Derpsch, R. 2010. Conservation Agriculture in the 21st Century: A Paradigm of Sustainable Agriculture. European Congress on Conservation Agriculture, 4-6 October 2010, Madrid, Spain.

Kassam, A., Friedrich, T. and Derpsch, R. 2019. Global spread of Conservation Agriculture. *International Journal of Environmental Studies* **76**(1): 29-51.

Kassam, A., Friedrich, T., Shaxson, F. and Pretty, J. 2018. The spread of Conservation Agriculture: Justification, sustainability and uptake. *International Journal of Agricultural Sustainability* **7**(4): 292-320.

Lal, R. 2013. Climate-resilient agriculture and soil Organic Carbon. *Indian Journal of Agronomy* **58**(4): 440-450.

Meena, M. S. and Singh, K. M. 2012. Decision process innovations, constraints and strategies for adoption of conservation agriculture. http://dx.doi.org/10.2139/ssrn.2088710

Meena, M. S. and Singh, K. M. 2013. Information and communication technologies for sustainable natural resource management. http://dx.doi.org/10.2139/ssrn.2244751

Pretty, J. 2008. Agricultural sustainability: concepts, principles and evidence. Philosophical Transactions of the Royal Society of London B **363**(1491): 447-466.

Pretty, J., Toulmin, C. and Williams, S. 2011. Sustainable intensification in African agriculture. *International Journal of Agricultural Sustainability* **9**(1): 5-24.

Raina, R. S., Sulaiman, R., Hall, A. J. and Sangar, S. 2005. Policy and institutional requirements for transition to conservation agriculture: An innovation systems perspective. In: Abrol, I. P., Gupta, R. K. and Mallik, R. K. (eds.). Conservation Agriculture - Status and Prospects. Centre for Advancement of Sustainable Agriculture, New Delhi. pp. 224-232.

Sharma, A. R., Jat, M. L., Saharawat, Y. S., Singh, V. P. and Singh, R. 2012. Conservation agriculture for improving productivity and resource-use efficiency: prospects and research needs in Indian context. *Indian Journal of Agronomy* **57** (IAC Special Issue): 131-140.

Singh, K. M. and Meena, M. S. 2012. Conservation Agriculture: Economic Perspective and Future Challenges. Available at: http://dx.doi.org/10.2139/ssrn.2026336

Wassmann, R., Jagadish, S. V. K., Sumfleth, K., Pathak, H., Howell, G., Ismail, A., Serraj, R., Redona, E., Singh, R. K., and Heuer, S. 2009. Regional vulnerability of climate change impacts on Asian rice production and scope for adaptation. *Advances in Agronomy* **102**: 91-133.

5

Soil Management for Sustainable Agriculture

Soil is a biologically active porous medium that has developed in the uppermost layer of the earth's crust which serves as a natural medium for the growth of plants. It is one of the principal substrata of life on earth, serving as a reservoir of water and nutrients, and as a participant in the cycling of carbon and other elements through the global ecosystem. It has evolved through weathering processes driven by biological, climatic, geologic, and topographic influences. Soil is a mixture of minerals, dead and living organisms, air and water. These four ingredients react with one another, making soil one of most dynamic and important natural resources of earth.

Soil is an extremely complex ecosystem and a highly valuable resource from an ecocentric and anthropocentric perspective. Soil is one of the most essential and strategic resources, due to its following crucial functions (Yang et al., 2020).

1. Provision of food, fibre, and fuel
2. Decomposition of organic matter (e.g., dead plant and animal material)
3. Recycling of essential nutrients
4. Detoxification of organic contaminants
5. Carbon sequestration
6. Regulation of water quality and supply
7. Habitat provision for countless animals and microorganisms
8. Source of raw materials (clay, sand, gravel)

Thus, healthy soils are a basic prerequisite to meeting varied needs for food, biomass (energy), fibre, fodder, and other products, and to ensuring the provision of essential ecosystem services in all regions of the world. Soil health is the foundation of productive farming practices (Stivers, 2017). However, humankind is facing unprecedented pressures on soil resources. Soil is currently being rapidly degraded due to a range of invasive anthropic activities in intensive agriculture, with concomitant adverse effects on human and ecosystem health. This is concerning as soil is a non-renewable resource at a human temporal scale, i.e., soil loss and degradation are not recoverable within a human lifespan.

SUSTAINABLE SOIL MANAGEMENT

Soils are an essential and non-renewable natural resource hosting goods and services vital to ecosystems and human life. Soils are fundamental for producing crops, feed, fibre, fuel, and they filter and clean tens of thousands of cubic kilometres of water each year. As a major storehouse for carbon, soils also help regulate emissions of carbon dioxide and other greenhouse gases, which is fundamental for regulating climate. Sustainable soil management is an integral part of sustainable land management, as well as a basis for addressing poverty eradication, agricultural and rural development, promoting food security and improving nutrition (FAO, 2017). Soil is the world's largest terrestrial pool of carbon (FAO, 2004), and approximately 95% of global food is produced in soil (FAO, 2015a). Sustainable soil management (SSM) is a valuable tool for climate change adaptation and a pathway for safeguarding key ecosystem services and biodiversity.

Due to the incalculable value soils provide to society through ecosystem services, SSM ensures a high return on investment by supporting and increasing these services. Widespread adoption of SSM practices generates multiple socioeconomic benefits, especially for smallholder farmers and large scale agricultural producers worldwide whose livelihoods directly depend on their soil resources.

Definition of sustainable soil management

Soil management is sustainable if the supporting, provisioning, regulating, and cultural services provided by soil are maintained or enhanced without significantly impairing either the soil functions that enable those services or biodiversity. The balance between the supporting and provisioning services for plant production and the regulating services the soil provides for water quality and availability and for atmospheric greenhouse gas composition is a particular concern (FAO, 2015b).

The types of ecosystem services and the soil functions referred to in this definition can be elaborated as follows (FAO, 2017).

1. Supporting services include primary production, nutrient cycling and soil formation
2. Provisioning services comprise the supply of food, fibre, fuel, timber and water; raw earth material; surface stability; habitat and genetic resources
3. Regulating services imply the regulation of aspects such as water supply and quality, carbon sequestration, climate regulation, control of floods and erosion
4. Cultural services denote the aesthetic and cultural benefits derived from soil use

Sustainable soil management is associated with the following characteristics (FAO, 2017).

1. Minimal rates of soil erosion by water and wind
2. The soil structure is not degraded (e.g., soil compaction) and provides a stable physical context for movement of air, water, and heat, as well as root growth
3. Sufficient surface cover (e.g., from growing plants, plant residues, etc.) is present to protect the soil
4. The store of soil organic matter is stable or increasing and ideally close to the optimal level for the local environment
5. Availability and flows of nutrients are appropriate to maintain or improve soil fertility and productivity, and to reduce their losses to the environment
6. Soil salinisation, sodification and alkalinisation are minimal
7. Water (e.g., from precipitation and supplementary water sources such as irrigation) is efficiently infiltrated and stored to meet the requirements of plants and ensure the drainage of any excess
8. Contaminants are below toxic levels, i.e., those which would cause harm to plants, animals, humans and the environment
9. Soil biodiversity provides a full range of biological functions
10. The soil management systems for producing food, feed, fuel, timber, and fibre rely on optimised and safe use of inputs
11. Soil sealing is minimised through responsible land use planning

Two factors underlie the ability of the earth to support the ever-increasing human population (FAO, 1994).

1. The resilience of soils in response to increasing demands made on them
2. The increasing knowledge of farmers and scientists about how to manage soils productively and sustainability

Many attempts to introduce continuous arable cropping systems into the semiarid, subhumid and humid tropics have failed due to inadequate maintenance of the nutritional status of the soil. Other causes of failure include soil degradation in one or more of the following ways.

1. Physical degradation due to erosion, compaction and crusting
2. Chemical degradation associated with nutrient mining and acidification
3. Biological degradation associated with loss of organic matter
4. Deterioration of drainage conditions causing waterlogging or salinisation

While arable cropping has often been found to be non-sustainable, production of tree crops in the humid and subhumid tropics has presented relatively few problems with respect to sustainability. In the semiarid areas, animal production has sometimes been possible on a sustainable basis. Problems of sustainability in animal production systems have arisen as stock numbers have increased with no improvement in the carrying capacity of the land. In periods of drought or water shortage, overgrazing can lead to denudation of vegetation and, as a result, exposure of the soil to wind erosion. Thus, for arable cropping and pasture production, there is a great need to establish economically viable and socially acceptable methods of soil management which are both productive and sustainable.

Good soil management has always required that the soil be used in such a way that its productivity is maintained or preferably, enhanced. This requires that the chemical and physical conditions of the soil do not become less suitable for plant growth than when cultivation commences. Cultivation normally means that the soil will, in fact, deteriorate due both to nutrient removal when harvesting crops, and to physical damage to the soil structure. What is essential is that the deterioration is reversible, by chemical additions to the soil, mechanical manipulation, or natural processes of fertility restoration under pasture or trees. This implies that the soil must be resilient, i.e., after being subjected to the stresses involved in crop production, it must have the ability to return to its former condition, or an improved condition (Greenland and Szabolcs, 1994).

Sustainability and non-sustainability of current soil management systems

The distribution of production systems is determined by soil and climate conditions, and by social and economic factors. As one moves from drier to wetter areas, pastures and animals become less important, and trees more so. Population density is greatest where soils are most fertile, and management systems the most intense. Most production systems have evolved so that they are sustainable in terms of the environmental conditions prevailing at the time including the level of demographic pressure. However, over the years demographic pressure has increased dramatically and accordingly sustainability in relation to agriculture and soil management must be defined to include the need for increases in demand to be met (FAO, 1994).

FAO has defined a sustainable agricultural system as one which involves the management and conservation of the natural resource base and the orientation of technological and institutional change in such a manner as to ensure the attainment and continued satisfaction of human needs for present and future generations. Such sustainable development conserves land, water, plant and animal genetic resources, and is economically viable and socially acceptable.

There are many other definitions of sustainable agriculture and land use. From all these definitions one point is clear that sustainable soil management must not degrade the soil or contaminate the environment, while providing the desired support to the production of food, fuel, fibre and construction materials.

If a soil is to sustain the production of crops it must provide the nutrient requirements of the crop. It must provide a physical medium

1. In which the plant roots can grow adequately so that water and nutrients can be absorbed;
2. Which stores sufficient water for the crop; and
3. Which allows water to enter and move in the soil to maintain the water supply as it is transpired by the crop and evaporates from the soil.

The soil must also provide a medium in which soil organisms are able to

1. Decompose organic materials, releasing nutrients to the plants;
2. Assist the transport of nutrients to plant roots;
3. Compete successfully with pathogens which might otherwise infect roots and damage the plants; and
4. Form the soil organic compounds which will have a favourable effect on other soil properties.

A few soils contain sufficient nutrients to allow them to be mined for many years without significant loss of yield, but the majority of soils can only be exploited for a few years before their ability to supply nutrients falls to a low level. If yields are to be maintained, and the soils used to produce crops on a continuing basis, a method by which nitrogen, phosphorus, potassium and other nutrients can be replaced has to be found.

Soils under natural vegetation normally support an active population of soil animals. These live on the roots and on litter from the vegetation, and they dig and burrow, keeping the soil loose and friable. When the vegetation is cleared to grow a crop, the soil is exposed to the impact of rain, and to the effects of people, animals and machines, treading and compacting the soil. Compaction will make the soil less suitable for the proliferation of plant roots, and reduce its ability to retain the water that plants need to survive. Exposure and subsequent drying of the soil can also lead to surface crusting. This reduces the rate at which water can enter the soil, and can cause water to runoff the surface leading to soil erosion. In arid and semiarid regions, exposure of some soils by overgrazing, or by annual burning of the vegetation, may also lead to compaction and the formation of surface crusts, even when the soil is not subjected to cultivation. Thus, managing a soil's physical properties must therefore aim to preserve the structure of the soil where this is already favourable for sustainable crop production or to create a favourable structure by suitable tillage or other practices where such a structure does not exist (FAO, 1993).

The contribution of soil organic matter to soil productivity is immense. Organic matter stabilises soil aggregates and supports the soil animals which create the pores through which air and water move. In addition, soil organic matter plays a major role in the retention of cationic nutrients by the dominant soils of the tropics which have clays composed of kaolinite, and iron and aluminium oxides with only a weak ability to hold the nutrient cations. Under acid conditions, some of the organic compounds present in the soil form complexes with aluminium which would otherwise be toxic to plants. Finally, soil organic compounds hinder the formation of insoluble complexes of iron and aluminium with phosphate, thus avoiding a reduction in the amount of phosphate available to plants. In addition to these physical and chemical effects, organic matter provides the substrate that supports almost all biological life in the soil. An active soil population is known to contain organisms that are directly beneficial to the soil, such as those related to nitrogen fixation and the maintenance of a favourable soil structure. Under natural vegetation the amount of organic matter in the soil tends to be established at a relatively high level. The actual amount depends on the quantities of organic matter added from the vegetation, the soil characteristics and the moisture and temperature regime prevailing in the soil. When the soil is cultivated, the addition from the crop is usually much less than from the natural vegetation, and consequently the level of organic matter tends to fall. This is especially so if the soil cultivation method increases the openness of the soil and the rate at which the organic matter is decomposed (FAO, 1994).

SOIL DEGRADATION

Soil degradation refers to decline in the productive capacity of land due to decline in soil quality caused through processes induced mainly by human activities. It is a global problem. The Global Assessment of the Status of Human-induced soil Degradation (GLASOD) was the first worldwide comparative analysis focusing specifically on soil degradation. Worldwide around 1.96 billion ha are affected by human induced soil degradation, mainly caused by water and wind erosion (1094 and 548 million ha respectively). Chemical degradation accounted for 240 million ha, mainly nutrient decline (136 million ha) and salinisation (77 million ha), physical degradation occurred on 83 million ha, mainly as a result of compaction, sealing and crusting (Oldeman, 2000).

The major forms of soil degradation are due to the displacement of soil, and the deterioration of soil without displacement, which usually involves deterioration of both chemical and physical properties. Soil degradation may be controlled by good soil management. Some soils are more resistant to certain forms of degradation than others, and the ease with which degradation can be corrected also differs between soil types. Many soils in the subhumid and semiarid tropics show some properties which place them in the category of 'fragile soil', notably those with easily erodible surface soils. These soils require careful management if degradation is to be prevented.

Soil degradation is also a very important problem in India. The processes leading to soil degradation are generally triggered by excessive pressure on land to meet the competing demands of growing population for food, fodder, fibre and fuel. Therefore, the direct causes for soil degradation are unsustainable land use and inappropriate land management. The most common direct causes include

1. Deforestation of fragile lands,
2. Overcutting and grazing of vegetation,
3. Extension of cultivation onto lands of low capability/potential,
4. Improper crop rotations,
5. Unbalanced fertiliser use,
6. Non-adoption of soil conservation practices,
7. Inadequacies in planning and management of irrigation resources, and
8. Overdraft of groundwater in excess of capacity to recharge.

Deforestation, in strict sense of the term, refers to the transfer of forest land to non-forest uses and includes all land where the forest cover has been stripped off and the land converted to such uses as permanent cultivation, shifting cultivation, human settlements, mining, reservoirs, etc. Deforestation continues to threaten and erode the area under forest cover in several countries of the world. The forest area in India has shrunk from 33% in 1947 to 21.67% at present. As per the recommended norm of the National Forest Policy 1988, about 33% of the geographical area should be under the forest cover, which comes to about 110 million ha compared to 71.22 million ha presently under the forest cover in India (FSI, 2019). Deforestation, together with overgrazing, is modifying the climate, and accelerating the erosion of biodiversity, which in turn posing the threat to ecological sustainability and food security.

Chemical degradation

Forms of chemical degradation of soil include

1. Loss of nutrients and organic matter,

2. Acidification (mostly associated with the removal of soil nutrients or the misuse of fertilisers),
3. Increased leaching (when the vegetative cover is removed and the soil is exposed),
4. Increased temperatures and oxidation of the soil organic matter (due to exposure and cultivation),
5. Salinisation and sodication (often associated with inappropriate irrigation practices and inadequate drainage), and
6. Pollution (most commonly from improper management of industrial and mining wastes).

Chemical degradation due to nutrient removal in crops cannot be avoided for the majority of soils. It can, however, be readily corrected by the use of fertilisers or manures, or both. Continued crop removal, or the use of ammonium fertilisers, leads eventually to acidification. This in turn may be corrected by judicious use of lime. For many upland soils in the tropics, the most satisfactory methods of ensuring sustainability appear to be those which involve the use of both inorganic fertilisers and organic manures. It may be possible to maintain yields by correcting both nutrient losses and acidification with inorganic materials alone. However, in many soils of the tropics it is extremely difficult to maintain the correct balance of nutrients without the use of organic amendments to buffer the changes of nutrient ion concentrations in the soil solution. Loss of organic matter is a form of soil degradation, not so much because of the direct effects of organic matter, but more importantly because of the many indirect effects. In addition to the buffering effect on nutrient ion concentrations, and the complexing of potentially toxic ions, the indirect benefits include the stabilising effects on soil aggregates, and the support offered to the soil population, both faunal and microbial (FAO, 1994).

Another form of soil degradation is associated with the cultivation of soils which are poorly drained, or where the water table is close to the soil surface. Salts and sometimes potentially toxic ions such as boron may be carried into the irrigated fields with the irrigation water. Unless there is adequate drainage, and sufficient water is supplied to wash out excess salt and undesired ions, salt will accumulate until it reaches levels at which plant growth is reduced. This problem is commonly found in areas where dams have been constructed, and where areas below the dam receive irrigation water. The water stored behind the dam is also likely to raise water tables in areas below the dam. Where groundwaters are saline, this may not only induce salinity problems but will inevitably make drainage of the area receiving irrigation more difficult (FAO, 1985).

In addition to salinity, sodication or alkalinisation can be a serious problem. This is not necessarily associated with poor drainage, but arises when irrigation water with a high content of sodium is used, or when seawater flooding of the land occurs. Sodium displaces other ions in the exchange complex, and at about 15% of sodium saturation the clay will disperse. This makes the soil extremely difficult to manage for crop production as it becomes extremely sticky when wet, cloddy when dry and the smaller clay particles will tend to move down the soil profile, a process sometimes known as internal erosion. This leads to extremely compact subsoil, in which water movement and root development are largely inhibited. In most situations these soils are also alkaline, and as such are often referred to as alkaline soils. However, many saline soils are also alkaline due to the presence of sodium carbonate, but these do not show the features of sodic soils because the salts prevent dispersion of the clay. Management of sodic soils involves the addition of gypsum, which provides a concentration of calcium sulphate in the soil solution sufficient to flocculate the clay, thus giving rise to a more easily manipulated structure. The calcium will also gradually displace sodium from the exchange complex, so eliminating the cause of the condition (FAO, 1985 & 1988).

Physical degradation

Deterioration of soil structure is the commonest form of physical degradation which involves

1. Loss of stability of aggregates in surface soils, leading to crusting and compaction, and consequently poorer infiltration rates and greater runoff and erosion; and
2. Translocation of clay particles to subsurface layers and loss of porosity in the subsurface and deeper soil layers, with consequent loss of water transmission and storage capacity.

Some forms of physical soil degradation can be readily reversed. These include surface soil crusting and compaction, which may be reversed in many arable soils by cultivation, although such improvements may be rather temporary. Physical degradation of soil is most commonly found where heavy machinery is used to clear and cultivate the soil. Problems occur mostly in soils of intermediate texture and low organic matter content, particularly sandy and silty loams. This can be avoided by using no-till and mulch farming techniques, and by giving careful attention to the need to restrict cultivation times to those when the soil is not too wet as a wet soil is readily damaged by cultivation. Increasing soil organic matter content by resting the land or incorporating organic manures can help to increase soil aggregate stability (FAO, 1994).

ENVIRONMENTAL DAMAGE

Soil management practices affect not only the site where the crop is being produced but also areas remote from the site. The off-site effects include those associated with the deposition of eroded soil material, the pollution of water supplies due to inefficient use of fertilisers and pesticides, and the production of greenhouse gases. These problems can be avoided with good soil management practices.

Off-site damage due to soil erosion

In semiarid areas, periods of drought are often accompanied by dust storms as the vegetation dies and the soil is exposed to wind erosion. Deposition of the dust gives rise to many problems, and dunes may be formed, covering otherwise valuable land with non-productive fine sand and coarse silt. Deposition of sand particles on the agricultural fields due to natural calamities is termed as sand casting.

1. Where sand casting is about 15 cm, the sand can be incorporated into the soil by ploughing. This may be followed by normal cultivation.
2. Where the sand casting is between 15 to 60 cm, sand may be removed using bulldozers. This may be followed by normal cultivation.
3. Where sand casting is more than 60 cm, no reclamation or removal of sand is suggested, since this would be costly and difficult to dump the sand so removed in the nearby fields. In such cases, horticultural and tree crops like cashew nut, coconut and casuarinas, etc. can be planted *in situ.*

In semiarid areas, the planting of drought resistant trees as windbreaks can be of considerable help. Strips of drought resistant grasses such as vetiver grass (*Chrysopogon zizanioides*) can also be invaluable to trap blowing sand and stabilise moving dunes.

Deposition of water-borne, eroded soil material can also be a serious problem, as it may also bury good land beneath less fertile material. It also causes damage by blocking waterways, including streams and rivers. This may lead to further problems due to rising water tables and the associated problems of waterlogging and salinity. The sediment may also be deposited in reservoirs, reducing their storage capacity and, in extreme cases, blocking the outlets to hydroelectric generators. Failure to take correct measures to check soil erosion in the catchment areas of dams may reduce the life and total economic value of dams. Where the catchment is in hilly areas it is important that deforestation of the catchment must be avoided. Where the catchment includes cropped areas, it is necessary to maintain a soil cover to minimise runoff. The careful design of channels and canal systems to dispose of unavoidable runoff is also imperative.

In arid and semiarid regions, overgrazing is a notorious factor in land denudation. It occurs frequently above small reservoirs where animals have direct access for drinking. Control depends on proper stock management, and on the construction of well-distributed watering points rather than direct soil management measures (FAO, 1994).

Damage due to misuse of fertilisers

When farmers purchase fertilisers their intention is to use them to feed their crops as efficiently as possible. However, nitrogen recovery efficiencies are often less than 50%, and phosphorus recovery is still lower. Much of the unrecovered nitrogen ends up in groundwater in the form of nitrate, or in the wetlands as ammonia in the atmosphere. In contrast to nitrogen, phosphate is normally adsorbed to soil clay and iron oxides, and does not move out of the soil except where the soil is essentially devoid of them. When it is washed off the soil surface, it may cause eutrophication problems, proliferation of algal growth, and exhaustion of dissolved oxygen. These in turn, may lead to the death of fish and other aquatic species. The problems due to surface wash are best controlled by ensuring that the phosphate applied is incorporated in the soil. Sewage and industrial effluents discharged directly into waterways is a major cause of phosphate contamination of water.

In whatever form nitrogen is added to the soil, most will normally be converted to nitrate, except in anaerobic conditions. Nitrate is freely mobile in the soil and so may be leached into groundwater unless it is intercepted by plant roots. Nitrogen fertilisers should be used most efficiently to avoid potential contamination problems as well as to obtain best economic advantage from the fertiliser. A low basal dose of N is recommended. This would be followed by a further dressing at the time of maximum crop demand, so avoiding concentrations in the soil which exceed the ability of the crop to absorb N.

Usually higher amounts of nitrogen fertilisers are applied to low-land rice crop where significant losses of N occur as ammonia which escapes to the atmosphere. Ammonia is not a serious atmospheric pollutant. Because of its high solubility, much of the ammonia volatilised is returned to the soil in rainfall. If a paddy soil is allowed to dry so that nitrification occurs, nitrous oxide, a greenhouse gas will be formed in small amounts. If after drying and nitrate formation, the soil is flooded again, much larger amounts of nitrous oxide are normally formed by biological denitrification processes. When organic manures are used on wetland soils, and anaerobic decomposition occurs, another greenhouse gas, methane, is formed in large amounts. Methane, like nitrous oxide, has a greenhouse effect several times greater than carbon dioxide. Hence, for paddy, good soil management practice should avoid the use of organic manures, and rely on the residues of the rice crop to maintain soil organic matter levels; and nitrogen fixation by blue-green algae, azolla, loppings from leguminous trees, etc. to supplement nitrogen fertilisers.

GUIDELINES FOR SUSTAINABLE SOIL MANAGEMENT

FAO has developed the following ten technical guidelines to address soil threats that hamper sustainable soil management (FAO, 2017).

Minimisation of soil erosion

Soil erosion by water and wind is the most significant threat to global soils and the ecosystem services they provide. Soil erosion causes the loss of surface soil layers containing organic and mineral nutrient pools, partial or complete loss of soil horizons and possible exposure of growth-limiting subsoil, as well as reduced water quality and sedimentation. Soil erosion is accelerated by human activities through reduced plant or residue cover, tillage and other field operations, and reduced soil stability leading to soil creep and landslides. The following recommendations have been made by Food and Agriculture Organisation (2017) to keep the soil erosion at minimum.

1. Land-use changes such as deforestation or improper grassland-to-cropland conversion that cause removal of surface cover and loss of soil carbon should be avoided or carefully planned and appropriately implemented if unavoidable.
2. A cover of growing plants or other organic and non-organic residues that protects the soil surface from erosion should be maintained through implementation of appropriate measures such as mulching, minimum tillage, no-till by direct seeding with attention to reduced herbicide use, cover crops, agroecological approaches, controlled vehicle traffic, continuous plant cover and crop rotation, strip cropping, agroforestry, shelterbelts, and appropriate stocking rates and grazing intensities.
3. Erosion by water on sloping and relatively steep lands should be minimised by measures that reduce runoff rates and velocity such as strip cropping, contour planting, crop rotation, intercropping, agroforestry, cross slope barriers (e.g., grass strips, contour bunds and stone lines), terrace construction and maintenance, and grassed waterways or vegetated buffer strips.
4. Where appropriate, riparian buffers, buffer strips, wetlands, water harvesting and cover crops should be used/installed to minimise export of soil particles and associated nutrients and contaminants from the soil system and protect the downstream areas from damaging impacts.

5. Erosion by wind, including dust storms, should be minimised and mitigated through vegetative (trees and shrubs) or artificial (stone walls) windbreaks to reduce wind velocity.

Enhancement of soil organic matter content

Soil organic matter (SOM) plays a central role in maintaining soil functions and preventing soil degradation. Soils constitute the largest organic carbon pool on the earth and play a critical role in regulating climate and mitigating climate change through trade-offs between greenhouse gas emission and carbon sequestration. For this reason, SOM is strategic for climate change adaptation and mitigation, and global stores of SOM should be stabilised or increased. A loss of soil organic carbon (SOC) due to inappropriate land use or the use of poor soil management or cropping practices can cause a decline in soil quality and soil structure, and increase soil erosion, potentially leading to emissions of carbon into the atmosphere. On the other hand, appropriate land use and soil management can lead to increased SOC and improved soil quality that can partially mitigate the rise of atmospheric CO_2. The following recommendations are offered by Food and Agriculture Organisation (2017) to enhance the organic matter in soil.

1. Biomass production should be increased by increasing water availability for plants using methods (e.g., irrigation with drippers or micro-sprinklers, irrigation scheduling, monitoring of soil moisture or loss of water via evapotranspiration) that maximise water-use efficiency and minimise soil erosion and nutrient leaching, using cover crops, balancing fertiliser applications and effective use of organic amendments, improving vegetative stands, promoting agroforestry and alley cropping, and implementing reforestation and afforestation.
2. Organic carbon-rich soils in peatlands, forests, pasturelands, etc. should be protected by taking appropriate measures.
3. Organic matter content in soil should be increased through practices such as managing crop residues, using forage by grazing rather than harvesting, practising organic farming, applying integrated soil fertility management and integrated pest management, applying animal manure or other carbon-rich wastes, using compost, and applying mulches or providing the soil with a permanent cover.
4. Fire should preferably be avoided, except where fire is integral to land management, in which case the timing and intensity of burning should aim to limit losses of soil functions. Where fire is a naturally occurring event, steps to minimise erosion and encourage revegetation after fire should be considered.

5. All sources of organic inputs, such as animal manure and properly processed human wastes should be used in the most favourable way.
6. Management practices such as cover crops, improved fallow plant species, reduced- or no-tillage practices, or live fences should be adopted to ensure the soil has a sufficient organic cover.
7. Decomposition rates of soil organic matter should be minimised by practising minimum or no-tillage without much increase in the use of herbicides.
8. Crop rotations should be practised with legumes including pulses to improve the crop mix.

Maintenance of soil nutrient balance and cycles

The concepts of sufficiency and utilisation efficiency apply especially to nutrient dynamics in the soil-water-nutrients-plant root continuum. Plant nutrition should be based on crop needs, local soil characteristics and conditions, and weather patterns. Plant nutrition can be enhanced through nutrient recycling or additions including chemical fertilisers, organic fertilisers and other soil amendments including primary sources (e.g., rock phosphate) and secondary sources (e.g., phosphorus from sewage sludge). It is crucial to select an appropriate plant nutrient management system and approach alongside assessing the suitability of the land for a given land use.

The benefits of sufficient and balanced nutrient supply for plant needs include production of food, feed, fibre, timber and fuel; reduced need for pest control measures, external application of organic and inorganic amendments, and mineral fertilisers; less pollution resulting from inappropriate use of agrochemicals; and enhanced soil carbon sequestration through biomass production and restoration to the soil. The lack of basic nutrients leads to the underdevelopment of plants and decrease in yields and crop nutritional value. Similarly, there are several adverse effects of excess nutrients in soils which include the loss of excess nutrients (especially nitrogen and phosphorus) from agricultural fields, causing eutrophication and deterioration of water quality and terrestrial and aquatic ecosystems; increased release of the greenhouse gas nitrous oxide from soils to the atmosphere; leaching of mobile forms of nitrogen to water used for human consumption, with potential human health impacts; and crop failure. The following recommendations are based on the guidelines given by Food and Agriculture Organisation (2017) to foster soil nutrient balance and cycle.

1. Natural soil fertility and natural nutrient cycles should be improved and maintained through the preservation or enhancement of soil organic matter. Improved soil fertility can be attained through soil conservation practices such as the use of crop rotations with legumes, green- and animal manures,

and cover crops in combination with reduced- or no-tillage with attention to reduced herbicide use, as well as agroforestry. Nutrient cycles are best managed in integrated systems such as crop-livestock systems or crop-livestock-forest systems.

2. Nutrient use efficiency should be optimised by adopting measures such as applying balanced soil organic and inorganic amendments (e.g., compost and liming agents, respectively) and/or innovative products (e.g., slow and controlled release fertilisers), as well as the recycling and reuse of nutrients.
3. Fertiliser application methods, types, rates and timing should be appropriate to limit losses and promote balanced crop nutrient uptake. This should be based on soil and plant analyses and is a long-term endeavour rather than short-term action.
4. The addition of soil micronutrients should be considered when planning soil fertilisation.
5. Practical sources of plant nutrients should be used, including the precise and judicious use of organic and mineral amendments, inorganic fertilisers, and agricultural bio-products. These amendments and bio-products include liquid, semisolid or solid manures, crop residues, composts, green manures, household refuse, clean ash generated during bio-energy production, soil amendments and inoculants. In order to increase their efficiency, such measures should be combined with the mitigation of other limiting factors (such as water deficiency). Safe use (including tolerable levels of contaminants and pollutants, and worker health) of the amendments should be ensured.
6. Soil and plant-tissue testing and field assessments should be adopted and used. This provides valuable guidance in diagnosing and correcting limiting factors in crop production related to plant nutrients, salinity, sodicity, and extreme pH conditions. Such guidance is key for making informed decisions and monitor progress.
7. Livestock movement and grazing should be managed to optimise manure and urine deposition.
8. Application of liming agents in acid soils is a prerequisite for optimal nutrient use efficiency in such soils, while application of organic amendments such as compost, as well as appropriate soil-crop management should be considered for alkaline and other soils.
9. Naturally occurring mineral fertiliser resources like rock phosphate or potash should be allocated efficiently and strategically to ensure the continued availability of adequate amounts of mineral inputs for future generations.

Prevention, minimisation and mitigation of soil salinisation and alkalinisation

Salinisation is the accumulation of water-soluble salts of sodium, magnesium and calcium in the soil. It is the consequence of high evapotranspiration rates, inland seawater intrusion, and human-induced (e.g., improper irrigation) processes. Salinisation reduces crop yields and, above certain thresholds, completely eliminates crop production. The following recommendations are offered by Food and Agriculture Organisation (2017) to prevent, minimise and mitigate soil salinisation and alkalinisation.

1. Surface cover should be optimised to reduce evaporation losses.
2. Efficiency of irrigation water use should be increased through improved conveyance, distribution, and field application methods. Application methods should be used that operate with low pressure and apply the water directly to the soil. Automisation of water supply and application of water on top of crops should be avoided to reduce evaporation losses.
3. Irrigation management should ensure sufficient water for plant growth and efficient drainage to avoid problems of salinisation.
4. Irrigation water quality should be tested and monitored and when feasible, water desalinisation should be performed.
5. Surface and subsurface drainage systems should be installed and maintained to control rising groundwater tables and control soil salinity. The design of these systems needs to be based on a thorough understanding of the water balance in these areas.
6. If soils are already degraded and prevention is no longer an option, reclamation of saline soils can be achieved using a variety of techniques such as direct leaching of salts, planting salt tolerant varieties, domestication of native wild halophytes for use in agripastoral systems, chemical amelioration and the use of organic amendments.

Prevention and minimisation of soil contamination

Soil may filter, fix and neutralise, but also release pollutants when conditions change (e.g., heavy metal release with lowering *p*H). Therefore, prevention of soil contamination remains the best way to maintain healthy soils and food safety in accordance to the Sustainable Development Goals. Contaminants can enter soils from a variety of sources including agricultural inputs, land application of by-products, atmospheric deposition, flood and irrigation water, accidental spills, inappropriate urban waste and wastewater management, and other means. Accumulation and contamination occur if the rate of addition of a given contaminant exceeds its rate of removal from the soil system. Negative consequences may include plant toxicities and subsequent productivity declines, contamination of water and off-site areas through sediment transport, and increased human and animal health risks through accumulation in the food-chain. The following recommendations have been made by Food and Agriculture Organisation (2017) to prevent and minimise soil contamination.

1. Governments should establish and implement regulations to limit the accumulation of contaminants below established levels to safeguard human health and well-being, and facilitate remediation of contaminated soils that exceed these levels.
2. Management of local soil contamination requires establishing background levels, followed by testing, monitoring and assessing contaminant levels to identify sites that are likely to be contaminated. Risk assessment, including total cost assessment, and remediation should be applied to reduce risks to humans and ecological systems.
3. Identification of soils that are the most susceptible to the harmful effects of diffuse pollutants is needed. Appropriate attention should be given to reduce contaminant loads in these soils.
4. Information on contaminated soil sites should be available to the public.
5. Contaminated soils should not be used for food and feed production.
6. Recycled nutrients originating from treated wastewater or other waste materials that are used as soil amendments should be properly processed and tested to ensure they contain safe levels of contaminants and plant available nutrients.
7. Outflows of floodwater from paddy rice cultivation after applying fertilisers and pesticides should be minimised to avoid off-site effects.

Prevention and minimisation of soil acidification

Human-induced acidification of agricultural and forest soils is primarily associated with removal of base cations and loss of soil buffering capacity or increases in nitrogen and sulphur inputs. Soils with low *p*H buffering capacity and/or high aluminium content are most prevalent when they have a low content of weatherable minerals (e.g., ancient, strongly weathered soils, and soils developed from quartz-rich parent materials). The following guidelines have been issued by Food and Agriculture Organisation (2017) to prevent and minimise soil acidification.

1. Monitoring soil acidity and minimising surface and subsurface soil acidity by using proper amendments such as lime, gypsum and clean ash.

2. Balanced fertiliser and organic amendment applications.
3. Appropriate use of acidifying fertiliser types.

Preservation and enhancement of soil biodiversity

Soils provide one of the largest reservoirs of biodiversity on earth, and soil organisms play key roles in the delivery of many ecosystems services. Little is known about the degree of biodiversity required to maintain core soil functions, but new tools for biochemical techniques and DNA analysis suggest significant progress in this area is possible. The following recommendations are offered by Food and Agriculture Organisation (2017) to preserve and enhance soil biodiversity.

1. Monitoring programmes for soil biodiversity, including biological indicators and *in situ* early warning signals, should be undertaken.
2. Soil organic matter levels supporting soil biodiversity should be maintained or enhanced through the provision of sufficient vegetative cover through cover crops and multiple crops, optimal nutrient additions, addition of diverse organic amendments, minimising soil disturbance, avoiding salinisation, and maintaining or restoring vegetation such as hedgerows and shelterbelts.
3. The authorisation and use of pesticides in agricultural systems should be based on the recommendations included in the International Code of Conduct on Pesticide Management and relevant national regulations. Integrated or organic pest management should be encouraged.
4. The use of nitrogen fixing leguminous species, microbial inoculants, mycorrhizae, earthworms and other beneficial micro-, meso- and macro-soil organisms should be encouraged where appropriate, with attention to limiting the risk of invasive processes by promoting the use of local biodiversity and avoiding the risk of disturbance in soil services.
5. Plant biodiversity should be restored in ecosystems. This would favour soil biodiversity.
6. Crop rotation, intercropping, and preservation of field margins, hedges and biodiversity refuges should be encouraged.
7. Any land use change in areas with high biodiversity should be subject to land use planning and in line with the relevant international instruments and national law.

Minimisation of soil sealing

Land conversion and subsequent soil sealing for settlements and infrastructure affect all soils, but are of particular concern on productive, arable soils because of their importance for food production and food security and nutrition. In many places, urban sprawl affects the most productive soils adjacent to the cities and settlements. Soil sealing and land conversion causes a largely irreversible loss of some or all soil functions and the ecosystem services they provide. While arable land is converted for non-agricultural uses the following points should be considered as suggested by Food and Agriculture Organisation (2017).

1. Considering the total value of soils and to ensure the preservation of productive, arable soils, existing policies, relevant laws and land use planning procedures for the development of settlements and infrastructure should be reviewed as appropriate.
2. Where policy and legislation aim to minimise land conversion, measures should be implemented to encourage densification and re-use of existing urban or industrial areas such as abandoned areas and brown fields, and restoring degraded neighbourhoods after appropriate reclamation measures have been implemented. Ecological restoration of quarries and mining sites should be encouraged.
3. Soils with significant ecosystem services including high soil carbon stocks, high biological diversity or high agricultural suitability should be protected from land conversion for settlements and infrastructure by special legislation.

Prevention and mitigation of soil compaction

Soil compaction is related to the degradation of soil structure due to imposed stresses by machinery and livestock trampling. Soil compaction causes reduced or disrupted pore continuity which in turn reduces soil aeration by destroying soil aggregates and collapsing macro-pore density, and reduces water drainage and infiltration, generating higher runoff. Compaction limits root growth and seed germination by high mechanical impedance, affecting soil biodiversity and causing surface soil crusting. The following suggestions are offered by Food and Agriculture Organisation (2017) to prevent and mitigate soil compaction.

1. Deterioration of soil structure due to inappropriate or excessive tillage should be prevented.
2. Vehicular traffic should be minimised to the absolutely essential, particularly on bare soils, by reducing the number and frequency of operations, creating controlled traffic systems, and by performing agriculture/forestry operations only when the soil moisture content is suitable down to deeper depth.
3. Machines and vehicles used in the field should be adjusted to soil strength and should be equipped with tyre pressure control systems or other means to reduce surface pressure, and use of heavy machinery should be avoided. During forestry operations, machine traffic should be restricted and brush

mats used to help protect exposed soils from physical damage; on agricultural soils, controlled traffic and drive rows should be established, where possible.

4. Cropping systems should be selected that include crops, pasture plants and, where appropriate, agroforestry plants with strong tap roots able to penetrate and break up compacted soils.
5. An adequate amount of soil organic matter should be maintained to improve and stabilise soil structure.
6. Macro-fauna and microbial (especially fungal) activity should be promoted to improve soil porosity for soil aeration, water infiltration, heat transfer and root growth.
7. In grazing systems, a sufficient cover of growing plants should be maintained to protect the soil from trampling and erosion. The livestock management should take into account grazing intensity and timing, animal types and stocking rates.

Improvement in soil water management

A sustainably managed soil has rapid water infiltration, optimal soil water storage of plant available water and efficient drainage when saturated. However, when these conditions are not met, waterlogging and water scarcity problems arise. On one hand, waterlogging creates rooting problems for many plants, thereby reducing yields, and can cause contaminants such as arsenic and methyl mercury to become mobile in the soil. On the other hand, water scarcity occurring in areas where water is lost by evaporation, surface runoff and percolation, can cause crop failure. The following recommendations are given by Food and Agriculture Organisation (2017) to improve soil water management.

1. In humid areas where precipitation exceeds evapotranspiration, additional drainage systems are needed to provide aeration for root functions like nutrient uptake. This is a concern especially in fine-textured soils which have high water retention capacity.
2. Surface and subsurface drainage systems should be installed and maintained to control rising groundwater tables in order to mitigate potential waterlogging.
3. The efficiency of irrigation water use by plants should be increased through improved conveyance, distribution, and field application methods (e.g., scheduled drip or micro-sprinkler irrigation) that reduce evaporation and percolation losses of irrigation water, as well as through better soil water reserve estimation, better species or variety choices, and better computing of water loading periods and amounts.

4. In dryland cropping systems, measures should be taken to optimise water-use efficiency such as the management of soil cover and water harvesting to increase soil water availability at sowing; reduction of runoff and evaporative losses from the soil surface; and ensuring that there is adequate water available at each stage of crop development.
5. Optimal soil water extraction by the crop through the selection of appropriate cultivars and careful timing of agronomic operation should be promoted.
6. Irrigation water quality should be regularly monitored for nutrients and potential harmful substances.

SUSTAINABLE AGRICULTURAL SYSTEMS FOR GOOD SOIL MANAGEMENT

The principles of good soil management are universally applicable. The practices which embody those principles vary quite widely according to soil types, climate and other environmental conditions.

Sustainable systems in the humid tropics

Perennial crops, such as oil palm, rubber, cocoa and banana have been grown for many years throughout the humid tropics. The crops provide a cover for the soil and usually return sufficient residues to the soil to maintain a satisfactory organic matter level. Nutrient replenishment is necessary if the system is to sustain productivity. Where nitrogen fertilisers are used, acidity may need to be regulated by liming. Maintaining a leguminous cover under the canopy of the perennial crop can provide nitrogen to the crop and complement the protection against erosion afforded by the tree canopy. The greatest risk of erosion arises during the initial establishment phase, when it is normally necessary to use a cover crop to protect the soil until the canopy of the perennial closes. The system is only sustainable as long as it is economically viable and the value of the produce is sufficient to cover the costs of essential inputs. The size of the world market for many of the perennial crops of the humid tropics is limited, and this in turn limits the extent to which land is used for such crops (FAO, 2014).

In arable cropping systems, the dominant soils of the humid tropics are often simple to manage as far as their physical properties are concerned. They have relatively stable aggregation and are free draining. However, efficient drainage and high rainfall means that they are often severely leached off nutrients, and are strongly acid. Hence, management of their chemical properties becomes critically important. Productivity of arable cropping systems can be maintained by careful management of fertilisers and lime (Sanchez et al., 1982). Combining tree crops with arable cropping systems is widely practised, as is the rotation of forest plantations with food crop production. These systems can be sustainable, depending on the efficiency of the tree crop in recycling nutrients, and on the effects of the trees on soil organic matter and acidity. Such systems are potentially of great importance for the humid tropics.

Sustainable systems in the subhumid tropics

In the wetter parts of the subhumid tropics, plantation agriculture with perennials has sometimes been sustainable, usually with coffee or tea at moderate elevations where the potential for erosion is not too great. Arable crop production has posed considerable problems for sustainable soil management. While the chemical condition of the soil is usually better than in the more humid areas, the dominant soils are usually physically weak, and therefore readily eroded. Long-fallow shifting cultivation systems were sustainable, but broke down when demographic pressure forced a reduction in the length of the fallows, and an extension of the cropping periods. These changes led to deterioration of organic matter levels and structural condition, which in turn led to erosion and the associated problems.

Agroforestry offers the best route to a sustainable system (Nair, 1989; Young, 1989), but requires that the trees in the system offer an economic return. Alley farming, in which rows of trees are planted along the contour, and crops grown between them, has been widely promoted as a sustainable system for the subhumid tropics. Growing maize or cowpeas between the rows of the fast-growing legume *Leucaena leucocephala* is a sustainable system, where loppings from the trees are used as a mulch to protect the soil and provide nitrogen. To obtain economic yields, phosphorus and other fertilisers including supplemental nitrogen are needed. Although it is inevitable that the trees will compete with the crop for water and nutrients, they too have the potential to provide an economic return. In addition to an economic return from the trees and arable crops, this system provides its value in controlling erosion and building soil fertility (FAO, 1994).

Sustainable systems in the semiarid tropics

Uncertain rainfall and long dry seasons make sustainable crop production difficult in the semiarid tropics. Animal husbandry, rather than crop production, offers the greatest prospect of sustainability, but subject to an important consideration, that stocking rates are kept below the carrying capacity of the land. This ensures that the vegetation persists, and continues to provide the essential ground cover. In most parts of the semiarid tropics the demand for food has meant that stocking rates have often exceeded sustainable limits, and arable cropping has continued to invade traditional grazing areas. The cropping practices have been based on shifting cultivation systems in which fertility restoration depends on the regrowth of native grasses. The extent to which fertility is restored is often limited. This has lead to falling productivity. Shorter duration crop varieties may help to overcome water shortages by reducing the length of time that the crop is using water. However, the demand for nutrients is not reduced. Legumes always require an adequate level of soil phosphorus, and so fertiliser phosphorus has normally to be used together with a suitable legume. In order to maintain the productive capacity of land in the semiarid tropics, a cropping system is required where the rate at which soil fertility can be restored is much improved, compared with the changes under native grassland (FAO, 1994).

Sustainable systems in the wetlands

The rice-based farming systems of the wetlands are the most sustainable system. The following factors contribute to this system.

1. Nutrients are washed into rather than out of the soil.
2. Water supplies are normally assured by irrigation and bunding of the fields to retain water.
3. Erosion is not a problem as the bunds prevent runoff and the water held on the surface prevents soil displacement by raindrop impact.
4. Acidity is not a problem as flooded soils will always approach the near neutrality of the flood water.
5. Nitrogen fixation in the flooded system is relatively high.
6. Phosphate availability in flooded soils is relatively high because iron is present in the ferrous rather than ferric state.
7. Weed problems are less in flooded than in dryland conditions.

The system is most sustainable where the rice crop is produced with fully controlled water supplies from irrigation systems, and with necessary inputs of fertilisers. Although organic manures, including green manures, have been widely advocated for use on rice, actual use has decreased considerably as rice yields and the use of nitrogen fertilisers have been increasing. Higher levels of organic manure may well be harmful due to the formation of phytotoxic by-products formed when the decomposition of fresh organic materials occurs in the soil under anaerobic conditions (FAO, 1994). The evolution of the highly active greenhouse gas methane is also considered as a threat to sustainability. The greatest threats to the sustainability of irrigated rice production are those relating to the water supply system. The problems of poor soil management in the catchment areas of dams feeding the irrigation are often the greatest threat. Eroded soil can both reduce the storage capacity of the dam and choke distribution canals. Where saline waters exist beneath the service area of the dam rising water tables can bring salt into the root zone and reduce yields. Poor management of the water supply can mean overwatering at the upper end of the distribution system, and droughts at the lower end. With less water becoming available from the dam, the disruption of water supplies can lead to occasional water shortage in all of the command area, with drastic effects on yields.

REFERENCES

FAO. 1985. Guidelines: Land evaluation for irrigated agriculture. Food and Agriculture Organisation of the United Nations, Rome, Italy.

FAO. 1988. Salt-affected soils and their management. Food and Agriculture Organisation of the United Nations, Rome, Italy.

FAO. 1993. Soil tillage in Africa: needs and challenges. Food and Agriculture Organisation of the United Nations, Rome, Italy.

FAO. 1994. Cherish the earth: Soil management for sustainable agriculture and environmental protection in the tropics. Land and Water Development Division, Food and Agriculture Organisation of the United Nations, Rome, Italy.

FAO. 2004. Carbon sequestration in dryland soils. Food and Agriculture Organisation of the United Nations, Rome, Italy.

FAO. 2015a. Healthy soils are the basis for healthy food production. Food and Agriculture Organisation of the United Nations, Rome, Italy.

FAO. 2015b. Revised World Soil Charter. Food and Agriculture Organisation of the United Nations, Rome, Italy.

FAO. 2017. Voluntary Guidelines for Sustainable Soil Management. Food and Agriculture. Organisation of the United Nations, Rome, Italy.

FSI. 2019. India State of Forest Report 2019. Forest Survey of India, Ministry of Environment, Forest & Climate Change, Govt. of India, Dehradun, India.

Greenland, D. J. and Szabolcs, I. (eds.). 1994. Soil resilience and sustainable land use. CAB International, Wallingford, UK.

Nair, P. K. R. 1989. Agroforestry Systems in the Tropics. Kluwer Academic Publishers, Dordrecht and ICRAF, Nairobi.

Oldeman, L. R. 2000. Impact of soil degradation: A Global scenario. International Soil Reference and Information Centre, Wageningen, the Netherlands.

Sanchez, P. A., Bandy, D. E., Villchica, J. H. and Nicholaides, J. J. 1982. Amazon basin soils: management for continuous crop production. *Science* **216**: 821–827.

Stivers, L. 2017. Introduction to Soils: Soil Quality. https://extension.psu.edu/introduction-to-soils-soil-quality

Yang, T., Siddique, K. H. M. and Liu, K. 2020. Cropping systems in agriculture and their impact on soil health-A review. *Global Ecology and Conservation* **23**: 1-13.

Young, A. 1989. Agroforestry for Soil Conservation. CAB International, Wallingford, UK.

6

Nutrient Management for Sustainable Agriculture

Sustainability in agricultural production is guided by the overriding philosophy of 'feed the soil to feed the plant'. Adequate supply of plant nutrients is required to ensure efficient crop production. Plants use individual chemical elements and ions as their food. In 1939, Arnon and Stout laid down three criteria for the essentiality of a plant nutrient.

1. A deficiency of the element makes it impossible for a plant to complete its life cycle.
2. The deficiency is specific for the element in question.
3. The element plays a direct role in metabolism.

According to these criteria, 16 plant elements have been recognised as being essential to all plants for their normal growth and development. Out of these, carbon (C), hydrogen (H) and oxygen (O) are derived from air and water while the remaining 13 are supplied by the soil (Chesworth, 1991). On the basis of the amounts in which these 13 elements are taken up by the crop plants, they are classified as primary nutrients, secondary nutrients or micronutrients. Nitrogen (N), phosphorus (P) and potassium (K) are used in large quantities by the plants and are called as major or primary nutrients. Calcium (Ca), magnesium (Mg) and sulphur (S) are required in relatively smaller but in appreciable quantities and thus named as secondary nutrients. Iron (Fe), manganese (Mn), copper (Cu), zinc (Zn), molybdenum (Mo), boron (B) and chlorine (Cl) are required in small quantities and are called as trace elements or micronutrients.

ESSENTIAL PLANT NUTRIENTS

Regardless of the class to which they belong, all essential nutrients are equally important for plant growth and need to be available in sufficient amounts in soil, within the soil volume in which crop roots are active. Plant nutrient concentrations may be inadequate, insufficient, deficient, toxic or excessive. When plant growth is satisfactory and optimum yields are produced, the amounts are adequate. When the optimum yields cannot be reached because of nutrient deficiencies, the amounts are insufficient. It is difficult to detect deficiencies if they are small or close to normal. On the other hand, if the deficiency is considerable, the symptoms are visible. When the situation is reverse, when the amount of the nutrients is excessive, it may cause deficiency of other nutrients or may result in environmental pollution. At excessive amounts, the nutrient becomes toxic to the plant. Different crops have different nutrient needs and whether the nutrients are sufficient, insufficient or excessive varies with the type of plant (Nicolas, 1961). Whether the land is plentiful or in short supply, efficient soil fertility management is the key to sustainable agriculture (Prasad and Power, 1997). Important functions and deficiency symptoms of essential plant nutrients are given in table 6.1.

Table 6.1. Functions in plants and deficiency symptoms of plant nutrients

Nutrient (forms absorbed)	Functions in plants	Deficiency symptoms
Nitrogen (NH_4^+, NO_3^-)	Synthesis of amino acids, proteins, chlorophyll, nucleic acids, coenzymes	Light-green coloured leaves, lower leaves turn yellow and die in severe deficiency
Phosphorus ($H_2PO_4^-$, HPO_4^-)	Metabolic transfer processes, ATP, ADP, photosynthesis, and respiration; component of phospholipids	Purplish leaves, especially on the margins
Potassium (K^+)	Involved in sugar and starch formation; lipid metabolism and nitrogen fixation; neutralises organic acids	Marginal burning of leaves, curling of leaves
Calcium (Ca^{2+})	Component of cell walls; cell growth and cell division; cofactor for some enzymes	Failure in the development of terminal bud, young leaves become small, distorted, cup-shaped; root growth impaired
Magnesium (Mg^{2+})	Component of chlorophyll, hence essential for food synthesis in plant	Light green leaves and yellowing of leaves similar to N. In rapeseed the leaves are cupped inward
Sulphur (SO_4^{2-})	Synthesis of the S-containing amino acids like cystine, cysteine and methionine; structural formation of chlorophyll; constituent of proteins; oil synthesis	Pale yellow or light green leaves in younger leaves (similar to N deficiency); short & slender stalks, retarded growth; suppressed fruit development
Zinc (Zn^{2+})	Formation of auxins and chloroplasts; carbohydrate metabolism; stabilising and structural orientation of membrane proteins	Stunted growth, pale to white colouration of young leaves, brownish red (rusty) discolouration of leaves in rice known as *khaira* disease

Nutrient (forms absorbed)	Functions in plants	Deficiency symptoms
Manganese (Mn^{2+})	Nitrogen assimilation, photosynthesis-evolution of oxygen, oxidation reduction processes, reactions	Intervenous discoloration - green veins against a pale background; whitening and abscission of leaves
Iron (Fe^{2+})	Structural component of cytochromes, perrichrome, and leghemoglobin and thus involved in oxidation-reduction reactions in respiration and photosynthesis	Yellowing or whitening of young leaves; in severe deficiency in rice nurseries, or direct seeded rice or sorghum fields the entire plants may turn pale or white; pale yellow interveinal chlorosis in stem
Copper (Cu^{2+})	Constituent of chlorophyll, catalyst for respiration, carbohydrate, and protein metabolism	Stunted growth, terminal leaf buds die, leaf tips become white, and leaves are narrowed and twisted
Boron (BO_3^-)	Involved in germination and pollen tube growth; fruiting, cell division, nitrogen metabolism	Terminal buds die, rosette formation, flower or fruit shedding
Molybdenum (MoO_4^-)	Essential component of nitrate reductase and nitrogenase, thus important in N fixation by legumes	Resembles N-deficiency symptoms, whip tail disease of cauliflower
Chlorine (Cl^-)	Involved in the oxygen evolution in the primary photosynthesis, raises cell osmotic pressure	Chlorotic leaves, some leaf necrosis

Source: Grundon et al. (1987); Katyal and Randhawa (1983); Mengel and Kirkby (1987); Tisdale et al. (1985); Tanka and Yoshida (1970)

In addition to these 16 essential elements, some elements are needed by a particular group or species of plants at very low concentrations. These are sodium (Na), silicon (Si), cobalt (Co), vanadium (V), iodine (I), selenium (Se), gallium (Ga), aluminium (Al), rubidium (Rb), strontium (Sr), nickel (Ni), chromium (Cr) and arsenic (As). These are often referred as beneficial elements.

NUTRIENT DYNAMICS IN SOIL

Soil is a heterogeneous mixture of organic and inorganic materials. It varies greatly in space and also in time; a single farmer's field may contain many identifiably different types of soil and the nutrient content, especially of nitrate (Clark et al., 2005) owing to the many processes that produce and consume nitrate (Milne et al., 2004). Other plant nutrients are also dynamic in time and space, but the degree of variation varies with their chemical and biological properties. The variation in the weather parameters like rainfall and temperature, and the changing nutrient requirements of crop plants as they grow, makes matching the supply of nutrients from the soil to the needs of the crop a complex task (Barraclough, 1986). Achieving high nutrient use efficiency (NUE) by manipulating inputs alone is difficult.

Soil structure affects nutrient supply in two main ways

1. The complex nature of the network of pores in soil influences the transport of soluble nutrients. This can be advantageous in the case of highly soluble forms of nutrients such as nitrate-N, which is then less able to leach out of the rooting zone (Addiscott and Whitmore, 1987). On the other hand, sufficient quantities of poorly soluble nutrients such as phosphate that diffuse only slowly as a result of a complex structure may not be available or accessible to plant roots (Barraclough, 1986).
2. Soil structure results from a complex relationship between organic matter and the mineral matrix, with a degree of physical protection from decomposition of soil organic matter (SOM) often are conferred by its adsorption onto mineral surfaces. SOM contains nutrients that, when decomposed by microorganisms, are transformed into mineral forms that are available to plants. However, the physical protection of SOM can cause the supply of N to plants to be smaller from clay soils, which protect SOM, than from sandy soils containing identical amounts of SOM (Hassink and Whitmore, 1997).

In addition, the very structure of the compounds mineralised confers a degree of protection by preventing decomposer organisms or their exoenzymes from accessing the nutrients in them. Soil structure can be altered by management such as compaction by vehicles or 'poaching' by livestock can close pores and makes nutrients less accessible to plants. Compacted soils are more prone to waterlogging and hence to anaerobic conditions and the loss of N by denitrification. On the other hand, ploughing loosens the soil and can improve the transport of nutrients, at the same time as ameliorating the physical condition of soil (Goulding et al., 2008).

Soil structure also influences the fate of excess nutrients in soils. Thus, well-drained grassland soils, particularly if they are grazed, have a much higher propensity for excess nitrate to be lost by leaching. Poorly drained soils, under most circumstances, have a lower leaching potential and N loss by denitrification is likely to be greater than by leaching. The provision of drainage in an inherently poorly drained soil also alters the pattern of phosphate loss (Haygarth et al., 1998).

Biology of soil affects nutrient dynamics and NUE in all farm systems (Abbott and Murphy, 2003). SOM contains most of the N and much of the P and S contained in and eventually supplied to crops by soils. The soil microbial biomass, comprising all of the living microorganisms in soil, plays a critical role by both mineralising nutrients in SOM and immobilising them in their bodies for later release and recycling (Powlson, 1994).

APPROACHES FOR NUTRIENT MANAGEMENT

The effects of declining soil fertility on yield growth are being realised in developing countries. The low level of chemical fertiliser use, decline in soil organic matter, and insufficient attention to crop nutrient studies contribute the most to the loss of soil fertility in these regions (Kumwenda et al., 1996). Gruhn et al. (2000) summarised the possible reasons for apparent diminished returns from increased fertiliser applications in Asia.

1. More fertilisers are being used on lands with poorer soils or uncertain water supply.
2. The increased intensity of cropping, especially changes in crop sequences, makes current management practices, including fertiliser use, less effective.
3. There is an imbalance in the supply of N, P and K, with applications of the latter two nutrients often being too low.
4. Deficiencies of secondary nutrients and micronutrients are beginning to appear.
5. There is an overall decrease of soil organic matter and an increase in soil degradation in general.
6. Adverse effects from pests and diseases are increasing in the region.

Nutrient use efficiency (NUE) represents a key indicator to assess progress towards better nutrient management. The efficient use of any nutrient depends on its balanced supply with other nutrients, i.e., all nutrients should be available at the right amount and time. Nutrient interactions can also be synergistic, i.e., benefits to yield and NUE are multiplicative, not just additive.

Goulding (2000) summarised best management practices for N but the principles set out can be applied to all nutrients.

1. The highest-yielding variety of a crop is to be chosen to maximise the use of the available nutrients.
2. A green cover is maintained as much as is practicable to retain N. A cover crop may be practised if necessary and autumn crops may be sown early, but this must be balanced against the risk of carry-over of pests and diseases and the need for effective weed, pest and disease control.
3. Regular soil analyses for pH, P, K and Mg and if possible, trace elements should be done.
4. Lime should be used to maintain the appropriate *p*H for optimum nutrient supply (6.5 for arable crops; 6.0 for grassland).

5. The fertiliser requirements are to be calculated using a recommendation system. P, K and Mg should be applied to maintain recommended levels of these nutrients in soils.
6. Timely applications of N is to be ensured to provide N when the crop is growing quickly.
7. The fertilisers and manures should be applied evenly and well away from watercourses, with a properly calibrated spreader.
8. Timely control measures are to be taken to minimise pest, disease and weed infestation.
9. Irrigation to the crops should be given only to support crop yield using a scheduling system that takes account of the weather.

Strategic approaches

The economic pressures and uncertainty experienced by farming encourage economic analyses of various nutrient management strategies. The fertiliser-based system gives the larger profit for most grain crops, but the alternative systems like organic, cover crop and manure-based systems are remunerative and sustainable, especially if the manure is not to be purchased or transported. With the increasing price of oil and, therefore, of fertilisers and need to effectively recycle manures, such alternative systems will become increasingly attractive (Gareau, 2004).

Nutrient conservation and uptake

Soil conservation technologies prevent the physical loss of soil and nutrients through leaching and erosion. Terracing, alley cropping, and low-till farming practices alter the local physical environment of the field and thereby prevent soil and nutrients from being carried away. Mulch application, cover crops, intercropping and biological nitrogen fixation act as physical barriers to wind and water erosion and help to improve soil characteristics and structure. Organic manures such as animal and green manures also aid soil conservation by improving soil structure and replenishing secondary nutrients and micronutrients (Kumwenda et al., 1996). Improved application and targeting of both inorganic and organic fertiliser not only conserves nutrients in the soil, but makes nutrient uptake more efficient. Most crops make inefficient use of nitrogen.

Nutrient recycling

Soil nutrient availability changes over time. The continuous recycling of nutrients into and out of the soil is known as the nutrient cycle. The cycle involves complex biological and chemical interactions. A simplified version of this cycle of plant growth, suggested by Smaling (1993), is shown in Fig. 6.1. The simplified cycle has two parts, 'inputs' that add plant nutrients to the soil and 'outputs' that export them from the soil largely in the form of agricultural products. Important input sources include inorganic fertilisers; organic fertilisers such as manure, plant residues, and cover crops; nitrogen generated by leguminous plants; and atmospheric nitrogen deposition. Nutrients are exported from the field through harvested crops and crop residues, as well as through leaching, atmospheric volatilisation, and erosion. The difference between the volume of inputs and outputs constitutes the nutrient balance. When nutrient additions to the soil are greater than the nutrients removed from the soil, the nutrient balance is positive and it indicates that farming systems are inefficient and, in the extreme, that they may be polluting the environment. On the other hand, negative balance indicates that soils are being mined and that farming systems are unsustainable over the long term. In the latter instance, nutrients have to be replenished to maintain agricultural output and soil fertility into the future (Gruhn et al., 2000).

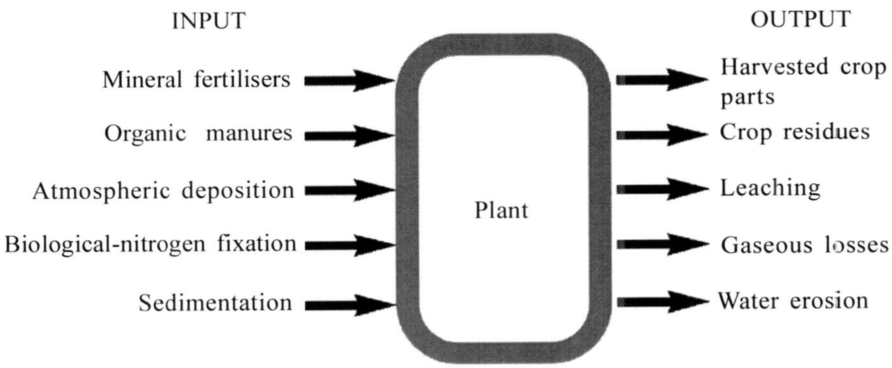

Fig. 6.1. The plant nutrient balance system
Source: Smaling, 1993

Plant nutrient application

Fertilisers need to be applied at the level required for optimal crop growth based on crop requirements and agroclimatic considerations. Over-application of fertilisers induces neither substantially greater crop nutrient uptake nor significantly higher yields (Smaling and Braun, 1996). Rather, excessive nutrient applications are economically wasteful and can damage the environment. Under-application, on the other hand, can retard crop growth and lower yields in the short-term and in the long term jeopardise sustainability through soil mining and erosion. The wrong kind of nutrient application can also be wasteful. The correction of nutrient imbalances can have a tremendous effect on yields. Balanced fertilisation should also include secondary nutrients and micronutrients, both of which are often most readily available from organic fertilisers such as animal and green manures. Application of balanced plant nutrient is necessary for sustainability in agriculture over time.

Precision farming

The spatial variation in yield across a field offers the prospect of variable-rate N and other nutrient applications, but yield maps are confounded by many potential causes of yield variability as well as errors (Stafford et al., 1999). Using yield maps alone to predict crop production for nutrient management without measuring other potential and often transient yield-limiting factors such as pest incidence may be a waste of effort and resource. However, variable-rate application maps can result in a 60% increase in the area correctly fertilised compared with a fixed-rate application (Ferguson et al., 1996). The precision farming is suitable in the areas where multisource data of high spatial and temporal resolution for decision making are available in the management of crop production. Farmers are more likely to be convinced by practices that increase yields as well as NUE (Olesen et al., 2004).

Nitrogen inhibitors

Nitrogen fertiliser management is challenging due to the many factors that influence fertiliser N after it is applied. Primary concern is the potential for fertiliser N to be lost to the environment and thus unavailable to the crop, either through ammonia volatilisation, denitrification, or leaching. Fertiliser N is mostly available in three chemical forms, ammonium, nitrate or urea.

Urease inhibitors: Urease inhibitors are the chemical compounds that block the activity of the enzyme urease. Urease catalyses the hydrolysis of urea, forming ammonium and carbon dioxide. It is found in soil as well as in plant residues. With high *p*H soil, or soil/residue environments that are poorly buffered against an increase in *p*H, the rapid hydrolysis of urea results in an accumulation of ammonia (NH_3) rather than ammonium (NH_4^+). Ammonia can be lost to the atmosphere as a gas. More than 50% of applied fertiliser N may be lost to the atmosphere under these conditions. By keeping urea from hydrolyzing, urease inhibitors protect against ammonia volatilisation, keeping fertiliser N in the urea form. Since urea is water soluble, such inhibition allows time for rainfall or irrigation to move urea into the soil. Once incorporated into the soil, to a depth of 5-7 cm, urea-N is protected from ammonia volatilisation, because ammonium that is released through hydrolysis is quickly held on soil cation exchange sites. Urease inhibitors break down with time, but typically can protect against ammonia loss from surface applied fertilisers for a couple of weeks, depending on temperature and moisture conditions (Ferguson et al., 2019).

Nitrification inhibitors: Bacteria *Nitrosomonas* is responsible for converting ammonium to nitrite and *Nitrobacter* is responsible for converting nitrite to nitrate. Nitrification inhibitors temporarily reduce populations of *Nitrosomonas* and *Nitrobacter* bacteria in soil. These chemical compounds slow the microbial conversion of ammonium-N to nitrate-N (nitrification), reducing the risk of loss through leaching or denitrification and thereby increasing the NUE of fertilisers. Nitrification inhibitors, like urease inhibitors, will break down over time and will no longer be effective (Ferguson et al., 2019).

Untapped nutrient sources

The recycling of organic waste from urban and rural areas is a potential, largely untapped, source of plant nutrients. Environmentally undesirable wastewater may be used to irrigate fields and this returns nutrients and organic matter to the soil (Tandon, 1992). Urban waste sludge is also a source of primary nutrients. Stabilised municipal waste sludge typically contains about 3.3% nitrogen, 2.3% phosphorus, and 0.3% potassium. Actual nutrient content, however, varies widely and depends on the source of the waste. Urban waste also has a number of other benefits. Like other organic manures, it helps improve soil structure by adding organic matter to the soil. It is also a source of the secondary nutrients and micronutrients. In addition, urban waste transforms material that would otherwise be slated for costly disposal into a useful farm product (Gruhn et al., 2000). However, urban waste needs to be treated carefully because it may contain heavy metals, parasites, and other pathogens. While trace amounts of some heavy metals play a critical role in plant metabolism, excessive amounts have reduced crop yields and could be dangerous to public and grazing livestock (Conway and Pretty, 1991). To minimise these risks the continuous application of urban waste needs to be monitored in order to ensure that heavy metal and overall nutrient concentrations do not reach toxic levels and do not damage the environment through leaching and eutrophication.

Knowledge transfer

Nitrogen use efficiency on cereal crops in farmers' field is usually 20-50%, whereas on research plots it is 60-90% (Dobermann and Cassman, 2004). The variation is caused by several factors like crop variety, crop rotation, tillage practices, pest, disease and weed management, irrigation and drainage, and the weather. Except weather factors over which there is little or no control, all other factors can be modified that can increase the nitrogen use efficiency, simply by improving knowledge transfer. However, the availability of skilled expertise, time and the application of resources are likely to be much higher in research stations than in farmers' field, and it may not be possible to increase

NUE on farms to the level obtained in experiments. There are still opportunities to improve NUE simply by applying the necessary nutrients in the correct amounts at the correct times (Krupnik et al., 2004). Soil analysis should be seen as an essential aid to best management practice for optimum NUE, including N in many cases, augmented by various techniques, such as chlorophyll-metres and remote sensing in high-cost systems and colour charts in low-cost systems.

Computer-based decision support system

Kitchen and Goulding (2001) have suggested to include computer models and interactive, internet based systems, with remote sensing, yield mapping and crop canopy measurements to fine-tune fertiliser applications. The computer-based systems are useful for marshalling the relevant information and presenting data in a fashion that is helpful to a farmer or farm adviser. Many of such decision support systems aim at N owing to the importance and complexity of the dynamics of this element in soil. Once decision support migrates to a computer-based environment, it becomes easier to integrate nutrient advice with support for pest, disease and weed control scheduling, financial and general farm planning.

NUTRIENT MANAGEMENT IN FARM SYSTEMS

Intensive farming

The intensive farming systems of developed countries seek to maximise yield through the best management practices that involves the efficient use of all inputs, including fertilisers and plant protection chemicals, crops and crop rotations, livestock breeding programmes and often precision agricultural techniques. Fertilisers have been central to this approach, which has resulted in a very rapid increase in productivity. The yield potential of high yielding varieties and hybrids of various crops are fully exploited through application of plant nutrients coupled with efficient water management. For sustainable food supply to the ever-increasing world population with a declining land area available for farming, the best land should be used for farming as intensively as possible to maximise the yields, and the poorer land should be farmed more extensively or used for environmental good or amenity. This would make best use of the most fertile soils and beneficial climates, but would result in large areas of relatively uniform landscape (Goulding et al., 2008).

Integrated farming

The aim of integrated farming is to use fewer inputs, especially less fertiliser and pesticide, and accept slightly smaller yields and gross returns, but to maintain net profitability through the reduced costs of inputs. Minimum tillage and integrated pest management, using strategies other than chemical sprays, are often practised. Integrated farmers are strong advocates of nutrient management plans and auditing systems. However, the distinction between 'intensive' and 'integrated' is not clear as intensive farmers adopt what they regard as useful integrated practices. Integrated systems have no specific mechanisms that optimise nutrient efficiency and minimise environmental impacts other than the careful and suboptimal use of inputs. The suboptimal applications cannot guarantee better NUE due to the complexity of soil nutrient dynamics (Goulding, 2000). In this case, the skill of the land manager is critical.

Extensive and organic farming

In extensive systems, inputs are used at low rates to compensate the offtakes through the products. In some situations, for example upland livestock production, inputs can be minimal or even zero (Goulding et al., 2000). Specialised local production system uses more extensive systems that can provide added value either in terms of product quality or the delivery of environmental benefits such as less pollution and greater biodiversity, which may provide an economic advantage. The organic farm systems may be considered an amalgam of integrated and extensive farming principles, i.e., minimising and optimising inputs, but with specialised and precisely defined requirements as to their origin. Recycling of animal manures and composted wastes is central to organic management systems.

Livestock production systems

In animal production system, the efficiency of N conversion into product is very poor and a large amount of N in animal diets is converted into excreta. At best, the theoretical maximum efficiency of conversion of N into milk is 40-45% and in many cases it is less than 25% (Jarvis, 1998). Use of higher energy feeds such as maize silage is one means of reducing concentrates fed to dairy animals and hence increasing N use efficiency. Much of the excess P in intensive livestock systems can be accounted for by the imported P in concentrates (Haygarth et al., 1998a), and there are opportunities to reduce this. The use of manures and slurries generated in livestock managements increasingly becomes a key part of nutrient management.

INTEGRATED NUTRIENT MANAGEMENT

Chemical fertilisation has enhanced the crop productivity, but to larger extent they have also contributed to soil deterioration. Organic manures are a vital resource not only for supplying plant nutrients, but also for the replenishing organic matter content of most agricultural soils. However, large-scale use of only organic manure as the sole source of nutrients has also several problems (Paarlberg, 1994; Rodriguez, 2011).

1. It is difficult to provide the proper nutrient balance to meet crop requirements with bulky organic matter because of low, variable and generally unbalanced nutrient contents.
2. Animal manure creates unique problems in application that increase farm workload and management requirements.
3. Bulky, organic manure can be applied only during a few weeks or months each year, usually before seeding or after harvesting a crop.
4. Nutrient deficiencies in crops that arise during the crop growth period can seldom be corrected with bulky, organic manure.

Thus, neither inorganic fertilisers, nor organic manures alone can sustain productivity. So judicious uses of organic manures and inorganic fertilisers are essential to safeguard soil health and augment productivity and input use efficiency. Integration of different sources of nutrients has a promising strategy for soil health management and sustained productivity (Jat et al., 2015). Integrated nutrient management (INM) involves the use of manures, chemical fertilisers and biological agent achieve sustainable crop production and improved soil health.

Concept of INM

Integrated nutrient management is a practice that combines old and modern techniques of fertiliser use and nutrient management. The idea of INM depends on a number of factors, including harmony in nutrient properties, a balance between crop nutrient demands, what sort of nutrient, in general, is available in soil and in the farmer's hand, information and skills about the most suitable nutrient can be harmonised in combination, and which materials can be safely used that lead to increase nutrient-use efficiency (Selim, 2020). Additionally, it is a method and a way of disposing organic wastes safely and also an effective method of recycling wastes into good quality compost (Roy et al., 2006; Selim and Al-Owied, 2017). It optimises all aspects of nutrient cycling, including macro- and micronutrient inputs and outputs, with the aims of synchronising nutrient demand by the crop and its release into the environment. Under INM practices, the losses through leaching, runoff, volatilisation, emissions and immobilisation are minimised, while high nutrient-use efficiency is achieved (Zhang et al., 2012).

INM also aims to optimise the soil conditions by improving its physical, chemical, biological and hydrological properties to enhance farm productivity and minimise land degradation (Saikia et al., 2015). There is now a greater awareness that INM cannot only increase crop productivity, but also simultaneously and almost imperceptibly preserve soil resources. Its practices use farmyard manures, farm wastes, soil amendments, crop residues, natural and chemical fertilisers, green manures, cover crops, intercropping, crop rotations, fallows, conservation tillage, irrigation and drainage to conserve available water and to boost plant nutrients (Wu and Ma, 2015). This strategy also includes new techniques, such as deep placement of fertilisers and the use of inhibitors or urea coatings that have been developed to minimise nutrient losses and improve plant uptake (Saikia et al., 2015). Such practices encourage farmers to focus on long-term planning and have greater consideration of environmental impacts, rather than only focusing on yield-scaled profit.

Principles of INM

The following points are the main principles of INM to be considered in planning the integrated nutrient management strategies (Cassman et al., 2002; Chen et al., 2011; Li et al., 2015; Wu and Ma, 2015; Selim, 2020).

1. Adopting INM practices must be compatible with the local farming system, for example, field biological condition (weeds, insects, and diseases), soil texture, irrigation services, and equipment's availability and climatic conditions.
2. Using both sources of nutrients optimises fertiliser input, increases the use efficiency by crops and then lessens their nutrient requirements and reduces food contamination, environmental pollution, and total costs.
3. Corresponding soil nutrient supplies must be ensured with crop demand spatially and temporally in order to conserve native soil contents.
4. Applying INM practice on root zone, where the most important process between plant and soil occur, is a 'bottleneck' controlling nutrient conversions, solubility, availability, and release from soil to plant roots and absorption.
5. Considerable attention must be focused on all possible nutrient sources that can be used as a tool of nutrient sources in planning nutrient input programmes for optimising nutrient-use efficiency and high yield production.
6. Nutrient losses, especially in the intensive agriculture system should be minimised.

Key components of INM

1. Integration of soil fertility restoring crops like green manures, legumes, etc.
2. Recycling of crop residues
3. Use of organic manures like FYM, compost, vermicompost, biogas slurry, poultry manure, biological composts, pressmud cakes, phosphocompost, etc.

4. Utilisation of biological agent
5. Use of efficient genotypes
6. Balanced use of fertiliser nutrients as per the requirement and target yields

Advantages of INM

The main objective of the integrated nutrient management is to maintain economic yield for a long period with little effect on native soil fertility and environmental pollution, making some changes in farmers' awareness toward the eco-friendly technique for producing healthy food free from contaminants and insuring satisfactory economic returns. INM provides following benefits (Prasad et al., 2002; Zhang et al., 2012; Selim and Al-Owied, 2017; Selim, 2018).

1. Systems can improve the soil nutrient natives and increase the solubility and availability of fertilisers to be used
2. Use the harmonious behaviour of nutrient supplies and making them match with the crop requirements
3. Offer the nutritional balance to the crops and lessen the aggressive effects resulting from the opposite impact between nutrient fractions and nutrient imbalance
4. Advance and sustain the physiochemical and biological functions of soil properties
5. Reduce the rate of soil degeneration, water, and ecosystem by enhancing carbon confiscation and decreasing nutrient losses to ground and surface water forms and/or to environment pollution
6. Minimise higher total costs of production and raise the farmer's returns (increasing profitability)
7. Improve the resistance to both biotic and abiotic stresses
8. An effective method of agricultural practices to ensure healthy food, covering population food demands with many soil and environmental impacts
9. Save the total costs at the satisfactory level with an increase in crop production
10. INM can have positive effects on the susceptibility or plant resistance against many types of biotic and abiotic stresses
11. INM enables to explore a larger volume of soil in order to access water and nutrients; improved root development enables the plant to absorb water from deeper soil layers and then reflect an increase in the ability of crops toward drought resistance

12. Changes in awareness of farmers towards the climate changes, which have greater ecological impacts in order to produce safe food rather than achieving higher yield aiming at attaining higher profit
13. INM practice reduces the fertiliser loss, improves the crop potential, and increases the profitability
14. In long-term, INM improves the physiochemical properties of the soil (biological and hydrological properties)

REFERENCES

Abbott, L. K. and Murphy, D. V. (eds.). 2003. Soil biological fertility. A key to sustainable land use in agriculture. Kluwer Academic Publishers, Dordrecht, the Netherlands.

Addiscott, T. M. and Whitmore, A. P. 1987. Computer simulation of changes in soil mineral nitrogen and crop nitrogen during autumn, winter and spring. *Journal of Agricultural Science* **109**: 141-157.

Arnon, D. I. and Stout, P. R. 1939. The essentiality of certain elements in minute quantity for plants with special reference to copper. *Plant Physiology* **14**: 371-375.

Barraclough, P. B. 1986. The growth and activity of winter wheat roots in the field: nutrient inflows of high-yielding crops. *Journal of Agricultural Science* **106**: 53-59.

Cassman, K. G., Dobermann, A. and Walters, D. T. 2002. Agro ecosystems, nitrogen use efficiency and nitrogen management. *AMBIO: A Journal of the Human Environment* **31**: 132-140.

Chen, X. P., Cui, Z. L., Vitousek, P. M., et al. 2011. Integrated soil crop system management for food security. *Proceedings of the National Academy of Sciences* **108**(16): 6399-6404.

Chesworth, W. 1991. Geochemistry of Micronutrients. In: Mortvedt, J. J., Cox, F. R., Shuman, L. M. and Welch, R. M. (eds.). Micronutrients in Agriculture. Soil Science Society of America, Madison. p. 1-30.

Clark, L. J., Gowing, D. J. G., Lark, R. M., Leeds-Harrison, P. B., Miller, A. J., Wells, D. M., Whalley, W. R. and Whitmore, A. P. 2005. Sensing the physical and nutritional status of the root environment in the field: a review of progress and opportunities. *Journal of Agricultural Science* **143**: 347-358.

Conway, G. R. and Pretty, J. N. 1991. Unwelcome harvest: agriculture and pollution. Earthscan Publications Ltd., London.

Dobermann, A. and Cassman, K. C. 2004. Environmental dimensions of fertiliser nitrogen: what can be done to increase nitrogen use efficiency and ensure global food security? In: Mosier, A. R., Syers, J. K. and Freney, J. R. (eds.). Agriculture and the nitrogen cycle. Assessing the impacts of fertiliser use on food production and the environment. Island Press, Washington, D.C. pp. 261-278.

Ferguson, R. B., Gotway, C. A., Hergert, G. W. and Peterson, T. A. 1996. Soil sampling for site-specific nitrogen management. In: Robert, P. C., et al. (eds.). Precision agriculture. American Society of Agronomy, Madison, Wisconsin. pp. 13-22.

Ferguson, R., Maharjan, B., Wortmann, C. and Krienke, B. 2019. Nitrogen inhibitors for improved fertiliser use efficiency. https://cropwatch.unl.edu/2019/nitrogen-inhibitors-improved-fertiliser-use-efficiency

Gareau, S. E. 2004. Analysis of plant nutrient management strategies: conventional and alternative approaches. *Agriculture and Human Values* **21**: 347-353.

Goulding, K. W. T. 2000. Nitrate leaching from arable and horticultural land. *Soil Use and Management* **16**: 145-151.

Goulding, K. W. T., Stockdale, E. A., Fortune, S. and Watson, C. 2000. Nutrient cycling on organic forms. *Journal of the Royal Agricultural Society of England* **161**: 65-75.

Goulding, K., Jarvis, S. and Whitmore, A. 2008. Optimising nutrient management for farm systems. *Philosophical Transactions of the Royal Society B* **363**: 667-680.

Gruhn, P., Goletti, F. and Y,udelman, M. 2000. Integrated nutrient management, soil fertility, and sustainable agriculture: current issues and future challenges. International Food Policy Research Institute, Washington, D. C. pp. 31.

Grundon, N. J., Edwards, D. G., Taakar, P. N., Asher, C. J. and Clark, B. R. 1987. Nutritional disorders of grain sorghum. Australian Centre for International Agricultural Research, Canberra. 99 pp.

Hassink, J. and Whitmore, A. P. 1997. A model of the physical protection of organic matter in soils. *Soil Science Society of America Journal* **61**: 131-139.

Haygarth, P. M., Hepworth, L. and Jarvis, S. C. 1998. Forms of phosphorus transfer in hydrological pathways from soil under grazed grassland. *European Journal of Soil Science* **49**: 65-72.

Haygarth, P. M., Jarvis, S. C., Chapman, P. and Smith, R. F. 1998a. Phosphorus budgets for two contrasting grassland farming systems in the UK. *Soil Use and Management* **14**: 160-167.

Jarvis, S. C. 1998 Nitrogen management and sustainability. In: Cherney, J. H. and Cherney, D. J. R. (eds.). Grass for dairy cattle. CABI Publishing, Wallingford, UK. pp. 161-192.

Jat, L. K., Singh, Y. V., Meena, S. K., Meena, Sunita. K., Parihar, M., H.S. Jatav, H. S., Meena, R. K. and Meena, V. S. 2015. Does integrated nutrient management, enhance agricultural productivity? *Journal of Pure and Applied Microbiology* **9**(2): 1211-1221.

Katyal, J. C. and Randhawa, N. S. 1983. Micronutrients. Food and Agricultural Organisation of the United Nations, Rome. 82 pp.

Kitchen, N. R. and Goulding, K. W. T. 2001. On-farm technologies and practices to improve nitrogen use efficiency. In: Follett, R. F. and Hatfield, J. L. (eds.). Nitrogen in the environment: sources, problems, and management. Elsevier, Amsterdam, the Netherlands. pp. 335-369.

Krupnik, T. J., Six, J., Ladha, J. K., Paine, M. J. and van Kessel, C. 2004. An assessment of fertiliser nitrogen recovery efficiency by grain crops. In: Mosier, A. R., Syers, J. K.

and Freney, J. R. (eds.). Agriculture and the nitrogen cycle. Assessing the impacts of fertiliser use on food production and the environment. Island Press, Washington, D.C. pp. 193-207.

Kumwenda, J. D. T., Waddington, S. R., Snapp, S. S., Jones, R. B. and Blackie, M. J. 1996. Soil fertility management research for the maize cropping systems of smallholders in southern Africa: a review. Natural Resources Group Paper 96-02. International Maize and Wheat Improvement Centre (CIMMYT), Mexico City.

Li, Y., Simunek, J., Zhang, Z., Jing, L. and Ni, L. 2015. Evaluation of nitrogen balance in a direct-seeded rice field experiment using Hydrus-1D. *Agricultural Water Management* **148**: 213-222.

Mengel, K. and Kirkby, E. A. 1987. Principles of Plant Nutrition, International Potash Research Institute, Berne, Switzerland. 687 pp.

Milne, A. E., Lark, R. M., Addiscott, T. M., Goulding, K. W. T., Webster, C. P. and O'Flaherty, S. 2004. Wavelet analysis of the scale- and location-dependent correlation of modelled and measured nitrous oxide emissions from soil. *European Journal of Soil Science* **56**: 3-17.

Nicholas, D. J. D. 1961. Minor mineral nutrients. *Annual Review of Plant Physiology* **11**: 314-324.

Olesen, J. E., Sørensen, P., Thomsen, I. K., Eriksen, J., Thomsen, A. G. and Berntsen, J. 2004. Integrated nitrogen input systems in Denmark. In: Mosier, A. R., Syers, J. K. and Freney, J. R. (eds.). Agriculture and the nitrogen cycle: Assessing the impacts of fertiliser use on food production and the environment. Island Press, Washington, D.C. pp. 129-140.

Paarlberg, R. L. 1994. Countrysides at risk: the political geography of sustainable agriculture. Overseas Development Council, Washington, DC.

Powlson, D. S. 1994. The soil microbial biomass - before, beyond and back. In: Ritz, K., Dighton, J. and Giller, K. E. (eds.). Beyond the biomass: compositional and functional analysis of soil microbial communities. Wiley, Chichester, U.K. pp. 3-20.

Prasad, P. V. V., Satyanarayana, V., Murthy, V. R. K. and Boote, K. J. 2002. Maximising yields in rice-groundnut cropping sequence through integrated nutrient management. *Field Crops Research* **75**(1): 9-21.

Prasad, R. and Power, J. F. 1997. Soil fertility management for sustainable agriculture. CRC Press, Lewis Publishers, New York.

Rodriguez, R. M. 2011. Soil fertility: An important aspect of sustainable agriculture. https://www.researchgate.net/publication/215828150_Soil_Fertility_-_An_Important_Aspect_of_ Sustainable_Agriculture

Roy, R. N., Finck, A., Blair, G. J. and Tandon, H. L. S. 2006. Plant nutrition for food security. FAO, Rome.

Saikia, P., Bhattacharya, S. S. and Baruah, K. K. 2015. Organic substitution in fertiliser schedule: impacts on soil health, photosynthetic efficiency, yield and assimilation in wheat grown in alluvial soil. *Agriculture, Ecosystems & Environment* **203**: 102-109.

Selim, M. M. 2018. Potential role of cropping system and integrated nutrient management on nutrients uptake and utilisation by maize grown in calcareous soil. *Egyptian Journal of Agronomy* **40**(3): 297-312.

Selim, M. M. 2020. Introduction to the integrated nutrient management strategies and their contribution to yield and soil properties. *International Journal of Agronomy*. https://doi.org/10.1155/2020/2821678

Selim, M. M. and Al-Owied, A. J. A. 2017. Genotypic responses of pearl millet to integrated nutrient management. *Bioscience Research* **14**(2): 156-169.

Smaling, E. M. A. 1993. Soil nutrient depletion in Sub-Saharan Africa. In: Van Reuler, H. and Prims, W. H. (eds.). The role of plant nutrients for sustainable food crop production in Sub-Saharan Africa. Dutch Association of Fertiliser Producers, Vereniging van Kunstmest Producenten, Leidschendan, the Netherlands.

Smaling, E. M. A. and Braun, A. R. 1996. Soil fertility research in Sub-Saharan Africa: new dimensions, new challenges. *Communications in Soil Science and Plant Analysis* **27**(3 & 4): 365-386.

Stafford, J. V., Lark, R. M. and Bolam, H. C. 1999. Using yield maps to regionalise fields into potential management units. In: Robert et al. (eds.). Precision agriculture. American Society of Agronomy, Madison, Wisconsin. pp. 225-237

Tanaka, A. and Park, Y. D. 1966. Significance of the absorption and distribution of silica in the growth of the rice plant. *Soil Science and Plant Nutrition* **12**: 23-28.

Tandon, H. L. S. 1992. Fertilisers, organic manures, recyclable wastes and biofertilisers: components of integrated plant nutrition. Fertiliser Development and Consultation Organisation, New Delhi.

Tisdale, S. L., Nelson, W. L. and Beaton, J. D. 1985. Soil Fertility and Fertilisers, 4th ed.. Macmillan, New York. 756 pp.

Wu, W. and Ma, B. 2015. Integrated nutrient management for sustaining crop productivity and reducing environmental impact: a review. *Science of the Total Environment* **512**: 415-427.

Zhang, F., Cui, Z., Chen, X., Ju, X., Shen, J., Chen, Q., Liu, X., Zhang, W., Mi, G., Fan, M. and Jiang, R. 2012. Integrated nutrient management for food security and environmental quality in China. *Advances in Agronomy* **116**: 1-40.

7

Water Management for Sustainable Agriculture

Water is a critical input for agricultural production and plays an important role in sustainable agriculture and food security. Agriculture is the principal user of water, accounting for around 70% of total withdrawals. Producing each calorie of food requires approximately one litre of water. Thus, to provide each consumer with 1,800 calories per day, Asia-Pacific region, the major rice-growing area would need an additional 2.4 billion cubic meters of water per day by 2050. With demand and competition for water on the rise, the planet's water resources are under increasing stress due to climate change, poor management and pollution (FAO, 2015).

The uses within the agricultural sector are very diverse and include mainly irrigation, pesticide and fertiliser application, and sustaining livestock. Further, along the value chain, water is used for food preservation and processing. Irrigated agriculture represents 20% of the total cultivated land and contributes 40% of the total food produced worldwide. Irrigated agriculture leads to more intensification of agricultural production and can be up to twice as productive as rainfed cultivation systems. Thus, it is certain that water consumption for agriculture will keep growing. Water use in agriculture not only consumes resources quantitatively, but also pollutes the valuable resource with pesticides and fertilisers.

Due to population growth, urbanisation, and climate change, competition for water resources is expected to increase, with a particular impact on agriculture. The future demand on water by all sectors will require as much as 25 to 40% of water to be reallocated from lower to higher productivity and employment activities, particularly in water stressed regions. In most cases, such reallocation is expected to come from agriculture due to its high share of water use. Inter-sectoral water reallocations and significant shifts of water away from agriculture will also need to be accompanied by improvements in water use efficiency and improvements in water delivery systems. Improving agriculture's water management is therefore essential to a sustainable and productive agro-food sector. Resolving the challenges of the future requires a thorough reconsideration of how water is managed in the agricultural sector, and how it can be repositioned in the broader context of overall water resources management and water security (World Bank, 2020).

WATER AND AGRICULTURAL PRODUCTION

Water requirement is the quantity of water, regardless of its source, required by a crop or diversified pattern of crops in a given period of time of its normal growth under field conditions at a given place. Water is required to meet the demands of evapotranspiration and the metabolic activities of the plants. These two demands are together known as consumptive use. Normally the water required for the metabolic activities is negligible and hence water requirement is usually equal to evapotranspiration. In addition to evapotranspiration demands, water requirement also includes losses during the application of irrigation water and the quantity of water required for special operations such as land preparation, transplanting, leaching, etc.

Water plays a vital role in crop plant life. It is essential to plants in the following ways.

1. Water is a structural constituent of plant cells and it maintains the cell form through turgor pressure. When plenty of water is available, cells are turgid and plants retain their structural form. Water accounts for the largest part of the bodyweight of an actively growing plant and it constitutes 85-90% of bodyweight of young plants and 20 to 50% of older or mature plants.
2. Water is a source of two essential elements, oxygen and hydrogen required for synthesis of carbohydrate during photosynthesis.
3. Water serves as a solvent of substances and a medium in plants allowing metabolic reactions to occur. It also acts as a solvent of plant nutrients and helps in uptake of nutrients from soils. Plants also absorb nutrients through leaves from nutrient sprays.
4. Water acts as a carrier of food materials synthesised in plants.
5. Transpiration is a vital process in plant and it occurs at a potential rate as long as water is available in adequate amount.
6. Adequate supply of water maintains the turgor pressure of guard cells helping stomata to open fully. Water deficit, on the other hand, closes stomata partially or completely reducing water loss through transpiration.
7. Water deficit slows down the growth processes.
8. Leaves get heated up with solar radiation. Plants dissipate heat by increased transpiration. Water acts as a buffer against high or low temperature injury as it has high heat by vapourisation and high specific heat.
9. Water, when available in plenty, encourages good growth, development and yield of plants. Conversely, plants die when water supply is curtailed down.

In rainfed agriculture, gaps persist between actual and potential yields and opportunities exist to improve yields and water productivity without irrigation. Rainwater harvesting, as well as supplemental irrigation, can substantially improve rainfed agriculture. On-farm water conservation, particularly the adoption of agricultural practices that reduce runoff and increase the infiltration and storage of water in the soil in rainfed agriculture, is the most relevant local supply enhancement option that farmers have to increase production. On a slightly larger scale, small, decentralised water harvesting and storage systems contribute to increasing water availability and agricultural production at the household and community levels. These small-scale measures promote local economic development and increase the climate resilience of local communities. Increased rainfall intensity due to climate change may increase the availability of surface water, but will result in more runoff, higher soil erosion and lower soil infiltration causing more moisture stress on plants and reduced recharge of groundwater. More intense and longer droughts will expose crops to moisture stress, and reduce rainfed yields and quality.

Agriculture currently uses about 70% of the total water withdraw, mainly for irrigation. The intensive irrigation could provide for the growth of irrigated areas and guarantee increased food production. In the 1980s, the global rate of increase in irrigated areas slowed considerably, mainly due to very high cost of irrigation system construction, soil salinisation, the depletion of irrigation water-supplying sources, and the problems of environmental protection. However, as the population is growing at a rapid rate, irrigation is being given an important role in increasing land use and cattle-breeding efficiency. Thus, irrigated farming is expected to expand rapidly in the future with subsequent increase of water use for irrigation (Chartzoulakisa and Bertaki, 2015).

Irrigation is not sustainable if water supplies are not reliable. Especially in areas of water scarcity the major need for development of irrigation is to minimise water use. Effort is needed to find economic crops using minimal water, to use application methods that minimise loss of water by evaporation from the soil or percolation of water beyond the depth of root zone and to minimise losses of water from storage or delivery systems. Nowadays, during a period of dramatic changes and water resources uncertainty there is a need to provide some support and encouragement to farmers to move from their traditional high-water demand cropping and irrigation practices to modern, reduced demand systems and technologies.

Under scarcity conditions considerable effort has been devoted over time to introduce policies aiming to increase water efficiency based on the assertion that more can be achieved with less water through better management. Better management usually refers to improvement of allocative and/or irrigation water efficiency. The former is closely related to adequate pricing, while the latter depends on the type of irrigation technology, environmental conditions and on scheduling of water application.

Crop yield increases with water availability in the root zone, until saturation level, above which there is little effect (Hillel, 1997). Over-irrigation can cause among others temporal water shortage to other farmers, water-logging conditions for the crop, favourable environment for disease development, loss of nutrients due to leaching or deep percolation, contamination of the aquifers from agrochemicals, reduction of crop yield and deterioration of the quality and increase of production cost.

FACTORS AFFECTING WATER USE EFFICIENCY

Water use efficiency (WUE) is a concept introduced more than 100 years ago by Briggs and Shantz (1913) showing a relationship between plant productivity and water use. They introduced the term, WUE, as a measure of the amount of biomass produced per unit of water used by a plant. Water use efficiency (or transpiration efficiency) describes the intrinsic trade-off between carbon fixation and water loss that occurs in plants because water evaporates from the interstitial tissues of leaves whenever stomata open for CO_2 acquisition. The transpiration efficiency of crop plants is generally low as huge quantity of water is transpired in comparison with the small amounts of carbon that are fixed by photosynthesis (Bramley et al., 2013). Typical crop plants transpire 200-1000 g of water per g of assimilated carbon (Martin et al., 1976).

Climatic factors

Weather affects both crop yield and evapotranspiration. Evapotranspiration (ET) is an evaporative process largely controlled by climatic factors; it is influenced by the weather elements in the following four ways (Singh et al., 2014).

1. **Temperature:** An increase in temperature brings about an increase in the rate of evapotranspiration.
2. **Relative humidity:** When the relative humidity is high, the rate of evapotranspiration decreases. This is because the atmosphere is saturated with moisture.
3. **Solar radiation:** Incident solar radiation and adverted heat provide energy for the evaporative process.
4. **Wind velocity:** When the wind is stagnant the rate of ET remains normal, when it is blowing gently the rate increases because it removes moisture from the vicinity and when the wind is blowing violently the rate of transpiration decreases because it creates hindrances in the outward diffusion of water vapours from the transpiring parts due to closing of the stomata.

Climate change also brings a change in water use by plants. There are four factors changing in the climate such as increasing carbon dioxide concentrations, increasing temperatures, more variable precipitation, and variations in humidity will also affect the water use efficiency. Trenberth (2011) and Collins et al. (2013) projected the following changes in climate.

1. Global mean temperatures will increase throughout the 21st century if CO_2 concentrations continue to increase and under the highest CO_2 emission scenario, increases range from 2.6 to 4.8° C.
2. Temperature changes will not be uniform across regions with land surfaces warming more than over the oceans.
3. Increasing global temperatures will result in more hot extremes and fewer cold extremes at both daily and seasonal time scales.
4. Precipitation will increase with increases in global mean surface temperature and could increase 1-3% per degree C; however, there will be substantial spatial variation in these changes.
5. The water holding capacity of air increases by 7% per degree C. The air can take up more water, and water vapour inclines. That leads to higher intensity of precipitation, i.e., higher amount of rainfall per rain event.
6. Annual surface evaporation will increase with temperature increases; however, over land, evaporation will be linked to precipitation.

These changes in climate will increase atmospheric water demand by crops and increase the potential for limitations in soil water availability, because of the increased variation in precipitation during the growing season and even more so in soils with limited water holding capacity (Hatfield and Dold, 2019).

Edaphic factors

Edaphic factors are the major detriment of the water use efficiency. Soil texture is the combination of relative proportion of sand, silt and clay, and has a direct role in the water holding capacity of soil. Water use efficiency in field crops is generally higher under loam textured soils. Soil depth represents the effective root zone depth. Another very important factor affecting water use efficiency is the soil structure. Soil aggregates influences the ratio of soil macropores and micropores (Singh et al., 2014).

Nature of the plant

There are considerable differences between plant species to produce a unit of dry matter per unit amount of water used resulting in widely varying values of WUE. There are also differences in WUE between varieties of the same crop. These variations are due to variations in their genetic build-up which affects both morphological traits controlling the rate of transpiration and water absorption by roots from the soil and the physiological functions responsible for photosynthesis, respiration, translocation of photosynthates to economically harvested plant parts. Varieties also differ in their adaptation to environment, resistance to pests and diseases and management levels. Selection of properly adopted crops, with good rooting habits, low transpiration rates and improved energy consumption in photosynthesis will increase WUE (Singh et al., 2008).

Agronomic practices

Planting pattern: Planting patterns have a direct effect on yield, solar energy capture, and evaporation and thus an indirect effect on water use efficiency. Planting pattern influences crop yield through its influence on light interception, rooting pattern and moisture extraction pattern. Usually square arrangement of plants is more efficient in the utilisation of light, water and nutrients available to the individual plants than in a rectangular arrangement. Close row-spacing generally gives more uniform distribution of plants over a given area, thus making the plant canopy more effective in intercepting radiant energy and shading weeds. It also reduces the raindrop impact on soil structure in the surface layer. Different planting patterns are followed to suit different cropping systems. Suitable rotational cropping and intercropping techniques should be adopted in order to maintain the crop yield (Meena et al., 2013).

Planting date: Sowing/planting dates should be planned in such a way that the stress periods does not coincide with anthesis of the crop. In arid and semiarid regions planting date is an extremely important cultural practice in efficient water use.

Plant population: Higher yield potential of the crops is achieved by the favourable water regime, the higher soil fertility level and the genetic potential of new varieties coupled with appropriate adjustments in plant population. To obtain maximum possible yield, it is essential for the crop to utilise as efficiently as possible all the available production factors such as water, nutrients, light and CO_2. Higher yield and WUE is possible only through optimum levels of soil moisture regime, plant population and proper fertilisation (Singh et al., 2013).

Fertilisation: When soil moisture tension is low, the abilities of plant to absorb nutrients and the soil to supply them are optimal and therefore, nutrient availability is at its highest level.

TECHNIQUES TO INCREASE WATER USE EFFICIENCY IN RAINFED AGRICULTURE

Agronomic management practices

With diminishing fresh water supplies, the threat of more frequent droughts due to climate change, and the increasing demand for more food crops, the possibility of increasing agricultural water productivity through agronomic means has received much attention (Passioura and Angus, 2010). Conservation agriculture, improved irrigation management, and other agronomic practices have made significant contributions to improving yields (Turner and Asseng, 2005). Sustainable water management in agriculture aims to match water availability and water needs in quantity and quality, in space and time, at reasonable cost and with acceptable environmental impact.

Tillage practices for moisture conservation: Tillage influences crop yields and water use efficiency. The principal effects of tillage are the preparation of seedbed conducive to the germination of seed and growth of seedling, conservation of soil moisture in unirrigated/ rainfed areas that influence infiltration characteristics of the soil and providing adequate soil depth for optimum root growth, proper placement of seeds and fertilisers in the soil and inter-cultivation for weed control. Deep tillage (deep ploughing and sub-soiling) is considered beneficial to conserve the monsoon rainfall (Meena et al., 2013). The general objectives of deep tillage are to break through plough soles and layers compacted by excessive use of heavy implements, impermeable soil horizons or other barriers to the movement of moisture and growth of roots through the soil profile.

Mulching: About 70% of the rainfall is lost through evaporation. Mulching restricts evaporation loss and conserves water in the field. Different types of mulches are stubble mulch, soil/dust mulch, straw mulch, plastic mulch and vertical mulching. Stubble mulching is based on stirring the soil with implements that leave considerably effective part of the vegetative material, crop residues or vegetative litter on the surface as a protection against erosion and for conserving moisture by favouring infiltration and reducing evaporation (Rana et al., 2003).

Intercropping: Intercropping systems are generally recommended for rainfed crops to get stable yields. The total water used in intercropping system is almost the same as for sole crops, but yields are increased, thus water use efficiency is higher than sole crops.

Direct seeded rice: It is an efficient method of water saving in rice. If adequate plant population is maintained and effective weed management practices are followed in direct seeded rice, there will be absolutely no low yield than the transplanted rice.

Seeding equipment: The moisture in the surface layer of the seed bed in dry areas is evaporated quickly. Germination of seeds may be adversely affected if the seeds are sown at shallow depths of 4 to 5 cm. Thus, the seeding equipment should be designed for rainfed areas to place the seed in firm moist soil.

Selection of crop and variety: Selection of crops and their varieties adapted to the total amount and distribution of water available and adoption of suitable agronomic practices are important factors in the success of agriculture in the rainfed areas. With the development of short-duration varieties of pulses, oilseeds and other crops, several double-cropping and mixed cropping patterns can be standardised for different regions.

Planting techniques/methods: Adoption of proper planting techniques/methods increases water use efficiency. Broad-bed and furrow method developed by ICRISAT for groundnut increases WUE.

Weed control: Weeds compete with crops for soil nutrients, water and light. A weed free crop uses water most efficiently.

Rainwater harvesting: Management techniques that increase infiltration and soil water storage, and decrease water losses by runoff, evaporation and evapotranspiration by weeds would lead to increase the amount of water retained in the soil for subsequent use by crops, as considered *in situ* water harvesting. Rainwater harvesting will not only conserve the soil, its fertility and vegetation, but also could be utilised as supplementary irrigation that will be advantageous in enhancing total water supply available to crop plants during low rainfall period (Singh et al., 2014). For *ex situ* rainwater harvesting, farm ponds are made either by constructing an embankment across water coarse or by excavating a pit. Runoff water is collected from a catchment and stored in farm pond. The stored water is utilised for supplemental irrigation during long dry spells at critical stages of crop growth. Farm ponds hold a great promise as a life-saving device for rainfed crops in low and erratic rainfall areas (Pandey et al., 2012).

Mechanical measures

Contour farming: Contour farming involves ploughing, planting and weeding along the contour, i.e., across the slope rather than up and down. Contour lines are lines that run across a slope such that the line stays at the same height and does not run uphill or downhill. Experiments show that contour farming alone can reduce soil erosion by as much as 50% on moderate slopes. However, for slopes steeper than 10%, other measures should be combined with contour farming to enhance its effectiveness. Contour ridges are used mainly in semiarid areas to harvest water, and in higher rainfall areas for growing potatoes. Trash lines made by laying crop residues or trash in lines along the contour. Grass barrier strips planted along the contour with fodder grass such as Napier, or are left with natural grass. They are effective soil conservation measures on soils that absorb water quickly, and on slopes as steep as 30%.

Terracing: Terraces are used in farming to cultivate sloped land. Graduated terrace steps are commonly used to farm on hilly or mountainous terrain. Terraced fields decrease erosion and surface runoff, and are effective for growing crops requiring much water, such as rice. Natural terracing, the result of small-scale erosion, can occur where cattle are grazed for long periods on steep sloping pasture

Chemical methods

Reducing transpiration is the most effective means of increasing the amount of water available to the crop. Antitransparants are the materials or chemicals that are used to reduce the transpiration. These chemicals reduce the transpiration either by closing the stomata or by forming the film on the leaf surface. Nearly 99% of the water absorbed by the plant is lost in transpiration. The common types of antitransparants are as follows.

1. **Stomatal closing type:** Most of the transpiration occurs through the stomata on the leaf surface. Some fungicides like phenyl mercuric acetate (PMA) and herbicides like atrazine in low concentration serve as antitranspirants by inducing stomatal closing.
2. **Film forming type:** Plastic and waxy material which form a thin film on the leaf surface and result into physical barrier that reduce transpiration.
3. **Reflectance type:** These are white materials which form a coating on the leaves and increase the leaf reflectance. By reflecting the radiation, vapour pressure gradient is reduced and thus transpiration rate is reduced. Application of 5% kaolin spray reduces transpiration losses.
4. **Growth retardant:** These chemicals reduce shoot growth and increase root growth and thus enable the plants to resist drought. They may also induce stomatal closure. Cycocel is useful for improving water status of the plant.

TECHNIQUES TO INCREASE WATER USE EFFICIENCY IN IRRIGATED AGRICULTURE

Irrigation is the artificial supply of water to the soil and the standing crop for enhanced growth and production. Irrigation supplements the quantity of water received from precipitation and the contribution of the soil from groundwater. It is applied for preparation of land before sowing or planting, for leaching of salts of saline soils, to regulate soil-plant temperature and also to restrict the population of soil borne pests. The application of water to a growing crop is essential for maintenance of optimum soil moisture balance suitable for the crop and its cultural practices. The entire amount of water released from the source does not reach the crop field and the volume of water entering the field is not utilised by the crop. A considerable amount of irrigation water is lost during conveyance, by deep percolation and seepage in the field. Thus, to increase the irrigation water use efficiency several protective measures have been suggested.

Usually, much attention is given to irrigation scheduling, i.e., when to irrigate and how much water to apply, giving the least attention to irrigation methods, i.e., how to apply the water in the field. Many parameters like crop growth stage and its sensitivity to water stress, climatic conditions, soil type and water availability in the soil determine when to irrigate or the irrigation frequency. However, this frequency depends upon the irrigation method and therefore, both irrigation scheduling and the irrigation method are interrelated and equally important for sustainable water management.

Localised irrigation

Localised irrigation is widely recognised as one of the most efficient methods of watering crops (Keller and Bliensér, 1990). Localised irrigation systems (drip irrigation, micro-sprayers) apply the water to individual plants by means of plastic pipes, usually laid on the ground surface. With drip irrigation water is slowly applied through small emitter openings from plastic pipes with discharge rate ≤ 12 l/h. With micro-sprayer (micro-sprinkler) irrigation water is sprayed over the part of the soil surface occupied by the plant with a discharge rate of 12 to 200 l/h. The main aims of localised irrigation are the application of water directly into the soil at root zone of crop plants, the reduction of water losses during or after water application and the reduction of the labour cost. Improvements in localised irrigation systems aiming to reduce the volumes of water applied and increase the water use efficiency include the use of a single drip line for a double row crop, the use of micro-sprayers in high infiltration soils, the adjustment of duration of water application and timing to soil and crop characteristics, the control of pressure and discharge variations, the use of appropriate filters to the water quality and the emitter characteristics used, the adoption of careful maintenance, automation, fertigation and chemigation.

Irrigation scheduling

Irrigation scheduling is the decision making process for determining when to irrigate the crops and how much water to apply. It forms the sole means for optimising agricultural production and for conserving water and it is the key to improving performance and sustainability of the irrigation systems. It requires good knowledge of the crops' water requirements and of the soil water characteristics that determine when to irrigate, while the adequacy of the irrigation method determines the accuracy of how much water to apply. With appropriate irrigation scheduling deep percolation and transport of fertilisers and agrochemicals out of the root-zone is controlled and waterlogging is avoided, water and energy are saved, optimum soil water conditions are created for plant growth, higher yields and better quality are obtained and rising of saline water table is avoided. In water scarce regions, irrigation scheduling is more important than under conditions of abundant water, since any excess in water use is a potential cause for deficit for other users or uses (Chartzoulakisa and

Bertaki, 2015). Irrigation scheduling techniques and tools varies greatly and has different characteristics relative to their applicability and effectiveness. Timing and depth criteria for irrigation scheduling can be established by using several approaches based on soil water measurements, soil water balance estimates and plant stress indicators.

Soil water estimates and measurements: Soil water affects plant growth directly through its controlling effect on plant water status. There are two ways to assess the availability of soil water for plant growth; by measuring the soil water content and by measuring how strongly that water is held in the soil (soil water potential). The accuracy of the information depends on the sampling methods adopted. Soil water estimates and measurements used for irrigation scheduling include soil appearance and feel, soil water content measurement, soil water potential measurement by tensiometers or soil spectrometers, remotely sensed soil moisture.

Crop stress parameters: Instead of measuring or estimating the soil water parameters, it is possible to get a signal from the plant itself indicating the time of irrigation but not defining the irrigation depths. This message can either come from individual plant tissues, where a correct sampling is required, or from the canopy as a whole. Therefore, crop stress parameters are useful when irrigation depths are predefined and kept constant during the irrigation season. Crop water stress parameters include leaf water content and leaf water potential, changes in stem or fruit diameter, sap flow measurement, canopy temperature, remote sensing of crop stress.

Climatic parameters: Climatic parameters are widely used for local or regional irrigation scheduling. Weather data and empirical equations are once locally calibrated which provide accurate estimates of reference evapotranspiration (ETo) for a given area. Then, crop evapotranspiration (ETc) is estimated using appropriate crop coefficients. Information may be processed in real time or, more often, using historical data. These techniques include evaporation measurements for ETo calculation, and assessment of crop evapotranspiration using climatic data like air temperature, relative humidity, wind speed and sunshine hours.

Soil-water balance: The aim of soil water balance approach is to predict the water content in the rooted soil. Soil water holding characteristics, crop and climate data are used by sophisticated models to produce typical irrigation calendars. This approach can be applied from individual farms to large regional irrigation schemes. Its effectiveness is very high, but depends on farm technological development and/or support services. Several commercial software for irrigation scheduling based on soil-water balance approach are available.

The effective application of any irrigation scheduling method and effective implementation of the corresponding delivery schedule are subject to the physical capability of the collective system for delivering water according to this schedule and to the capacity of the management for operating the system properly. One of the major obstacles to effective implementation of crop-based and water-saving irrigation scheduling is the inability of most conveyance and delivery systems to deliver water at the farm gates with the reliability and flexibility required. In modern pressurised irrigation networks water is available on demand, although discharges may be limited due to technical or economic reasons (Chartzoulakisa and Bertaki, 2015).

Fertigation

The application of fertilisers through the irrigation system is now a common practice in modern irrigated agriculture. Localised irrigation systems, which could be highly efficient for water application, are also suitable for fertigation. Thus, the soluble fertilisers at concentrations required by crops are applied through the irrigation system to the wetted volume of the soil. Possible disadvantages include the non-uniform chemical distribution when irrigation design or operation are inadequate, the over-fertilisation in case that irrigation is not based on actual crop requirements and the excessive use of soluble fertilisers.

Deficit irrigation practices

In arid and semiarid regions water is a scarce resource. Higher demands for water by municipal and industrial establishments gradually reduce the water allocation to agriculture in these areas, and water availability is not enough to achieve maximum yields. Thus, the irrigation strategies should be adopted for more effective and rational use of water. Such management practices include regulated deficit irrigation, partial root drying and subsurface irrigation.

Regulated deficit irrigation: Regulated deficit irrigation is an optimising strategy under which crops are allowed to sustain some degree of water deficit and yield reduction. During regulated deficit irrigation the crop is exposed to certain level of water stress either during a particular period or throughout the growing season. The main objective of regulated deficit irrigation is to increase water use efficiency of the crop by eliminating irrigations that have little impact on yield. The resulting yield reduction may be small compared with the benefits gained through diverting the saved water to irrigate other crops for which water would normally be insufficient under conventional irrigation practices. Regulated deficit irrigation is a sustainable issue to cope with water scarcity since the allowed water deficits favour water saving and the optimisation approach leads to economical viability. To achieve the best results from deficit irrigation, the farmer should have sound knowledge of crop evapotranspiration, crop response to water deficits including the identification of critical crop growth stages, and the economic impact of yield reduction strategies (Chartzoulakisa and Bertaki, 2015).

Partial root drying: Partial root drying is a new irrigation technique, where one half of the root system is allowed to drying conditions by withholding irrigation and the other half is irrigated. Wetted and dried sides of the root system alternate on a 7-14 day cycle. Partial root drying uses biochemical responses of plants to water stress to achieve balance between vegetative and reproductive growth. During water stress, the first line of defence of the plant is to close its stomata to conserve moisture. One of the principal compounds that elicit this response is the abscisic acid. As soil water availability falls following the cessation of irrigation, the abscisic acid is synthesised in the drying roots and transported to the leaves through the transpiration stream (Loveys et al., 1999). Stomata respond by reducing aperture, thereby restricting water loss. Improvement of WUE results from partial stomatal closure. However, an inevitable consequence is reduced photosynthesis. Partial root drying has been successfully applied with drip irrigation in grapevines (Dry et al., 2000), and with furrow irrigation in pear, citrus and grapevines (Clancy, 1999).

Subsurface drip irrigation

Subsurface drip irrigation is a low-pressure, low-volume irrigation system that uses buried tubes to apply water. The applied water moves out of the tubes by soil matrix suction. Wetting occurs around the tube and water moves out in the soil all directions. The potential advantages of subsurface drip irrigation are

1. water conservation,
2. enhanced fertiliser efficiency,
3. uniform and highly efficient water application,
4. elimination of surface infiltration problems and evaporation losses,
5. flexibility in providing frequent and light irrigations,
6. reduced problems of disease and weeds, and
7. lower pressure required for operation.

However, the major disadvantages are the high cost of initial installation and the increased possibility for clogging, especially when poor quality water is used. Subsurface irrigation is suitable for high value fruits and vegetables, turfs and landscapes. The depth of installation depends upon soil characteristics and crop species ranging from 15-20 cm for vegetables and 30-50 cm for tree crops.

Agricultural practices

Agricultural practices, such as soil management, fertiliser application, and disease and pest control are related with the sustainable water management in agriculture and the protection of the environment. Agricultural practice today is characterised by the abuse of fertilisers. Farmers very rarely carry out soil and leaf analyses in order to clarify the proper quantity and type of fertiliser needed for each crop and they apply them empirically. This practice increases considerably the cost of agricultural production and is potentially critical for the deterioration of the groundwater quality and the environment. Agrochemicals are also excessively used, endangering the quality of the surface water and negatively affecting the environment. Pesticides are often used preventively, even when there is not real threat in the area (Chartzoulakisa and Bertaki, 2015).

The following traditional and modern soil and crop management practices for water conservation such as runoff control, improvement of soil infiltration rate, increase soil water capacity, control of soil water evaporation, etc. can be followed (Pereira et al., 2002).

1. Soil surface tillage, which concerns shallow tillage practices to produce an increased roughness on the soil surface permitting short time storage in small depressions of the rainfall in excess to the infiltration.
2. Contour tillage, where soil cultivation is made along the land contour and the soil is left with small furrows and ridges that prevent runoff. This technique is also effective to control erosion and may be applied to row crops and small grains provided that field slopes are low.
3. Bed surface profile, which concerns cultivation of wide beds and is typically used for horticultural row crops.
4. Conservation tillage, including no-tillage and reduced tillage, where residuals of the previous crop are kept on the soil at planting. Mulches protect the soil from direct impact of raindrops, thus controlling crusting and sealing processes. Conservation tillage helps to maintain high levels of organic matter in the soil, thus it is highly effective in improving soil infiltration and controlling erosion.
5. Mulching with crop residues on soil surface which shades the soil, slows water overland flow, improves infiltration conditions, reduces evaporation losses and also contributes to control of weeds and therefore of non-beneficial water use.
6. Increasing or maintaining the amount of organic matter in the upper soil layers, since it provides for better soil aggregation, reduced crusting or sealing on soil surface and increased water retention capacity of the soil.

7. Addition of fine material or hydrophilic chemicals to sand/coarse soils increases the water retention capacity of the soil and controls deep percolation. Thus, water availability in soils with low water holding capacity is increased.
8. Control of acidity by liming, similarly to gypsum application to soils with high pH favours more intensive and deep rooting, better crop development and contributes to improved soil aggregation, thus producing some increase in soil water availability.
9. Adoption of appropriate weed control techniques alleviates competition for water and transpiration losses by weeds.

Wallace and Batchelor (1997) suggested four options for enhancing water use efficiency in irrigated agriculture (table 7.1) and pointed out that focusing on only one category will likely be unsuccessful.

Table 7.1. Options for improving irrigation efficiency at a field level

Improvement category	**Options**
Agronomic	1. Crop management to enhance precipitation capture or reduce water evaporation, e.g., crop residues, conservation till, and plant spacing
	2. Improved varieties
	3. Advanced cropping strategies that maximise cropped area during periods of lower water demands and/or periods when rainfall may have greater likelihood of occurrence
Engineering	1. Irrigation systems that reduce application losses, improve distribution uniformity, or both
	2. Cropping systems that can enhance rainfall capture, e.g., crop residues, deep chiselling
Management	1. Demand-based irrigation scheduling
	2. Slight to moderate deficit irrigation to promote deeper soil water extraction
	3. Avoiding root zone salinity yield thresholds
	4. Preventive equipment maintenance to reduce unexpected equipment failures
Institutional	1. User participation in an irrigation district or scheme operation and maintenance
	2. Water pricing and legal incentives to reduce water use and penalties for inefficient use
	3. Training and educational opportunities for learning newer, advanced techniques

Source: Wallace and Batchelor, 1997

Food and Agriculture Organisation provided a summary of practical measures recommended in order to improve water use efficiency (FAO, 1997).

Conservation of water

1. Reduce conveyance losses by lining channels or, preferably, by using closed conduits.
2. Reduce direct evaporation during irrigation by avoiding midday sprinkling. Minimise foliar interception by under-canopy, rather than by overhead sprinkling.
3. Reduce runoff and percolation losses due to over-irrigation.
4. Reduce evaporation from bare soil by mulching and by keeping the inter-row strips dry.
5. Reduce transpiration by weeds, keeping the inter-row strips dry and applying weed control measures where needed.

Enhancement of crop growth

1. Select most suitable and marketable crops for the region.
2. Use optimal timing for planting and harvesting.
3. Use optimal tillage (avoid excessive cultivation).
4. Use appropriate insect, parasite and disease control.
5. Apply manures and green manures where possible and fertilise effectively (preferably by injecting the necessary nutrients into the irrigation water).
6. Practise soil conservation for long-term sustainability.
7. Avoid progressive salinisation by monitoring water-table elevation and early signs of salt accumulation, and by appropriate drainage.
8. Irrigate at high frequency and in the exact amounts needed to prevent water deficits, taking account of weather conditions and crop growth stage.

DEVELOPMENT OF GROUNDWATER

Groundwater is the water found underground in the cracks and pores in soil, sand and rock, called aquifer. It is naturally recharged by precipitation or by infiltration from other waterbodies. The underground water movement from areas of recharge to areas of aquifer discharge is called groundwater flow. This occurs generally at low velocities through pore spaces and fractures in rock materials and depends on the geological composition of the aquifer. Groundwater levels may vary seasonally and annually and are usually high after the wet and low at the end of the dry season. Different types of agricultural activities may affect the recharge process negatively, reaching from soil sealing to soil compaction due to the use of heavy machinery, which avoid water infiltration. Crop choice and vegetation cover can also influence infiltration patterns. Regulative measures can help overcoming groundwater shortages and allow sustainable water resource management.

Overexploitation of groundwater

Although groundwater is a renewable resource, it is not inexhaustible. If groundwater supplies are to be maintained perennially, the recharge must balance discharge. However, all that recharged is not necessarily recoverable. The groundwater withdrawal must be limited to safe yield, which is defined as the amount of water that can be withdrawn annually from a groundwater basin without producing an undesired result. Overdraft of aquifers occurs if groundwater extraction exceeds the sustained yield. Overdraft may be reduced or controlled by changing over to less water consuming crops and most importantly by artificial recharge.

In many arid and hard rock areas, overdraft and associated quality problems are increasingly emerging. The overdraft of groundwater has resulted in

1. Failure of shallow wells,
2. Shortage of water supplies,
3. Deepening of wells,
4. Increased pumping lights and pumping costs,
5. Increased energy requirements,
6. Seawater ingress in coastal areas,
7. Increased inland salinity in groundwater, and
8. Land subsidence due to compression of the aquifers.

Advantages of artificial recharge of groundwater

Rainwater harvesting is the technique of collection and storage of rainwater at surface or in subsurface aquifers, before it is lost as surface runoff. The augmented resource can be harvested in the time of need. Artificial recharge to groundwater is a process by which the groundwater reservoir is augmented at rate exceeding that under natural conditions of replenishment. Artificial recharge implies diverting of excess surface water resources to the areas of needs and it facilitates infiltration of surplus water into the groundwater zones. Artificial recharge of groundwater has the following advantages (CGWB, 2003).

1. Cost of recharge to subsurface reservoir is lower than surface reservoirs
2. The aquifer serves as distribution system
3. No land is wasted for storage purpose and no population displacement is involved
4. Groundwater is not directly exposed to evaporation and pollution
5. Storing water underground is environment friendly
6. It increases the productivity of aquifer
7. It reduces flood hazards
8. Effects rise in groundwater levels
9. Mitigates the effects of drought
10. Reduces soil erosion

METHODS OF RAINWATER HARVESTING TO AUGMENT GROUNDWATER RESOURCES

The Central Groundwater Board (of India) has designed and recommended several methods and techniques for rainwater harvesting and groundwater recharge for urban and rural areas (CGWB, 2003).

Urban areas

In urban areas, rainwater available from rooftops of buildings, paved and unpaved areas goes waste. This water can be recharged to aquifer and can be utilised gainfully at the time of need. The rainwater harvesting system needs to be designed in a way that it does not occupy large space for collection and recharge system. A few techniques of rooftop rainwater harvesting in urban areas are described below.

Rooftop rainwater harvesting through recharge pit

In alluvial areas where permeable rocks are exposed on the land surface or at very shallow depth, rooftop rainwater harvesting can be done through recharge pits. The technique is suitable for buildings having a roof area of 100 m^2 and is constructed for recharging the shallow aquifers. Recharge pits may be of any shape and size and are generally constructed 1 to 2 m wide and 2 to 3 m deep which are backfilled with boulders (5-20 cm), gravels (5-10 mm) and coarse sand (1.5-2 mm) in graded form. Boulders at the bottom, gravels in between and coarse sand at the top so that the silt content that will come with runoff will be deposited on the top of the coarse sand layer and can easily be removed. For smaller roof area, pit may be filled with broken bricks/cobbles. A mesh should be provided at the roof so that leaves or any other solid waste/debris is prevented from entering the pit and a desilting/collection chamber may also be provided at the ground to arrest the flow of finer particles to the recharge pit. The top layer of sand should be cleaned periodically to maintain the recharge rate. By-pass arrangement is provided before the collection chamber to reject the first showers.

Rooftop rainwater harvesting through recharge trench

Recharge trenches are suitable for buildings having roof area of 200-300 m^2 and where permeable strata are available at shallow depths. Trench may be 0.5 to 1 m wide, 1 to 1.5 m deep and 10 to 20 m long depending upon availability of water to be recharged. These are backfilled with boulders (5-20 cm), gravel (5-10 mm) and coarse sand (1.5-2 mm) in graded form; boulders at the bottom, gravel in between and coarse sand at the top so that the silt content coming with runoff will be deposited at the top of the sand layer and can easily be removed. A mesh should be provided at the roof so that leaves or any other solid waste/debris is prevented from entering the trenches and a desilting/ collection chamber may also be provided on ground to arrest the flow of finer particles to the trench. By-pass arrangement may be provided before the collection chamber to reject the first showers. The top layer of sand should be cleaned periodically to maintain the recharge rate.

Rooftop rainwater harvesting through existing tubewells

In areas where the shallow aquifers have dried up and existing tubewells are tapping deeper aquifer, rainwater harvesting through existing tubewell can be adopted to recharge the deeper aquifers. PVC pipes of 10 cm diameter are connected to roof drains to collect rainwater. The first roof runoff is let off through the bottom of drainpipe. After closing the bottom pipe, the rainwater of subsequent rain showers is taken through a 'T' pipe to an online PVC filter. The filter may be provided before water enters the tubewells. The filter is 1.0-1.2 m in length and is made up of PVC pipe. Its diameter should vary depending on the roof area, 15 cm if roof area is less than 150 m^2 and 20 cm if the roof area is more. The filter is provided with a reducer of 6.25 cm on both the sides. Filter is divided into three chambers by PVC screens so that filter material is not mixed up. The first chamber is filled up with gravel (6-10 mm), middle chamber with pebbles (12-20 mm) and last chamber with bigger pebbles (20-40 mm). If the roof area is more, a filter pit may be provided. Rainwater from roofs is taken to collection/desilting chambers located on ground. These collection chambers are interconnected as well as connected to the filter pit through pipes having a slop of 1:15. The filter pit may vary in shape and size depending upon available runoff and are backfilled with graded material, boulder at the bottom, gravel in the middle and sand at the top with varying thickness (0.30-0.50 m) and may be separated by screen. The pit is divided into two chambers, filter material in one chamber and other chamber is kept empty to accommodate excess filtered water and to monitor the quality of filtered water. A connecting pipe with recharge well is provided at the bottom of the pit for recharging of filtered water through well.

Rooftop rainwater harvesting through trench with recharge well

In areas where the surface soil is impervious and large quantities of roof water or surface runoff is available within a very short period of heavy rainfall, the use of trench/pits is made to store the water in a filter media and subsequently recharge to groundwater through specially constructed recharge wells. This technique is ideally suited for area where permeable horizon is within 3 m below ground level. Recharge well of 100-300 mm diameter is constructed to a depth of at least 3 to 5 m below the water level. Based on the lithology of the area well assembly is designed with slotted pipe against the shallow and deeper aquifer. A lateral trench of 1.5 to 3 m width and 10 to 30 m length, depending upon the availability of water is constructed with the recharge well in the centre. The number of recharge wells in the trench can be decided on the basis of water availability and local vertical permeability of the rocks. The trench is backfilled with boulders, gravels and coarse sand to act as a filter media for the recharge wells. If the aquifer is available at greater depth, i.e., more than 20 m, a shallow shaft of 2 to 5 m diameter and 3-5 m deep may be constructed depending upon availability of runoff. Inside the shaft a recharge well of 100-300 mm diameter is constructed for recharging the available water to the deeper aquifers. At the bottom of the shaft a filter media is provided to avoid choking of recharge well.

Rural areas

In rural areas, rainwater harvesting is taken up considering watershed as a unit. Surface spreading techniques are common since space for such systems is available in plenty and quantity of recharged water is also large. Following techniques may be adopted to save water going waste through slopes, rivers, rivulets, etc.

Rainwater harvesting through gully plug

Gully plugs are built using local stones, clay and bushes across small gullies and streams running down the hill slopes carrying drainage to tiny catchments during rainy season. Gully plugs help in conservation of soil and moisture. The sites for gully plugs may be chosen whenever there is a local break in slope to permit accumulation of adequate water behind the bunds.

Rainwater harvesting through contour bund

Contour bunds are effective methods to conserve soil moisture in watershed for long duration. These are suitable in low rainfall areas where monsoon runoff can be impounded by constructing bunds on the sloping ground all along the contour of equal elevation. Flowing water is intercepted before it attains the erosive velocity by keeping suitable spacing between bunds. Spacing between two contour bunds depends on the slope of the area and the permeability of the soil. Lesser the permeability of soil, the close should be spacing of bunds. Contour bunding is suitable on lands with moderate slopes without involving terracing.

Rainwater harvesting through gabion structure

This is a kind of check dam commonly constructed across small streams to conserve streamflows with practically no submergence beyond stream course. A small bund across the stream is made by putting locally available boulders in a mesh of steel wires and anchored to the stream banks. The height of such structures is around 0.5 m and is normally used in the streams with width of less than 10 m. The excess water overflows this structure storing some water to serve as source of recharge. The silt content of stream water in due course is deposited in the interstices of the boulders in due course and with growth of vegetation, the bund becomes quite impermeable and helps in retaining surface water runoff for sufficient time after rains to recharge the groundwater body.

Rainwater harvesting through percolation tank

Percolation tank is an artificially created surface waterbody, submerging in a highly permeable land so that surface runoff is made to percolate and recharge the groundwater storage. Percolation tank should be constructed preferably on second to third order streams, located on highly fractured and weathered rocks, which have lateral continuity downstream. The recharge area downstream should have sufficient number of wells and cultivable land to benefit from the augmented groundwater. The size of percolation tank should be governed by percolation capacity of strata in the tank bed. Normally percolation tanks are designed for storage capacity of 0.1 to 0.5 million cubic meters. It is necessary to design the tank to provide a ponded water column generally between 3 and 4.5 m. The percolation tanks are mostly earthen dams with masonry structure only for spillway. The purpose of the percolation tanks is to recharge the groundwater storage and hence seepage is permissible. For dams up to 4.5 m height, cut off trenches are not necessary and keying and benching between the dam seat and the natural ground is sufficient.

Rainwater harvesting through check dams/ cement plugs/ nala bunds

Check dams are constructed across small streams having gentle slope. The site selected should have sufficient thickness of permeable bed or weathered formation to facilitate recharge of stored water within short-span of time. The water stored in these structures is mostly confined to stream course and the height is normally less than 2 m and excess water is allowed to flow over the wall. In order to avoid scouring from excess runoff, water cushions are provided at downstream side. To harness the maximum runoff in the stream, series of such check dams can be constructed to have recharge on regional scale. Clay-filled cement bags arranged as a wall are also being successfully used as a barrier across small nalas. At places, shallow trench is excavated across the nala and asbestos sheets are put on two sides. The space between the rows of asbestos sheets across the nala is backfilled with clay. Thus, a low cost check dam is created. On the upstream side clay-filled cement bags can be stacked in a slope to provide stability to the structure.

Rainwater harvesting through recharge shaft

This is the most efficient and cost effective technique to recharge unconfined aquifer overlain by poorly permeable strata. Recharge shaft may be dug manually if the strata are of non-caving nature. The diameter of shaft is normally more than 2 m. The shaft should end in more permeable strata below the top impermeable strata. It may not touch water table. The unlined shaft should be backfilled, initially with boulders/cobbles followed by gravel and coarse sand. In case of lined shaft the recharge water may be fed through a smaller conductor pipe reaching up to the filter pack. These recharge structures are very useful for village ponds where shallow clay layer impedes the infiltration of water to the aquifer. It is seen that in rainy season village tanks are fully filled up but water from these tanks does not percolate down due to siltation and tubewell and dugwells located nearby remains dried up. The water from village tanks gets evaporated and is not available for the beneficial use. By constructing recharge shaft in tanks, surplus water can be recharged to groundwater. Recharge shafts of 0.5 to 3 m diameter and 10 to 15 m deep are constructed depending upon availability of quantum of water. The top of shaft is kept above the tank bed level preferably at half of full supply level. These are backfilled with boulders, gravels and coarse sand. In upper portion of 1 or 2 m depth, the brick masonry work is carried out for the stability of the structure. Through this technique all the accumulated water in village tank above 50% full supply level would be recharged to groundwater. Sufficient water will continue to remain in tank for domestic use after recharge.

Rainwater harvesting through dugwell recharge

Existing and abandoned dugwells may be utilised as recharge structure after cleaning and desilting the same. The recharge water is guided through a pipe from desilting chamber to the bottom of well or below the water level to avoid scouring of bottom and entrapment of air bubbles in the aquifer. Recharge water should be silt free and for removing the silt contents, the runoff water should pass either through a desilting chamber or filter chamber. Periodic chlorination should be done for controlling the bacteriological contaminations.

Groundwater dams or subsurface dykes

Subsurface dyke or underground dam is a subsurface barrier across stream which retards the base flow and stores water upstream below ground surface. By doing so, the water levels in upstream part of groundwater dam rises saturating otherwise dry part of aquifer. The site where subsurface dyke is proposed should have shallow impervious layer with wide valley and narrow out let. After selection of suitable site, a trench of 1-2 m wide is dug across the breadth of stream down to impermeable bed. The trench may be filled with clay or brick/concrete wall up to 0.5 m below the ground level. Since the water is stored within the aquifer, submergence of land can be avoided and land above the reservoir can be utilised even after the construction of the dam. No evaporation loss from the reservoir and no siltation in the reservoir take place. The potential disaster like collapse of the dams can also be avoided.

REFERENCES

Bramley, H., Turner, N. C. and Siddique, K. H. M. 2013. Water use efficiency. In: Kole, C. R. (ed.). Genomics and Breeding for Climate-Resilient Crops, Vol. 2. Springer-Verlag Berlin Heidelberg. pp. 226-268.

Briggs, L. J., and Shantz, H. L. 1913. The water requirement of plants. *Bureau of Plant Industry Bulletin*. US Department of Agriculture, Washington, DC: 282-285.

CGWB. 2003. Rainwater Harvesting Techniques to Augment Groundwater. Central Groundwater Board, Ministry of Water Resources, Govt. of India, Faridabad. pp. 30.

Chartzoulakisa, K. and Bertaki, M. 2015. Sustainable water management in agriculture under climate change. *Agriculture and Agricultural Science Procedia* **4**: 88-98.

Clancy A. 1999. Revenina has the capacity to deliver diverse requirements. *Australian Viticulture* **3**: 38-42.

Collins, M., Knutti, R., Arblaster, J., et al. 2013. Long-term climate change: projections, commitments and irreversibility. In: Climate Change 2013: The Physical Science Basis. Stocker, T. F., Qin, D., Plattner, G. K., Tignor, M. and Allen, S. K. (eds.). Cambridge University Press, Cambridge.

Dry, P. R., Loveys, B. R., Stoll, M., Stewart, D., and McCarthy, M. G. 2000. Partial root drying - an update. *Australian Grapegrower and Winemaker* **438**: 35-39.

FAO. 1997. Small-scale irrigation for arid zones: principles and options. Food and Agriculture Organisation, Rome. http://www.fao.org/docrep/W3094E/w3094e04.htm

FAO. 2015. Towards a water and food secure future: Critical perspectives for policy-makers. Food and Agriculture Organisation of the United Nations, Rome and World Water Council, Marseille. 76 p.

FAO. 2017. Water for sustainable food and agriculture. Food and Agriculture Organisation of the United Nations. Rome. pp. 27

Hatfield, J. L. and Dold, C. 2019. Water-use efficiency: advances and challenges in a changing climate. *Frontiers in Plant Science* **10**:1-14.

Hillel, D. 1997. Small-scale irrigation for arid zones: principles and options. FAO Technical Paper 2, Rome.

http://ecoursesonline.iasri.res.in/mod/page/view.php?id=1554

http://www.fao.org/assets/infographics/FAO-Infographic-water-thirsty-en.pdf

https://energypedia.info/wiki/Water_Use_in_Agriculture

https://vikaspedia.in/energy/environment/rainwater-harvesting-1/rain-water-harvesting-techniques-to-augment-ground-water

https://www.oecd.org/agriculture/topics/water-and-agriculture/

https://www.worldbank.org/en/topic/water-in-agriculture

Keller, J. and Blinser, R. D. 1990. Sprinkle and trickle irrigation. Chapman and Hall, New York, USA.

Loveys, B. R., Dry, P. R. and McCarthy M. G. 1999. Using plant physiology to improve water use efficiency of horticultural crops. *Acta Horticulture* **537**: 187-199.

Martin, J. H., Leonard, W. H., Stamp, D. L. 1976. Principles of field crop production. Macmillan, New York.

Meena, B. L., Singh A. K., Phogat, B. S. and Sharma, H. B. 2013. Effects of nutrient management and planting systems on root phenology and grain yield of wheat. *Indian Journal of Agricultural Science* **83**(6): 627-632

Pandey, A. K., Singh, A. K., Singh, S. K. and Kumar, A. 2012. Livelihood sustainability of small and marginal farmers of Bihar through multiple use of water for enhancing agricultural productivity. *Environoment and Ecology* **30**(4A): 1529-1533.

Passioura, J. B., Angus, J. F. 2010. Improving productivity of crops in water-limited environments. *Advances in Agronomy* **106**: 37-75.

Pereira, L. S., Cordery, I., Iacovides, I. 2002. Coping with water scarcity. UNESCO IHP VI, Technical documents in Hydrology No 58, UNESCO, Paris . p. 267.

Rana, N. S., Singh A. K., Kumar, Sanjay and Kumar, Sandeep. 2003. Effect of trash mulching and nitrogen application on growth yield and quality of sugarcane ratoon. *Indian Journal of Agronomy* **48**(2): 124-126.

Singh, A. K., Bhatt, B. P., Sundaram, P. K., Gupta, A. K. and Singh, D. 2013. Planting geometry to optimise growth and productivity faba bean (*Vicia faba* L.) and soil fertility. *Journal of Environmental Biology* **34**(1): 117-122.

Singh, A. K., Manibhushan, Chandra, N. and Bharati, R. C. 2008. Suitable crop varieties for limited irrigated conditions in different agro climatic zones of India. *International Journal of Tropical Agriculture* **26**(3-4): 491-496.

Singh, L., Mirza, K. A. B., Akhter, S., Qayoom, S., Lone, B. A., Singh, P. and Singh, P. 2014. Efficient techniques to increase water use efficiency under rainfed eco-systems. *Journal of AgriSearch* **1**(4): 193-200.

Trenberth, K. E. 2011. Changes in precipitation with climate change. *Climate Research* **47**: 123-138.

Turner, N. C. and Asseng, S. 2005. Productivity, sustainability, and rainfall-use efficiency in Australian rainfed Mediterranean agricultural systems. *Australian Jcurnal of Agricultural Research* **56**: 1123-1136.

Wallace, J. S. and Batchelor, C. H. 1997. Managing water resources for crop production. *Philosophical Transactions of the Royal Society B* **352**: 937-947.

8

Integrated Pest Management for Sustainable Agriculture

The integrated pest management is a set of effective and environmentally safe pest management practices that takes advantages of all appropriate pest management options including, but not limited to, the judicious use of pesticides. The comprehensive information on the life cycles of pests and their interaction with the environment, and the available pest control methods are used to manage pest damage by the most economical means, and with the least possible hazard to people, property, and the environment. Thus, the integrated pest management aims to balance three important factors of sustainable agriculture, i.e., economic viability, environmental safety and social acceptability, in achieving sustainable food production and food security.

Food and Agriculture Organisation defines 'integrated pest management' as careful consideration of all available plant protection methods and subsequent integration of appropriate measures that discourage the development of populations of harmful organisms and keep the use of plant protection products and other forms of intervention to levels that are economically and ecologically justified and reduce or minimise risks to human health and the environment. Integrated pest management emphasises the growth of a healthy crop with the least possible disruption to agroecosystems and encourages natural pest control mechanisms.

EVOLUTIONARY TRENDS IN PEST MANAGEMENT

Until the beginning of 20th century, farmers relied exclusively on cultural practices such as crop rotation, healthy crop variety, manipulations in sowing dates, etc. to manage the pests. Use of pesticides, although began in 1870s with the development of arsenical and copper-based insecticides, discovery of pesticidal properties of DDT during the World War II revolutionised the pest control. DDT was effective against almost all insect species and was relatively harmless to the humans, animals, and plants. It was effective at low application rates and was also less expensive. Farmers were amazed with its effectiveness and started to use it increasingly particularly during the green revolution era. As a result of rising demand, the pesticide industry rapidly expanded its research on synthetic organic insecticides as well as on much more toxic, organophosphates and pyrethroid insecticides (Singh and Sharma, 2004).

Most of the pesticides were originally based on the toxic heavy metals such as arsenic, mercury, lead and copper. Pesticides often kill the natural enemies along with the pests. With natural enemies eliminated, it is difficult to prevent recovered pest populations from exploding to higher and more damaging levels, and often developing resistance to chemical pesticides. Repeated applications of chemical pesticide only repeat this cycle. At low yields, benefits from pest control were not huge. However, as yields started increasing, pesticide use started becoming widespread. Their adverse effects on the environment and human health also soon became apparent. Indiscriminate, excessive and continuous use of pesticides acted as a powerful selection pressure for altering the genetic make-up of the pests. Naturally resistant individuals in a pest population were able to survive onslaughts of the pesticides, and the survivors could pass on the resistance traits to their generations. This resulted in a much higher percentage pest population being resistant to pesticides (Singh and Sharma, 2004).

The solution to the pesticide externalities lies in the implementation of integrated pest management (IPM), which combines the use of different pest control strategies such as cultural, resistant varieties, biological and chemical control. During 1960s when the IPM began to be promoted as a pest control strategy, there were fewer IPM technologies available for field application. During 1970s, research generated some novel products and knowledge for successful implementation of IPM in crops like rice, cotton, sugarcane and vegetables. Integration of multiple pest suppression techniques has the highest probability of sustaining long-term crop protection. Pesticides may be used to remove/ prevent the target organism, but only when assessment with the help of monitoring and scouting indicates that they are needed to prevent economic damage. In the context of crop protection, sustainability refers to the substitution of chemicals and capital with farm grown biological inputs and knowledge, aimed at reduction in the cost of production without lowering the yields (Swaminathan, 1995).

Early models of integrated pest management focused on ecological aspects of pest management. However, considering recent advances in crop production and protection, communication technology, and globalisation of agriculture and commerce, a new paradigm of IPM is presented with its management, business, and sustainability aspects having the following direct or indirect objectives (Dara, 2019).

1. To update the IPM concept as appropriate for modern times and encourage reevaluation of what is perceived as sustainable.

2. To build consumer confidence and education in an IPM-based production system that is ideal for all crops and situations, ensures global food security, and eliminates food-based social inequality.
3. To ensure profitability for the producers while allowing informed consumers, rather than special interest groups or retailers, to make their food choice.
4. To minimise potential negative impact of the non-IPM-based conventional practices or those perceived to be sustainable alternatives on the environment and challenges associated with managing certain pests.

GOAL OF AN IPM PROGRAMME

A well structured integrated pest management programme should have the following goals (Alston, 2011).

1. Optimum profits over the long-term
2. Sustained resource (agricultural or natural over the long-term)
3. Rational use of pesticides
4. Reduced environmental contamination of soil, groundwater, surface water, pollinators, wildlife, endangered species, etc.
5. Utilisation of natural biological controls, and use of selective pesticides with proper timing of applications
6. Minimisation of pesticide resistance problems
7. Minimisation of pest resurgence and secondary pest outbreaks
8. Food safety by ensuring reduced residues of pesticides on food products
9. Worker safety through adoption of safer pest management tactics

PRINCIPLES OF INTEGRATED PEST MANAGEMENT

Directive 2009/128/EC of the European Parliament prescribed eight general principles of integrated pest management establishing a framework for community action to achieve the sustainable use of pesticides (European Union, 2009; Barzman et al., 2015). These principles may be universally applicable in integrated pest management for sustainable agriculture.

Principle 1: Prevention and suppression

'Prevention is better than cure' is the first general rule in any production system. Prevention is adoption of measures to reduce the chance of occurrence of pest. Suppression is reducing the impact of the pests. Prevention complements prevention. This principle means that the aim is not to completely eliminate pests but to prevent any single one from becoming dominant or damaging in a cropping system (Barzman et al., 2015). Certain aspects of prevention dealing with the use of healthy and weed-free planting material and detection of pathogens in substrates deserve more attention, particularly in light of new technologies. Many pathogens associated with seed become the source of disease in the subsequent year. Weed seed contaminating harvest can become a major problem in the subsequent year. Certification of disease-free seed and planting materials, and new sorting technologies are very helpful in avoiding problems but it is important to apply measures early, prior to certification of harvested seed (Van derWolf et al., 2013). Soil substrates, manure, and other amendments should be screened to qualitatively and quantitatively assess the disease situation (Van Gent-Pelzer et al., 2010; Sikora et al., 2012). Plant breeding for pest resistance is recognised as an important contributor to the development of prevention strategies. The use of pest-tolerant and resistant cultivars will help to decrease dependence on pesticides in arable crops. The other options for prevention and/or suppression of harmful organisms include crop rotation, use of suitable cultivation techniques (such as stale seedbed technique, altering sowing dates and planting densities, conservation tillage, pruning, etc.), use of balanced fertilisation, liming and irrigation/drainage practices, adoption of hygiene measures (e.g., by regular cleansing of machinery and equipment).

Principle 2: Monitoring

Harmful organisms must be monitored at regular intervals by adequate methods and tools such as observations in the field, scientifically sound warning, forecasting and early diagnosis systems, as well as the use of advice from professionally qualified advisors. Farmers are often reluctant to monitor weeds because they look similar at the young growth stages when management decisions are needed. While the level of infestation of most pathogens and insect pests will significantly vary between years, shifts in weed flora occur slowly (Walter, 1996). The gradual year-to-year changes in weed populations makes it possible to generate weed maps by monitoring end-of-season weeds or by establishing small untreated plots in the field. The information collected in this way can then be used to plan weed management in the subsequent crops.

Principle 3: Decision-making

Decision making is done based on the results of the monitoring. It is to be decided whether and when to apply plant protection measures. The threshold values are essential components for decision-making. Threshold is the pest density or population level, which when exceeded, management is required. The pest life cycle, climatic condition, etc. should be considered before defining the threshold of any crop in a particular region or area. While it is true that sound intervention thresholds play an important role in IPM, they are, however, not always applicable, available, or sufficient. In many cases, thresholds have not been established for weeds (Sattin et al., 1992). This is also the case for pathogens, particularly those that switch from a saprophytic to a pathogenic lifestyle depending on environmental events and climatic conditions (Underwood et al., 2007). Developing thresholds for weeds is a challenge because they usually appear as a community of multiple species, typically have a patchy distribution, and have long-term impact through a persistent seed bank. Similarly, the pertinence of the threshold approach can be questioned in the case of polycyclic diseases, where it is often necessary to target the primary cycle while the inoculums level is very low and disease symptoms invisible. In such cases the observations, in general should be given importance before taking any decision. The present-day decision-support systems are usually based on real-time tactical decision making involving one crop, one pest, and one control technique. However, new systems could support strategic approaches encompassing the whole range of IPM options. Instead of 'spray/do not spray' guidance, the new systems would provide insights on desired varieties, cropping sequences, combinations of direct control methods, and relevant agronomic practices. They could also provide information on expected damage and economic consequences as well as on impacts on non-target and beneficial organisms (Barzman et al., 2015).

Principle 4: Non-chemical methods

Sustainable biological, physical and other non-chemical methods must be preferred to chemical methods if they provide satisfactory pest control. The highest level of control attainable by chemical measures is often not sustainable, creates new pest problems, and is not a proper standard against which single non-chemical tactics are evaluated, rather, a satisfactory while sustainable level of pest management can be achieved via a broad IPM strategy that includes an array of protection methods. There is a wide range of non-chemical pest control measures such as soil solarisation or pheromone-based mating disruption, but their availability, efficacy, or pertinence varies considerably. The use of live natural enemies represents a major non-chemical IPM tool. Whereas biological control agents are well developed in protected crops, significant opportunities for their use still exist in other systems such as arable crops. Beyond the production and marketing of live biological agents, the cropping system and the landscape into which these organisms are released also needs to be taken into account to optimise and sustain their efficacy (Barzman et al., 2015).

Though a few biological control options are available for weeds, there are many effective agronomic, mechanical and physical control methods are available. They can be incorporated into integrated weed management strategies to reduce impact of weeds and over-use of conventional herbicides over the long-term. A number of non-chemical direct methods such as suppressive winter cover crops, stale seedbed technique, pre-emergence cultivation, more dense crop stands, inter-row precision weeding, and use of hoes equipped with tools for intra-row weeding are feasible (Barzman et al., 2015).

Principle 5: Pesticide selection

Integrated pest management does not totally avoid the use of the pesticides and it seeks to reduce reliance on pesticides. When the prevention and alternative control methods are not adequate or available then the pesticides are used for pest control. However, the use of pesticides should be as specific as possible for the target. The pesticides should not have any threat to the health of human, non-target organisms and the environment. There are undesired non-target impacts of broad-spectrum insecticides on arthropod natural enemies. So the insecticides should be more compatible with beneficial arthropods to minimise disruption of biological control of pests and improve IPM. Selective biopesticides represent a particularly desirable alternative to chemical pesticides, but a wider range of such products needs to be made available (Barzman et al., 2015).

Principle 6: Reduced pesticide use

Reduced pesticides use means the reduction in doses and application frequency or partial applications. Reduced pesticide use, in terms of frequency, spot spraying, or dose reduction is a recognised tactic along the IPM that can be supported by the other means of intervention. Though reducing doses, application frequency, and resorting to partial application of pesticides contribute to the IPM goal of reducing or minimising risks to human health and the environment, applying reduced doses may have the potential influence on the risk of pesticide resistance developing in the pest population (Barzman et al., 2015).

Principle 7: Anti-resistance strategies

Unmanaged and haphazard use of the pesticides create the problem of resistance, i.e., pests develop the resistance and the use of pesticides have less effect on them. The number of pest species resistant to pesticides is increasing and jeopardising the efficacy of many products. There are many instances of resistance among all pest categories. This issue is particularly acute for weed management because very few new herbicidal modes of action remain available (Duke, 2012).

The new vision of sustainable pesticide use focuses on a desirable control level that is then related to the selection pressure due to the biological activity and persistence of the active ingredients concerned. The available anti-resistance strategies should be applied to maintain the effectiveness of the products. This may include the use of multiple pesticides with different modes of action. The occurrence of resistance is also to be monitored regularly to guide decision-making. Focusing on weeds, Owen et al. (2014) conclude that merely modifying herbicide use will not yield lasting solutions to herbicide resistance in weeds. But this may be true for the whole range of pests. To reduce the selection of resistant pest biotypes and lengthen the commercial lifespan of pesticides, farmers can strive for a higher level of IPM, consider the spatial distribution of tolerant varieties or non-host crops, and make full use of preventive measures. The root causes of increased risk of evolution of non-target resistance can be addressed by revisiting over-simplification and excessive intensification of cropping systems (Barzman et al., 2015).

Principle 8: Evaluation

Evaluation is necessary in studying the effectiveness of the crop protective measures and plan further. It is done based on the records of the use of the pesticides, its effects and many more. Farmers should be encouraged to assess the soundness of the crop protection measures they adopt, and this is an important aspect of sound management. IPM-compatible assessment covers multi-season effects, trade-offs with other compartments of production and economics, as well as human health and the environment. Many positive effects of IPM strategies are multi-year, and effective evaluation therefore covers all crops of the rotation over more than one season. This is particularly pertinent to the management of weed seed banks, accumulation of soil-borne pathogens, resistance development of pathogens, and unpredictable insect outbreaks. The level of short-term control attained by chemical measures alone is not the standard by which success can be measured. A process of rethinking and reassessing evaluation needs to be initiated. It would emphasise the evaluation of yield, yield stability and profit over multiple years at the cropping system level (Lechenet et al., 2014). Sustainability in pest management calls for new evaluation criteria that take into account multi-season effects and a diversity of trade-offs, and can be widely shared among the farmer community (Barzman et al., 2015).

BENEFITS AND LIMITATIONS OF IPM

Benefits of IPM

Adoption of IPM strategies provides economic benefits due to sustained development, increased productivity and reduced pest damage. The options that IPM can offer make sole reliance upon synthetic pesticides a thing of the past. In the long-term, everyone benefits through a healthier environment.

The benefits of integrated pest management can be listed under the following major heads (Adhikari, 2018).

1. **Lower cost intervention:** Traditionally, the use of the pesticides to control the pest invasion would account to lots of cost since some of these pesticides need to be imported. So the adoption of IPM would lessen the financial burden. It increases the cost-effectiveness of a pest management programme. Various techniques involved in IPM are more sustainable with long lasting benefits.
2. **Environmental benefits:** Use of the pesticides is often linked degradation of the environment causing some more additional problems to human and animal health. IPM is an eco-friendly approach and the effect on the environment is always considered before the application of any interventions. It reduces environmental risk associated with pest management by encouraging the adoption of more ecologically control tactics. IPM promotes sustainable bio-based pest management alternatives which reduces the air and groundwater contamination. It also protects non-target species through reduced impact of pest management activities.
3. **Reduced residue hazards:** In an IPM schedule the use of pesticides is considerably reduced, hence the pesticide residue hazards will also get automatically minimised.
4. **Anti-resistance:** The IPM model in itself is the anti-resistant mode for pest control. It discourages the use of chemicals and thus creates less cases of anti-resistance. Pesticides are used only when the other alternatives are not satisfying.

Limitations of IPM

Integrated pest management also has some limitations.

1. **Need of skill and technicalities:** Integrated pest management schedule needs to be planned well ahead of cropping season, starting from seed collection and field preparation. It demands more attention and dedication and also requires expertise of various fields. All those involved in the IPM needs to be educated and trained adequately which often requires much time. However, inadequate interaction between research and extension agencies results low awareness and innovativeness of extension personnel and target groups.
2. **Time and energy consuming:** Application of IPM takes more time than the chemical control of pests. Much time is needed in planning itself. As IPM strategies differ from region to region, a separate plan is required for

each region. The expected results of intervention may take long time to be achieved.

3. **Timely non-availability of inputs:** Location-specific IPM modules for all crops are not available. Adequate quantity of quality inputs, including biocontrol agents and biopesticides are often not available at the time of requirement.

TOOLS OF IPM

Monitoring

Crop monitoring, that keeps track of the pests and their potential damage, is the foundation of IPM. This provides knowledge about the current pests and crop situation and is helpful in selecting the best possible combinations of the pest management methods. Pheromone traps have got advantage over other monitoring tools such as light and sticky traps. Being selective to specific pest, they have proven their usefulness in large scale IPM validations in cotton, basmati rice, chickpea and pigeonpea (Singh and Sharma, 2004).

Pest resistant varieties

Breeding for pest resistance is a continuous process. At the same time the pests also, particularly the plant pathogens, co-evolve with their hosts. Thus, gene transfer technology is useful in developing cultivars resistant to insects, plant pathogens and herbicides. An example of this is the incorporation of genetic material from *Bacillus thuringiensis* (Bt), a naturally occurring bacterium, in cotton, corn, and potatoes, which makes the plant tissues toxic to the insect pests. However, due to ethical, scientific and social considerations, this potential technology has been surrounded by controversies (Singh and Sharma, 2004).

Cultural pest control

It includes crop production practices that make crop environment less susceptible to pests. Crop rotation, fallowing, manipulation of planting and harvesting dates, manipulation of plant and row spacing, and destruction of old crop debris are a few examples of cultural methods that are used to manage the pests. Planting of cover crops, nectar-producing plants and inter-planting of different crops to provide habitat diversity to beneficial insects are important management techniques. Cover crops suppress weeds. Some cover crops of the *Brassica* family when incorporated in the soil have the ability to suppress nematode pests and wilt diseases. Wheat crop residues left in the field provide more than 90% weed suppression. Cultural controls are selected based on knowledge of pest biology and development (Singh and Sharma, 2004).

Physical or mechanical controls

These are based on the knowledge of pest behaviour. Placing plastic-lined trenches in potato fields to trap migrating Colorado potato beetles is one example of the physical control. Shaking of the pigeonpea plant to remove *Helicoverpa* larvae is a common practice in pigeonpea growing areas. Hand picking of insect pests is perhaps the simplest pest control method. Installation of dead as well as live bird perches in cotton and chickpea fields has proved effective in checking the bollworm infestation. Using mulches to smother weeds and providing row covers to protect plants from insects are other examples (Singh and Sharma, 2004).

Biological controls

Biological control is defined as applied natural control wherein man intervenes to improve the efficiency of natural enemies including parasites, predators, and pathogens of pest species by introductions, conservation or augmentation to maintain pest populations below economically injurious levels. In IPM programmes, native natural enemy populations are conserved, and non-native agents may be released with utmost caution. *Trichogramma* spp. are the most popular parasitoids being applied on a number of host crops. A number of microorganisms such as *Trichoderma* spp., *Verticillium* spp., *Aspergillus* spp., *Bacillus* spp. and *Pseudomonas* spp. that attack and suppress the plant pathogens have been exploited as biological control agents (Singh and Sharma, 2004).

Biopesticides: Biopesticides also known as biological pesticides are certain types of pesticides derived from such materials as animals, bacteria, and certain minerals. These are an important group of pesticides that can reduce pesticide risks. Biopesticides have a narrow target range and a very specific mode of action. These have relatively critical application times. Biopesticides are slow acting and have limited field persistence and short shelf life. These suppress, rather than eliminate a pest population. The following advantages are achieved by using biopesticides.

1. Biopesticides are inherently less harmful than conventional pesticides.
2. Biopesticides are designed to affect only one specific pest or, in some cases, a few target organisms, in contrast to broad-spectrum, conventional pesticides that may affect beneficial organisms.
3. Biopesticides often are effective in very small quantities and they decompose quickly thereby resulting in lower exposures and largely avoiding the pollution problems caused by conventional pesticides. These do not have residual problems.

4. When used as a component of integrated pest management programmes, biopesticides can greatly decrease the use of conventional pesticides while crop yields remain high.

Microbial pesticides: Microbial pesticides contain a microorganism (bacterium, fungus, virus, protozoon or algae) as the active ingredient. They suppress pests by producing a toxin specific to the pest, causing a disease or preventing establishment of other microorganisms through competition. An example of microbial pesticide is *Bacillus thuringiensis*. It is a naturally occurring soil bacterium that is toxic to the larvae of several species of insects but non-toxic to non-target organisms. *Bacillus thuringiensis* can be applied to plant foliage or incorporated into the genetic material of crops. It is toxic to the caterpillars of moths and butterflies. Microbial pesticides need to be continuously monitored to ensure they do not become capable of harming non-target organisms, including humans. Other examples of microbial pesticides include

1. *Pseudomonas fluorescens* against *Pythium* spp., *Rhizoctonia* spp., *Fusarium* spp.
2. Nematodes like green commandoes and soil commandoes against caterpillars and grubs
3. Nuclear polyhedrosis virus (NPV)
4. *Trichoderma viride* against many common diseases of vegetables and spices
5. Weevils *Neochetina eichhorniae* and *Neochetina bruchi* against water hyacinth
6. Beetle *Zygogramma bicolorata* against weed *Parthenium*

Biochemical pesticides: These are naturally occurring substances that control pests by non-toxic mechanisms. Biochemical pesticides include substances that interfere with growth or mating, such as growth regulators, or substances that repel or attract pests, such as pheromones. Pheromones are often used to detect or monitor insect populations, or in some cases, to control them.

Botanical pesticides: Botanical pesticides can be prepared in various ways. They can be as simple as raw crushed plant leaves, extracts of plant parts, and chemicals purified from the plants. Pyrethrum, neem, tobacco, garlic, and pongamia formulations are some examples of botanicals. Some botanicals are broad-spectrum pesticides. Botanicals are generally less harmful to the environment, because of their quick degrading property. They are less hazardous to transport. The major advantage is that these can be formulated on-farm by the farmers themselves (Singh and Sharma, 2004).

Bioherbicides: The biological control of weeds involves the use of living organisms such as plant pathogens, insects and mites, herbivorous fish, nematodes, other animals and competitive plants to limit their infestation. The objectives of biological control are not eradication, rather the reduction and, regulation of weed population below the levels of economic injury. A successful bio-agent is host specific, rapid destroyer of the target weed, effective on several species of the weed in question, adjustable to new environment and easy to multiply. A bioherbicide can often live in the environment and wait for the next growing season when there will be more weeds to infect. This reduces the farmer's cost of applying herbicides year after year. A new range of bioherbicides may essentially allow farmers to replace or reduce the expensive chemical herbicides.

Table 8.1. Plants commonly used as biopesticides

Plant	Parts used	Major mode of action	Target pest
Acorus calamus	Rhizome	Contact poison or repellent	Ants
Allium sativum	Pulp	Fungicide	*Pseudoperonospora cubensis*
Alpinia galanga	Rhizome	Repellent/contact poison	*Bactrocera dorsalis* , *Nilaparvata lugens*, aphids, brown spot
Annona muricata	Leaf	Contact poison	*Phyllotreta slimata, Phyllotreta punctata*
Annona squamosa	Leaf, seed	Contact poison	*Phyllotreta sinuata*
Azadirachta indica	Leaf, seed	Repellent, ovicidal	Sucking insects, butterflies, aphids larvae
Curcuma domestica	Rhizome	Contact poison, repellent, fungicide	Red-mite, thrips, ants, *Xanthomonas citri, Alternaria porri*
Derris elliptica	Stalk, root	Contact poison	Various insects
Dioscorea hispida	Tuber	Contact poison	Various insects
Euphorbia trigona	Whole plant	Contact poison	Various insects
Nicotiana tabacum	Whole plant	Contact poison	Various insects
Stemona tuberosa	Leaf, root	Contact poison	Various insects
Tinospora crispa	Vine	Anti-feedant	Diamond back moth
Toona tomentosa	Leaf	Contact poison	*Nilaparvata lugens*, aphids, butterflies
Zingiber serumber	Rhizome	Contact poison	Aphids, butterflies

Source: Chejew et al. (1988); Kuye et al. (2006)

Chemical controls

Pesticides are used to keep the pest populations below economically damaging levels when the pests cannot be controlled by other means. Pesticides include both the synthetic pesticides and plant-derived pesticides. Synthetic pesticides include a wide range of man-made chemicals. These are easy to use, fast-acting and relatively inexpensive. Ideally, pesticides should be used as a last resort in IPM programmes because of their potential negative effect on the environment. Pesticides with the least negative impacts on non-target organisms and the environment are most useful. Economic threshold assessment is based on the concept that most plants can tolerate at least some pest damage. In an IPM programme where the economic threshold is known, chemical controls are applied only when the pest's damaging capacity is nearing to the threshold, despite application of other alternative management practices (Singh and Sharma, 2004).

FACTORS AFFECTING ADOPTION OF IPM

Although, IPM has been accepted as the most attractive option for crop protection, its adoption at the farmers' level remains very low. Pesticides continue to dominate and their injudicious use represents the greatest threat to IPM. The following factors have been identified for low adoption of IPM by the farmers.

Available technologies and their efficacy

Sustainability of any IPM programme depends upon the compatible technologies available. A lot of technologies have been generated using naturally occurring enemies of insect pests (parasitoids, predators and pathogens) for use in IPM. However, the suitability of these technologies should be evaluated under different agroecological situations before their recommendation. Some important commercially available products include *Trichogramma, Bracons, Chrysoperla carnea, Cryptolaemus montrouzieri, Bacillus thuringiensis, Bacillus sphaericus,* nuclear polyhedrosis viruses (NPV) and *Trichoderma*. In addition, a number of plant products such as azadirachtin, pyrethrum, nicotine, etc. are also valuable as biopesticides. For IPM to be a success, it must be sound on technical parameters. Technical feasibility of IPM is judged on two criteria, i.e., change in the pesticide use, and yield change over the conventional chemical control (Birthal, 2004).

Economic feasibility

Technical feasibility may not be the sole condition for commercialisation and adoption of a technology. The economic feasibility in terms of net benefits over the conventional technology counts much for adoption of an IPM technology. The magnitude of net benefits would depend on the type of input used in IPM package, its application rate and price. Some technically feasible IPM packages may not be popular due to their economic infeasibility because of higher prices of some of its constituents.

Socioeconomic and policy issues

Despite its techno-economic superiority over conventional chemical control, adoption of IPM remains restricted to a very less area as compared to chemical control. The structure of agrochemical market suggests a low level of adoption; biopesticides share only 2% of the agrochemical market in India (Saxena, 2001). There could be a number of technological, socioeconomic, institutional and policy factors restricting large scale adoption of IPM.

Technology characteristics

The characteristics of technology have an important role in farmers' adoption decisions (Adesina and Zinnah, 1993; Lapar and Pandey, 1999). IPM draws heavily on complementarities and interactions of different methods of pest control (chemical, biological, cultural and mechanical), and each of the components has its own specific characteristics and requirements for application. This makes IPM a complex technology. Generally, the farmers adopt those components that show immediate effect, and are easily available. Biopesticides comprise a major component of IPM. Most of the biopesticides are host-specific, slow in action and have short shelf-life. Besides, application of some of the components is labour intensive compared to conventional chemical control (Birthal et al., 2000). Efforts should be made for development of broad-spectrum biological pesticides and improvements in their efficacy and shelf-life. Biopesticides based on predators, parasites, viruses, fungi, etc. are sensitive to chemical pesticides. Thus, biopesticides having better compatibility with chemical pesticides should be developed. Biotechnology has tremendous potentialities for developing biopesticides.

Technology dissemination

IPM is very much knowledge intensive. Its effective implementation requires extension workers to have a sound understanding of the characteristics of the technology, its target host and relationship with natural enemies, and its method of application before the technology is delivered to the farmers. Lack of understanding of any of these would adversely affect its adoption. The extension workers should act as a collaborator, consultant and facilitator in dissemination of the knowledge. Extension system needs to be overhauled in knowledge about IPM inputs, methodology of IPM and timely delivery of services to the farmers to accelerate the adoption of IPM (Birthal, 2004).

Community participation

Pest has the characteristics of a detrimental common property resource. It does not recognise spatial boundaries. Thus, successful pest control requires collective efforts. However, pest control efforts are more or less individualistic, giving rise to a number of pest control related problems, such as pest resistance, resurgence and secondary outbreak, destruction of natural enemies of insect pests and other beneficial insects. Collective pest management assumes greater significance in the context of IPM. There are a number of management practices such as observance of synchronicity in sowing dates, use of resistant varieties, crop rotations, etc. that require close cooperation among farmers to achieve maximum pest control efficiency. Further, IPM relies on inputs derived from living organisms, and the application of different control methods in a locality, in particular chemical pesticides, would adversely affect the activities of the biological inputs. Though, a majority of the farmers could be aware of the benefits of collective action, a number of socioeconomic factors act as a disincentive to participate in it. Therefore, institutional mechanisms should be developed that promote group action (Birthal, 2003).

Timely availability of biopesticides

Most of the production and distribution of biopesticides take place in public sector units. However, many of the laboratories producing biopesticides are small and cater to the location-specific needs only of a small area. The average gross cropped area per biocontrol laboratory is large. Some inherent technical characteristics of biopesticides however act as disincentive to the entry of private sector. Most of the biopesticides are not broad-spectrum and are slow in action. Many of these like *Trichogramma* and *Chrysoperla* have a short shelf-life. Thus, production of biopesticides is fraught with risk. Other constraints in expansion of biopesticide industry are uncertain demand, and lack of appropriate infrastructure for transportation, storage and marketing. Rural unemployed and educated youths should be encouraged to establish small-scale biopesticide production units. The registration requirements for biopesticide production units should be relaxed, though without reconciliation with quality standards. Appropriate measures such as training to the potential entrepreneurs, provision of institutional credit, subsidies and exemption from taxes would encourage stimulate production of biopesticides (Birthal, 2004).

Enforcement of pesticide regulations

Many pesticides that have been banned for use in agriculture are available in market. Lower prices of such pesticides induce farmers to use them. A number of spurious pesticides are available in the market because of lack of strict enforcement of regulations. Strict enforcement of the regulations governing production, use, distribution and quality of pesticides would help weed out spurious elements from the industry and would benefit the farmers.

Economic incentives

Price of a technology is an important determinant in farmers' decision for its adoption. An increase in the price of biopesticides due to cost considerations or withdrawal of subsidies would upset the economics of IPM. Provisions should be made for subsidies to biopesticides and no incentives or subsidies should be given on synthetic pesticides which would promote the IPM. Production and use of chemical pesticides would be unattractive through fiscal instruments of taxes, excise duties, sales taxes, etc. on intermediary inputs and final output. Withdrawal of subsidies on chemical pesticides and diversion of the same towards production and use of biopesticides, and linking institutional credit and insurance with IPM adoption would induce farmers switching over to IPM (Birthal, 2003).

Development of market for pesticide-free products

Economic incentives may not be sustained for a long. An alternative is to develop markets for pesticide-free or low-pesticide residue produce by creating consumer awareness about health benefits of such produce. Premium markets for pesticide-free products and standards for organic food should be developed. The certification procedures and labelling system for pesticide-free and organic products should be simplified. The cost of certification for an individual farmer should be low which will boost adoption of IPM.

REFERENCES

Adesina, A. A. and Zinnah, M. M. 1993. Technology characteristics, farmers' perceptions and adoption decisions. *Agricultural Economics* **9**: 297-311.

Adhikari, S. 2018. Integrated pest management (IPM): Principles, advantages and limitations. https://www.publichealthnotes.com/integrated-pest-management-ipm-principles-advantages-and-limitations/

Alston, D. G. 2011. The integrated pest management (IPM) concept. http://digitalcommons.usu.edu/extension_histall/766

Barzman, M., Barberi, P., Birch, A. N. E., et al. 2015. Eight principles of integrated pest management. *Agronomy for Sustainable Development* **35:** 1199-1215.

Birthal, P. S. 2003. Economic potential of biological substitutes for agrochemicals. NCAP Policy Paper 18. National Centre for Agricultural Economics and Policy Research, New Delhi.

Birthal, P. S. 2004. Integrated pest management in Indian agriculture: An overview. In: Birthal, P. S. & Sharma, O. P. (eds.). Proceedings 11, Integrated Pest Management in Indian Agriculture. ICAR-National Centre for Agricultural Economics and Policy Research, New Delhi & ICAR-National Centre for Integrated Pest Management, New Delhi, India. pp. 1-10.

Birthal, P. S., Sharma, O. P. and Kumar, S. 2000. Economics of integrated pest management: Evidence and issues. *Indian Journal of Agricultural Economics* **55**(4): 644-659.

Chejew, V. S., Hiranmusaphon and Srihakam, S. 1998. Alternative to Chemical Pest Control. ATTA and HDF. Bangkok, Thailand.

Dara, S. K. 2019. The new integrated pest management paradigm for the modern age. *Journal of Integrated Pest Management* **10**(1): 1-9.

Duke, S. O. 2012. Why have no new herbicide modes of action appeared in recent years? Pest *Management Science* **68**: 505-512.

European Union. 2009. Directive 2009/128/EC of the European Parliament and of the Council of 21 October 2009 establishing a framework for community action to achieve the sustainable use of pesticides. *Official Journal of the European Union* **52**: 71-86. http://eur-lex.europa.eu/legal-content/EN/TXT/?uri=OJ:L:2009:309:TOC

https://www.epa.gov/safepestcontrol/integrated-pest-management-ipm-principles

Kuye, O. O., Onya, V. E., Adinya, I. E., Oniah, M. O. and Inyang, N. N. 2006. Indigenous knowledge for cost minimisation in small holder crop production in Nigeria. Proceeding of the 20th Annual Conference of Farm Management Association of Nigeria. pp. 491-501.

Lapar, M. L. A. and Pandey, S. 1999. Adoption of soil conservation: The case of the Philippine uplands. *Agricultural Economics* **21**: 241-256.

Lechenet, M., Bretagnolle, V., Bockstaller, C., Boissinot, F., Petit, M. S., Petit, S. and Munier-Jolain, N. M. 2014. Reconciling pesticide reduction with economic and environmental sustainability in arable farming. *PLoS ONE* **9**(6): e97922. doi: 10.1371/journal.pone.0097922

Owen, M. D. K., Beckie, H. J., Leeson, J. Y., Norsworthy, J. K. and Steckeld, L. E. 2015. Integrated pest management and weed management in the United States and Canada. *Pest Management Science* **71**: 357-376.

Sattin, M., Zanin, G., Berti, A. 1992. Case history for weed competition/population ecology: velvetleaf (*Abutilon theophrasti*) in corn (*Zea mays*). *Weed Technology* **6**: 213-219.

Saxena, S. 2001. Pesticide industry in a climate of signal shifts. *Agriculture Today* **4**(4).

Sikora, K., Verstappen, E., Mendes, O., Schoen, C., Ristaino, J. and Bonants, P. 2012. A universal microarray detection method for identification of multiple *Phytophthora* spp. using padlock probes. *Phytopathology* **6**: 635-645.

Singh, A. and Sharma. O. P. 2004. Integrated pest management for sustainable agriculture. In: Birthal, P. S. & Sharma, O. P. (eds.). Proceedings 11, Integrated Pest Management in Indian Agriculture. ICAR-National Centre for Agricultural Economics and Policy Research, New Delhi & ICAR-National Centre for Integrated Pest Management, New Delhi, India. pp. 11-24.

Swaminathan, M. S. 1995. Agrochemicals in sustainable agriculture. In: Parmar, B. S. (ed.). Natural and man-made chemicals in sustainable agriculture in Asia

Underwood, W., Melotto, M., He, S. Y. 2007. Role of plant stomata in bacterial invasion. *Cell Microbiology* **9**: 1621-1629.

Van derWolf, J. M., Van derZouwen, P. S., Van derHeijden, L. 2013. Flower infection of *Brassica oleracea* with *Xanthomonas campestris* pv. campestris results in high levels of seed infection. *European Journal of Plant Pathology* **136**: 103-111.

Van Gent-Pelzer, M. P. E., Krijger, M. and Bonants, P. 2010. Improved real-time PCR assay for detection of quarantine potato pathogen *Synchytrium endobioticum* in zonal centrifuge extracts from soil and plants. *European Journal of Plant Pathology* **126**: 129-133.

Walter, A. M. 1996. Temporal and spatial stability of weeds. In: Brown, H. (ed.) Proceedings of Second International Weed Control Congress. pp. 125-130.

9

Climate Smart Sustainable Agriculture

Agriculture is the largest human activity in the world which depends on climatic parameters. Despite technological advances, such as improved crop varieties, genetically modified organisms, and irrigation systems, weather is still a key factor in agricultural productivity, as well as soil properties and natural communities.

Climate is changing naturally at its own pace since the beginning of evolution of the earth. But presently it has gained momentum due to inadvertent anthropogenic disturbances. Today, climate change is rapidly emerging as a global critical development issue affecting many sectors in the world and is considered to be one of the most serious threats to sustainable agricultural development. The earth is facing one of the greatest environmental, economic and social threats through climate change. Ecological imbalance, extreme climate variability, increasing temperature, limited food supply, increased occurrence of pests and diseases, more powerful cyclones, rise in sea levels, etc. are some of the most common evidences of climate change.

EFFECT OF CLIMATE CHANGE ON AGRICULTURE

Climate change and agriculture are interrelated processes, both of which take place on a global scale. However, the effect of climate change on agriculture is related to variability in local climate rather than in global climate patterns. Climate change has created challenges for the agricultural sector and will continue to do so. Climate change-induced increases in temperatures, rainfall variation and the frequency and intensity of extreme weather events are adding to pressures on global agricultural and food systems. Climate change is expected to negatively affect both crop and livestock production systems. The changing climate is also adding to resource problems, such as water scarcity, pollution and soil degradation. The growth, development and reproductive processes in both plants and animals are also temperature dependent. All these physiological processes show their maximum activity, hence good productivity, only at their optimum temperature. A genotype gives good yield only under suitable climatic conditions and other management practices. For each plant variety, there is an optimal temperature for vegetative growth, with growth dropping off as temperature increases or decreases. Similarly, there is a range of temperatures at which a plant will produce seed. Outside of this range, the plant will not reproduce. More than 50% differences in yield are due to climatic variations. Thus, climate change has a profound effect on agricultural practices. This will have negative impacts on all dimensions of food security. The impacts of climate change on agriculture will be the key channel through which climate change will affect food security. Climate change directly affects food availability through its increasingly adverse impacts on crop and animal productivity and health, and fish stocks (FAO, 2016). The Intergovernmental Panel on Climate Change Fifth Assessment Report (IPCC, 2014) also points out that changes in climate and carbon dioxide concentrations will enhance the distribution and increase the competitiveness of important invasive weeds. Thus, some cultivated areas may become unsuitable for crop production, and some tropical grassland may become more arid.

Impact of climate change on crop

Changes in temperature, amount of carbon dioxide, and the frequency and intensity of extreme weather could have significant impacts on crop yields. Many weeds, pests and fungi thrive under warmer temperatures, wetter climates, and increased CO_2 levels.

Warmer temperatures may make many crops grow more quickly, but warmer temperatures could also reduce yields. Crops tend to grow faster in warmer conditions. However, for some crops, faster growth reduces the length of time available for seeds to grow and mature. In the case of an annual crop, the duration between sowing and harvesting will shorten. The shortening of such a cycle could have an adverse effect on productivity because senescence would occur sooner. Thus, yield potential of the crops may not be fully exploited. In some areas, warming may benefit the types of crops that are typically planted there. However, if warming exceeds a crop's optimum temperature, yields can decline. Increase in temperature also increases crop respiration rate and alters partitioning of photosynthates to economic products.

Higher CO_2 levels can increase yields. The yields for some crops, like wheat and soybean, could increase by 30% or more under a doubling of CO_2 concentrations. However, some factors may counteract these potential increases in yield. For example, if temperature exceeds a crop's optimal level or if sufficient water and nutrients are not available, yield increases may be reduced or reversed.

More extreme temperature and precipitation can prevent crops from growing. Extreme events, especially floods and droughts, can harm crops and reduce yields. Dealing with drought could become a challenge in areas where summer temperatures are projected to increase and precipitation is projected to decrease.

As water supplies are reduced, it may be more difficult to meet water demands. Extreme climate conditions may affect quality of fruits, vegetables, tea, coffee, aromatic and medicinal plants.

Decrease in stratospheric ozone may increase biologically dangerous ultraviolet radiation on earth's surface. Ultraviolet radiation may damage nucleic acids creating long-term effects. Excess ultraviolet radiation can affect plant physiology either directly or indirectly through changed pollinator behaviour and cause massive amounts of mutations. Tropospheric ozone causes foliar injury. This may reduce productivity of many crops.

More detailed analysis of rice yields by the International Rice Research Institute forecasts 20% reduction in yields over the region per degree (in Celsius) rise in temperature. Rice becomes sterile if exposed to temperatures above 35° C for more than one hour during flowering and consequently produces no grain.

Impact of climate change on soil health

The warmer atmospheric temperature observed over the past decades is expected to lead to a more vigorous hydrological cycle, including more extreme rainfall events. Erosion and soil degradation is more likely to occur, creating sedimentation in streams and reservoirs. Soil fertility would also be affected by global warming. The increase in the temperature would induce a greater rate in the N mineralisation but there will be more volatilisation and denitrification loss. High soil temperature reduces soil organic matter both in quantity and quality. This may also decrease microbial population reducing soil productivity and nutrient cycling. Excess rainfall causes waterlogging, soil salinity problems and salt water ingression in coastal areas.

Impact of climate change on water resources

Climate change will modify rainfall, evaporation, runoff and soil moisture storage. In some areas, more frequent heavy downpours may cause localised flooding. Increased evaporation from the soil and accelerated transpiration in the plants will cause moisture stress. Thus, crops will need frequent irrigation. The water resources will come under increasing pressure and could become seasonally scarce as a result of regional changes in water supply due to changes in precipitation or snow and glacier melt. Climate change affects the availability of fresh water for irrigation. The overdraft and non-replenishment of groundwater will lower the groundwater level permanently.

Impact of climate change on pest

Global warming would cause an increase in rainfall in some areas, which would lead to an increase of atmospheric humidity and the duration of the wet seasons.

Combined with higher temperature, these could favour the development of fungal diseases. Similarly, because of higher temperature and humidity, there could be an increased pressure from insects and disease vectors. An increase in temperature may accelerate the growth and multiplication of insect pests. The polycyclic diseases become epidemic with climate change. The geographical range of insects is expanded with increased risk of infestation.

The effects of climate change on agriculture by 2050 as predicted by Ministry of Agriculture, Fisheries and Food, Govt. of United Kingdom are presented in table 9.1.

Table 9.1. Predicted effects of climate change on agriculture by 2050

Climatic element	Expected changes by 2050	Confidence in prediction	Effects on agriculture
CO_2	Increase from 360 ppm to 450 - 600 ppm	Very high	Good for crops; increased photosynthesis; reduced water use
Sea level rise	Rise by 10 -15 cm	Very high	Loss of land, coastal erosion, flooding, salinisation of groundwater
Temperature	Rise by 1-2° C; winters warming more than summers; increased frequency of heatwaves	High	Faster, shorter, earlier growing seasons, range moving north and to higher altitudes, heat stress risk, increased evapotranspiration
Precipitation	Seasonal changes by ± 10%	Low	Impacts on drought risk, soil workability, waterlogging, irrigation supply, transpiration
Storminess	Increased wind speeds, more intense rainfall events	Very low	Lodging, soil erosion, reduced infiltration of rainfall
Variability	Increases across most climatic variables; predictions uncertain	Very low	Changing risk of damaging events (heatwaves, frost, droughts, floods) which affect crops and timing of farm operations

Source: MAFF, 2000

EFFECT OF CLIMATE CHANGE ON FARM ANIMALS

Weather and climate have also great influence on farm animal production. The environment in which an animal lives is referred to as its habitat. A habitat includes both biotic (living) and abiotic (non-living) components of the animals' environment. The abiotic components of an animal's environment such as temperature, humidity, oxygen, wind, soil composition, day length and elevation have direct impact on livestock production. Climate change also affects feed production and nutrition of livestock. There will be less fodder production due to increased water scarcity. Rise in temperature causes more lignifications of plant tissues reducing its digestibility. Global warming increases the requirements for water, shelter and energy to meet the projected milk demands.

Drought may threaten pasture and feed supplies. Drought also reduces the amount of quality forage available to grazing livestock. Some areas could experience longer, more intense droughts, resulting from higher summer temperature and reduced precipitation. For animals that rely on grain, changes in crop production due to drought could also become a problem. Heatwaves, which are projected to increase under climate change, could directly threaten livestock. Heat stress affects animals both directly and indirectly. Over time, heat stress can increase vulnerability to disease, reduce fertility, and reduce milk production.

Climate change may increase the prevalence of parasites and diseases that affect livestock. In areas with increased rainfall, moisture-reliant pathogens could thrive. Vector borne diseases may have more impacts on livestock. Increases in carbon dioxide may increase the productivity of pastures and plants on which livestock feed, but may also decrease their quality.

Thornton, et al. (2008) examined some of the direct and indirect impacts of climate change on livestock and livestock systems. These are summarised in table 9.2.

Table 9.2. Direct and indirect impacts of climate change on livestock and livestock systems

Factor	Impacts
Water	Water scarcity is increasing at an accelerated pace and affects between 1 and 2 billion people. Climate change will have a substantial effect on global water availability in the future. Not only will this affect livestock drinking water sources, but it will also have a bearing on livestock feed production systems and pasture yield.
Feeds	As climate changes and becomes more variable, niches for different species alter. This may modify animal diets and compromise the ability of smallholders to manage feed deficits. Effects will depend significantly on location, system and species. In C_4 species, a rise in temperature to 30-35° C may increase the productivity of crops, fodder and pastures. In C_3 plants, rising temperature has a similar effect, but increases in CO_2 levels will have a positive impact on the productivity of these crops. For food-feed crops, harvest indexes will change, as will the

	availability of energy that can be metabolized for dry season feeding. In semiarid rangelands where the growing season is likely to contract, productivity is expected to decrease. As temperature and CO_2 levels change, optimal growth ranges for different species also change; species alter their competition dynamics, and the composition of mixed grasslands changes. For example, the proportion of browse in rangelands will increase in the future as a result of increased growth and competition of browse species due to increased CO_2 levels. Legume species will also benefit from CO_2 increases and in tropical grasslands the mix between legumes and grasses could be altered. Rising temperatures increase lignifications of plant tissues and thus, reduce the digestibility and the rates of degradation of plant species. The resultant reduction in livestock production may have an effect on the food security and incomes of smallholders. Interactions between primary productivity and quality of grasslands will require modifications in the management of grazing systems to attain production objectives.
Biodiversity (genetics and breeding)	In some places there will be acceleration in the loss of the genetic and cultural diversity already occurring as a result of globalization, in crops as well as domestic animals. A 2.5° C rise in global temperature would cause extinction of 20-30% of all plant and animal species assessed (IPCC, 2007). Ecosystems and species display a wide range of vulnerabilities to climate change, depending on the imminence of exposure to ecosystem-specific critical thresholds. Local and rare breeds could be lost as a result of the impact of climate change and disease epidemics. Biodiversity loss has global health implications and many of the anticipated health risks driven by climate change will be attributable to a loss of genetic diversity.
Livestock (and human) health	Vector-borne diseases could be affected by (i) the expansion of vector populations into cooler areas (in higher altitude areas, malaria and livestock tick-borne diseases) or into more temperate zones (such as bluetongue disease in northern Europe); and (ii) changes in rainfall pattern during wetter years, which could also lead to expanding vector populations and large-scale outbreaks of disease. Temperature and humidity variations could have a significant effect on helminth infections. Changes in crop and livestock practices could produce effects on the distribution and impact of malaria in many systems, and schistosomiasis and lymphatic filariasis in irrigated systems (Patz et al., 2005).

Source: Thornton et al., 2008

Effect of heat stress on cattle

All animals have a range of ambient environmental temperatures termed as the thermoneutral zone. This is the range of temperatures that are conducive to health and performance. The upper critical temperature is the point at which heat or thermal stress effects begin to affect the animal. The environmental factors that contribute to heat stress include high temperature, high humidity and sunlight. Heat stress can be simply defined as the point where the cow cannot dissipate an adequate quantity of heat to maintain body thermal balance.

The environmental conditions that induce heat stress can be calculated using the temperature humidity index (THI). A THI is a single value representing the combination of air temperature and humidity associated with the level of thermal stress. The THI was developed during early 1990s as a weather safety index to monitor and reduce heat stress-related losses. Utilising both humidity and temperature is important compared with utilising temperature or humidity alone to evaluate heat stress. THI is calculated from air temperature and relative humidity using the following equation.

THI = Dry bulb temperature in degree Celsius + (0.36 × dew point temperature in degree Celsius) + 41.2

The dew point is the temperature at which the water vapour in a sample of air at constant barometric pressure condenses into liquid water at the same rate at which it evaporates. Heat stress begins to occur in dairy cattle when the THI is > 72. The THI threshold values for beef cattle, swine and poultry are 72-75, 72-74 and 70-78, respectively.

Table 9.3. Effect of heat stress on dairy cattle based on temperature humidity index

THI	Stress level	Comments
< 72	None	–
72 - 79	Mild	Dairy cows will adjust by seeking shade, increasing respiration rate and dilation of the blood vessels. The effect on milk production will be minimal.
80 - 89	Moderate	Both saliva production and respiration rate will increase. Feed intake may be depressed and water consumption will increase. There will be an increase in body temperature. Milk reduction and reproduction will be decreased.
90 - 98	Severe	Cows will become very uncomfortable due to high body temperature, rapid respiration and excessive saliva production. Milk production and reproduction will be markedly decreased.
> 98	Danger	Death of cow may occur.

Source: Hahn, 1999; West, 2003

Impacts of thermal stress on animal health

Thermal stresses have both direct and indirect effects on animal health. Direct effects of thermal stress include simple physiological disturbances, organ dysfunction and death. Reduction in feed intake together with diversion of more energy to maintain normal body function creates negative energy balance which deteriorates animal health conditions. These animals with poor health are prone to infectious diseases. The high ambient temperature and moisture level creates the environment suitable for fungus growth in feed and feedstuffs which may lead to mycotoxicosis in animals due to mycotoxin produced by fungi.

Impacts of thermal stress on reproduction

Thermal stress causes imbalance in secretion of reproductive hormones. High ambient temperature increases incidence of ovarian cysts and reduces fertility of cattle. Conception rates are less in cattle under thermal stress as compared to those in thermoneutral zone. Thermal stress prior to and immediately after artificial insemination causes reduction in conception rate in high producing lactating cows. Similarly, climatic warming also affects reproductive performance in bulls.

Impacts of thermal stress on milk production

Reduction in milk production is one of the major economic impacts of climatic stress in dairy cattle. Reduction in milk yield is further intensified by decrease in feed consumption by the animals. However, reduced milk production due to thermal stress is attributable only partly to decrease in feed intake. Actually 35% of reduced milk production is due to decreased feed intake while remaining 65% is attributable to direct effect of thermal stress (Bajagai, 2011). Milk production in cow has been found to be reduced when ambient temperature and temperature humidity index increase above critical threshold. The yield and composition of milk are affected by the temperature of the environment. Environmental temperatures lower and higher than the comfort zone temperatures reduce yield and alter many of the components of milk. At all India level an estimated annual loss due to direct thermal stress on livestock is about 1.8 million tonnes of milk, that is, nearly 2% of the total milk production in the country. This decrease can be either transitory or long-term depending on the length and severity of heat stress. The decline in milk production will be higher in crossbreds followed by buffalo and indigenous cattle.

EFFECT OF CLIMATE ON POULTRY BIRDS

Hot weather can have a severe impact on performance of poultry. The climate in poultry houses influences the health of the birds. Respiratory, digestive and behavioural disorders are more likely to occur in houses in which the climatic conditions are not up to standard. Production efficiency can be affected long before the temperature reaches a level at which survival becomes a concern.

Table 9.4. Effect of ambient temperature on adult poultry birds

Temperature range (in degree Celsius)	Effect on adult poultry birds
13 - 24	Thermoneutral zone. The temperature range in which the bird does not need to alter its basic metabolic rate or behaviour to maintain its body temperature.
19 - 24	Ideal temperature range.
24 - 30	A slight reduction in feed consumption can be expected, but if nutrient intake is adequate, production efficiency is good. Egg size may be reduced and shell quality may suffer as temperatures reach the top of this range.
30 - 33	Feed consumption falls further. Weight gains are lower. Egg size and shell quality deteriorate. Egg production usually suffers. Cooling procedures should be started before this temperature range is reached.
33 - 35	Feed consumption continues to drop. There is some danger of heat prostration among layers, especially the heavier birds and those in full production. At these temperatures, cooling procedures must be carried out.
35 - 38	Heat prostration is probable. Emergency measures may be needed. Egg production and feed consumption are severely reduced. Water consumption is very high.
> 38	Emergency measures are needed to cool birds. Survival is the concern at these temperatures.

Source: Anderson and Carter, 2007

The layers are warm blooded with a body temperature of 41 - 42.2° C. Body temperature is kept quite constant and is regulated by part of the chicken brain. It takes some time before heat regulating mechanisms start functioning in newborn chicks and therefore, they need a higher ambient temperature than adult birds do. Furthermore, the ratio between the surface area and weight of young chicks is unfavourable and they do not have any fat reserves. The comfort zone is defined as the temperature zone in which the birds are able to keep their body temperature constant with minimum effort. Behaviour of birds will change when temperatures rise to above or decrease to below the comfort zone.

The thermoneutral zone is defined as the temperature zone in which the birds are able to keep their body temperature constant with the help of physical heat regulation. The lowest temperature in the thermoneutral zone is called the lowest critical temperature. If temperatures fall to under this temperature the bird will start to use feed energy to warm itself and will consequently consume more feed. The highest temperature in the thermoneutral zone is called the highest critical temperature. If the temperature rises above this temperature the birds can no longer dissipate their heat. They will start to consume less feed and as a result, production will drop. The highest and lowest critical temperature of a bird depends on its age and bodyweight, housing system, feeding level, relative humidity and air velocity. The critical temperature for layers is 20° C. For every 1° C lower than 20° C, the birds require an extra 1.5 g of feed per day. The most efficient temperature for layers lies between 20 and 24° C. When temperature rises above 24° C, shell quality and egg weight will reduce. The critical temperature for broilers and rearing birds is highly dependent on age.

Relative humidity in poultry houses is measured to determine its impact on respiratory disorders. If the relative humidity is too high, condensation may accumulate in the house which leads to wet litter. This has a direct effect on the growth of microorganisms. Humidity is controlled by the intense heating or cooling of air inside the house in response to the temperature outside the house. The ideal relative humidity for poultry is 60 - 80%.

IMPACT OF CLIMATE CHANGE ON FISHERIES

Fishery enterprises are already facing multiple stresses, including overfishing and water pollution. Climate change may still worsen these stresses. In particular, temperature changes could lead to significant impacts. Increased water temperature affects fish breeding, migration and harvest. Many marine species have certain temperature ranges at which they can survive. Increased temperature and tropical cyclonic conditions affect capture, production and marketing cost of marine fish. Many aquatic species migrate towards colder areas of streams, lakes or ocean to avoid warm water. However, moving into new areas may put these species into competition with other species over food and other resources. Changes in temperature and seasons could affect the timing of reproduction and migration.

Many steps within an aquatic animal's lifecycle are controlled by temperature and the changing of the seasons. The rising ocean temperature and increased acidity makes it more difficult for marine organisms such as shrimps, oysters, or corals to form their shells. The entire marine food web is being altered, and may result in 'cracks in the food chain'. As a result, the distribution, productivity, and species composition of global fish production is changing, generating complex and inter-related impacts on oceans, estuaries and sea grass beds that provide habitats and nursery areas for fish (Mohanty et al., 2010). Climate change has been implicated in mass mortalities of many aquatic species, including plants, fish, corals, and mammals (Battin et al., 2007). Some species will also shift away from shallow coastal waters and semi-enclosed areas, where temperatures will increase fastest, into deeper cooler waters. In general, fish tend to live near their tolerance limits of a range of factors, and as a result, increased temperature and acidity, lower dissolved oxygen and changes to salinity may have deleterious effects on their populations.

Some diseases that affect aquatic life may become more prevalent in warm water. The general effects of increased temperature on parasites include, rapid growth and maturation, earlier onset of spring maturation, increased parasite mortality, increased number of generations per year, increased rates of parasitism and disease, earlier and prolonged transmission, the possibility of continuous, year-round transmission (Marcogliese, 2008). There is much evidence to suggest that parasite and disease transmission, and possibly virulence, will increase with global warming. However, the effects of climate change will be superimposed on a multitude of other anthropogenic environmental changes. Climate change itself may exacerbate these anthropogenic effects. The repercussions of climate change are not limited solely to temperature effects on hosts and their parasites, but also have other possible effects such as alteration in water levels and flow regimes, eutrophication, stratification, changes in acidification, reduced ice cover, changes in ocean currents, increased ultraviolet light penetration, runoff, weather extremes (Cochrane et al., 2009). Increasing acidity could harm shellfish by weakening their shells, which are created from calcium and are vulnerable to increasing acidity. Acidification may also threaten the structures of sensitive ecosystems upon which some fish and shellfish rely.

CONTRIBUTION OF AGRICULTURE TO CLIMATE CHANGE

Modern agriculture, food production and distribution are major contributors to global greenhouse gas emissions. Agriculture contributes a significant share of the greenhouse gas (GHG) emissions that are causing climate change; 17% directly through agricultural activities and an additional 7-14% through changes in land use. In agriculture, the non-CO_2 sources (CH_4 and N_2O) are reported as anthropogenic GHG emissions. The CO_2 emitted is considered neutral, being associated to annual cycles of carbon fixation and oxidation through photosynthesis (IPCC, 2007). Soil respiration is roughly balanced by the net uptake of CO_2 through plant photosynthesis. The organic matter decomposition and microbial respiration are influenced by soil physicochemical and biological soil properties controlling the activity of soil microorganisms and fauna. Soils with higher soil organic carbon (SOC) content emit more CO_2 than soil with low SOC, thereby increasing concentration of CO_2 in the atmosphere. Increased productivity can supply more plant residues to the soil, possibly increasing storage of soil organic matter. But, higher level of atmospheric CO_2 is also coupled with temperature rise which would have both positive and negative effects on plant productivity. If the productivity decreases it would have a negative effect on soil carbon storage. Thus, accounting CO_2 emission and removal in agriculture should be considered in creating GHG inventory from agriculture (Lenka et al., 2015).

Methane emission from rice fields

The rice fields are a major source of emission of GHGs like CH_4 and N_2O. Methane is produced as the terminal step of the anaerobic breakdown of organic matter in wetland rice soils. The oxygen supply from the atmosphere to the soil of a flooded rice field is cut off which leads to anaerobic fermentation of soil organic matter. Methane is a major end-product of anaerobic fermentation. It is released from submerged soils to the atmosphere by diffusion and ebullition and through roots and stems of rice plants. Disturbance of wetland soil by cultural practices favours soil trapped methane to escape to atmosphere through ebullition. Methane is exclusively produced by methanogenic bacteria that can metabolize only in the strict absence of free oxygen. Though flooded paddy soils have a high potential to produce CH_4, but part of produced CH_4 is consumed by CH_4 oxidising bacteria, or methanotrophs. In rice fields, it is possible that a part of produced CH_4 in anaerobic soil layer is oxidised in aerobic layers such as surface soil-water interface and the rhizosphere of rice plants, and the net emission will be positive or negative depending on the relative magnitudes of methanogenesis and methanotrophy, respectively (Lenka et al., 2015).

In tropical flooded rice soils, where soil temperature is 25-30° C, methane production is rapid in alkaline and calcareous soils, may start hours after flooding and slow in acid soils, and may take five or more weeks after flooding. Methane production is negatively correlated with a soil redox potential and positively correlated with soil temperature, soil carbon content, and rice growth. Easily degradable crop residues, fallow weeds, and soil organic matter are the major source for initial methane production. At later growth stages of rice, root exudates, decaying roots, and aquatic biomass are more important for CH_4 emission. Use of organic amendments and chemical fertilisers also promotes CH_4 emission.

N_2O emission from agricultural soils

Most of the N_2O emissions take place in soils and are related with agricultural activities. The N_2O emitted from the soil represents some 50% of the total agricultural emissions. Even when it is not being cultivated, the soil naturally releases GHGs. The N_2O is generated as a by-product of microbial activities that convert ammonium into nitrate (nitrification) or nitrate into nitrogen gas (denitrification). Both processes are influenced and controlled by environmental conditions such as soil moisture, temperature, oxygen concentration, amount of available organic carbon and nitrogen and soil C/N ratio. Emissions also increase with agricultural activity, partly as a result of N input from manure, mineral fertilisers, or from symbiotic N fixation in legumes. Among these factors, those related to soil could be easily altered by management practices. Therefore, understanding the processes of N_2O formation in soils and the factors influencing these emissions is fundamental to develop efficient strategies to reduce N_2O emissions in agricultural soils (Signor and Cerri, 2013).

Burning of agricultural residues in field

The contribution of crop residue burning is the lowest 0.5% of the total agricultural emissions among different sources of GHG emission in the agriculture sector. In developing countries, farmers prefer crop residue burning as a quick and labour-saving process to dispose of the crop residues of rice, wheat, maize and sugarcane. Emissions of CO_2 during burning of crop residues are considered neutral, as it is reabsorbed during the next growing season. However, biomass burning is one of the significant sources of atmospheric aerosols and trace gas emissions, which has a major impact on human health. In addition to aerosol particles, biomass burning due to forest fires and crop residue burning are considered a major source of carbon dioxide (CO_2), carbon monoxide (CO), methane (CH_4), volatile organic compounds (VOC), nitrogen oxides, and halogen compounds. Carbon monoxide is a chemically active gas in the troposphere influencing the abundance of O_3 and the oxidising capacity (OH) of the troposphere. Thus, an increase in concentration of CO, VOC, and nitrogen oxides also increases concentration of GHGs in the atmosphere (Lenka et al., 2015).

Use of fossil fuels in agriculture

Use of fossil fuels in agriculture results in CO_2 emissions, and there are additional emissions associated with production and delivery of fuels to the farm. Non-traditional fuels sometimes used in processing agricultural materials include scrap tires and biomass. The CO_2 emission attributed to electricity consumption is based on the fuels used in power generation.

Use of fertilisers

Fertiliser production consumes about 1.2% of the world's energy and is responsible for approximately 1.2% of the total GHG emissions (Kongshaug, 1998). Carbon dioxide emissions result from the energy required for production of fertilisers and the energy required for their transport and application. The energy required per tonne of N and phosphate (P_2O_5) varies considerably with the form in which the nutrient is supplied. Carbon emissions from fossil fuels used in the production of fertilisers include emissions from mineral extraction and fertiliser manufacture. Postproduction emissions can include those from packaging, and transportation. Energy is also used during field application of fertilisers using farm machinery, thus the greater the fertiliser use, the greater are the emissions. Carbon emissions from agricultural lime are calculated from the fuel used for mining limestone and for grinding the stone into a usable product.

Pesticides

Modern pesticides are almost entirely produced from crude petroleum or natural gas products. The total energy input is thus both the material used as feedstock and the direct energy inputs. Carbon dioxide emissions from production of pesticides consist of both these contributions to manufacture the active ingredient. Postproduction emissions include those from formulation of the active ingredients into emulsifiable oils, wettable powders, or granules and those from packaging, transportation, and application of the pesticide formulation (Lenka et al., 2015).

Irrigation

The on-farm wells, on-farm surface reservoirs, and off-farm surface reservoirs are the major sources of water for farm activities. Fossil fuels used to power pumps, which distribute irrigation water. The energy use and carbon emissions from pumping water are applied to both on-farm wells and off-farm surface reservoirs.

Harvesting and threshing

Energy and CO_2 emissions during harvesting and threshing of agricultural produce are also important. The greater the productivity, the greater are the energy and emissions required for harvesting and threshing operations.

Farm machinery

Energy and CO_2 emissions associated with different tillage practices are a consequence of the fuel used by farm machines and the energy consumed in manufacture, transportation and repair of these machines.

Forestry and other land use changes

The expansion of agricultural land area has a major environmental impact. Deforestation and depletion of the humus releases large quantities of CO_2 from the carbon bound in the trees and the soil organic matter (SOM). Furthermore, deforestation has an immediate impact on the natural water cycle, resulting in a greater likelihood of flooding or drought. Some 24% of the total global GHG emissions can be attributed to agriculture and about 12% of these are due to change in land use and with extended agricultural production, this would rise considerably (FAOSTAT, 2013). Changes in land use have negatively affected the net ability of ecosystems to sequester C from the atmosphere. For instance, the C-rich grasslands and forests in temperate zones have been replaced by crops with much lower capacity to sequester C (Lenka et al., 2015).

CONTRIBUTION OF LIVESTOCK TO CLIMATE CHANGE

Ruminant animals, such as cattle, sheep, buffalo, and goats have special digestive systems that can convert otherwise unusable plant materials into nutritious food and fibre. This helpful digestive process, called enteric fermentation, however, produces methane (CH_4), a potent greenhouse gas that can contribute to global climate change. Enteric fermentation is a digestive process by which carbohydrates are broken down by microorganisms into simple molecules for absorption into the bloodstream of an animal. It is one of the factors in increased methane emissions. Enteric fermentation occurs when methane is produced in the rumen as microbial fermentation takes place.

Livestock production systems can also emit other greenhouse gases such as nitrous oxide (N_2O) and carbon dioxide (CO_2). In recent years, the increasing use of intensive livestock production systems has become a source of solid, liquid and airborne emissions that can be both a nuisance and environmentally harmful. In spite of the low amount of CH_4 in the atmosphere relative to that of CO_2, its importance as a pollutant is considered to be 21 times greater than that of CO_2, while that of N_2O is 310 times greater (Hartung, 2003).

It is estimated that nearly 20% of CH_4 and 35% of N_2O come from livestock production. According to the European Environment Agency, nearly 50% of the overall amount of CH_4 released in Europe originates from agriculture and stems mainly from ruminant animals (EEA, 2001).

Table 9.5. Relative contribution of various sources to the global emission of methane (CH_4) and nitrous oxide (N_2O)

Gas	Natural sources	Anthropogenic sources	
		Livestock production	Others
CH_4	30	20	50
N_2O	55	35	10

Source: Monteny, 2003

Manure of cattle is a much more important source of CH_4 emission than enteric fermentation. This problem is due to liquid manure storage in tanks or pits. The amount and surface area of the stored manure, and the ambient and core manure temperature are the major factors that determine the quantity of CH_4 gas emission (Hartung and Monteny, 2000).

Table 9.6. Methane emission from livestock production (kg/animal/year)

Species	Enteric fermentation	Manure
Dairy cows	100	345
Pigs	1	32
Poultry	0.1	2.4

Source: Hartung, 2003

Thus, the livestock sector which will be a sufferer of climate change is itself a large source of methane emissions. In India, although the emission rate per animal is much lower than the developed countries, due to vast livestock population the total annual methane emissions from Indian livestock ranged from 7.26 to 10.4 million tons/year. In India more than 90% of the total methane emission from enteric fermentation is being contributed by the large ruminants (cattle and buffalo) and rest from small ruminants and others.

CONCEPT OF CLIMATE SMART AGRICULTURE

Agricultural production systems are facing increasing competition from other sectors for limited natural resources. The availability of these resources and their quality are also being affected by unsustainable management practices and changing climatic and weather conditions. To respond to this situation, the agriculture sectors must improve their sustainability performance and adapt to the impacts of climate change in ways that do not compromise global efforts to ensure food security for all. These challenges are intimately and inextricably related, and need to be addressed simultaneously (FAO, 2021) by a technical and policy framework called climate smart agriculture.

Climate smart agriculture (CSA) is a relatively new concept which was launched in 2009 advocating for better integration of adaptation and mitigation actions in agriculture to capture synergies between them and to support sustainable agricultural development for food security under climate change. The rapid uptake of the concept after its launch indicates the tremendous demand for a framework to guide policy and technical interventions in agriculture that integrates the effects of change, the challenges of achieving sustainable agricultural development and the critical role of agriculture in attaining food security (Lipper and Zilberman, 2018). The concept of climate smart agriculture embodies the following three objectives.

1. To sustainably increase food security through increases in productivity and incomes
2. To build resilience and adapting to climate change
3. To reduce greenhouse gas emissions compared to a baseline scenario

Climate smart agriculture is not intended to provide a new set of sustainability principles, but rather a means of integrating the specificities of adaptation and mitigation into sustainable agricultural development policies, programmes and investments. CSA strategies and practices then should adhere to the principles that underpin sustainable agriculture and food systems. A set of guidelines and approach has been issued to achieve sustainable agriculture and food systems which meet the following criteria (FAO, 2016).

1. Improving the efficiency of resource use
2. Conserving, protecting and enhancing natural resources
3. Protecting and improving rural livelihoods
4. Enhancing resilience of people, ecosystems and communities
5. Responsible and effective governance mechanisms

Climate smart agriculture provides the means to help stakeholders at all levels identify agricultural strategies suitable to their local conditions. Climate smart agriculture is not a set of practices that can be universally applied, but rather an approach that involves different elements that are embedded in specific contexts and tailored to meet local needs. Climate smart agriculture builds on sustainable agricultural approaches, using principles of ecosystem and sustainable land and water management and landscape analysis, and assessments of the use of resources and energy in agricultural production systems and food systems (FAO, 2021).

ACTION POINTS FOR IMPLEMENTING THE CLIMATE SMART AGRICULTURE

Climate smart agriculture relates to actions in fields, pastures, forests, and oceans and freshwater ecosystems. It involves the assessment and application of technologies and practices, the creation of a supportive policy and institutional framework and the formulation of investment strategies. Climate-smart agricultural systems include the following four major elements (FAO, 2021).

1. The management of land, crops, livestock, aquaculture and capture fisheries to balance near-term food security and livelihoods needs with priorities for adaptation and mitigation;
2. Ecosystem and landscape management to conserve ecosystem services that are important for food security, agricultural development, adaptation and mitigation;
3. Services for farmers and land managers that can enable them to better manage the risks and impacts of climate change and undertake mitigation actions; and

4. Changes in the wider food system including demand-side measures and value chain interventions that enhance the benefits of climate-smart agriculture.

FAO (2021) has developed five action points for implementing the 'climate smart agriculture' approach.

Expand the evidence base

The first step in implementing a climate-smart agriculture approach is to develop a robust evidence base. The purpose of this step is to formulate strategies for increasing productivity and agricultural incomes, and estimate their potential mitigation co-benefits. An important part of building the climate-smart agriculture evidence base involves determining the current and projected effects of climate change on specific agricultural production systems and producers in the near and medium term, and pinpointing key vulnerabilities in the agricultural sectors and for food security. Another major component in developing the evidence base is the identification and evaluation of potential climate-smart options for adapting to the expected impacts of climate change while at the same time supporting sustainable agricultural development. In making the initial assessment, consideration needs to be given to the potential synergies and trade-offs for the proposed climate-smart agriculture interventions relative to the baseline activities. The expansion of evidence base is completed with estimating the costs and barriers to the adoption of different practices, identifying issues related to the sustainability of production systems and preparing the required policy and institutional responses.

Support enabling policy frameworks

The second step towards implementation of climate-smart agriculture to develop enabling policy frameworks. The development of supportive policies, plans and investments, and coordination in the policy-making processes and institutions responsible for agriculture, climate change, food security and land use are required to create this enabling policy framework. There may be a need to modify existing policy measures to exploit the synergies and minimise the trade-offs between the three objectives of climate-smart agriculture. Before designing new climate-smart agriculture policies, policy makers should systematically assess the intended and unintended effects of a wide range of current international and national agricultural and non-agricultural agreements and policies on climate-smart agriculture objectives, and take into account other national development priorities. New policies to stimulate the adoption of climate-smart agriculture systems should focus on filling policy gaps and contribute to a country-driven approach to capacity development in the short- and long-term. Understanding the socioeconomic and gender-differentiated barriers and incentive mechanisms that determine the adoption of climate-smart agriculture practices is also critical for designing and implementing supportive policies.

Strengthening national and local institutions

Enabling institutions (e.g., financial institutions, land tenure regimes, institutions regulating customary law, community-based organisations, insurance schemes, and information and extension services) are essential for harnessing the evidence base to empower, enable and motivate farmers to adopt climate-smart agriculture practices. Cross-sectoral dialogues, which form an important part of the climate-smart agriculture methodology, enhance coordination between institutions dealing with agricultural, climate change, social protection, food security and other issues at the local, national and international levels. These dialogues can take the form of dedicated workshops to consider emerging policies, or presentations and discussions among standing committees or public sector bodies involved in policy formation. In some cases, efforts also need to be made to build the capacities of national policy makers to participate in international policy fora on climate change and agriculture, and reinforce their engagement with local government authorities.

Enhancing financing options

Innovative financing mechanisms that improve the links between climate finance and agricultural investments from the public and private sectors are central to implementing climate-smart agriculture. Climate financing mechanisms may be a way of spurring sustainable agricultural development. Strong national sector budgets and official development assistance should continue to be the main sources of funding. Integrating climate change issues into sector planning and budgeting is a prerequisite for successfully addressing the impacts of climate change. Linking climate finance to agricultural investments requires the capacity to measure, report and verify the interventions that have received funding are indeed generating adaptation and mitigation benefits. An evidence base for climate-smart agriculture can provide much of the information needed for making this link.

Implementing practices in the field

Farmers, pastoralists, foresters and fisherfolk are the primary custodians of knowledge about their environment, agricultural ecosystems, crops, livestock, forests, fish and local climatic patterns. Efforts to adapt a climate-smart agriculture approach to a specific setting must take into account local producers' knowledge, requirements and priorities. Local project managers and institutions can engage with agricultural producers to identify suitable climate-smart agriculture options that can be easily adopted and implemented.

CLIMATE SMART CROP PRODUCTION

There are many climate change adaptation and mitigation options that can sustainably improve yields of a crop by minimising the harmful environmental impacts on production systems. These approaches may differ for each farming household, depending on its coping and adaptive mechanisms. Management practices and technologies for climate change adaptation and mitigation include practices with an explicit focus on adaptation and practices with a broader scope on reducing production risks and reducing emissions. The following five climate smart approaches to crop production have been suggested by FAO (2021).

1. Increasing diversity and complexity within the agricultural ecosystem, which can be done in many ways (e.g., expanding the diversity of crops or crop varieties), at many spatial scales (e.g., landscape level, within farms, and/or within the same crop) and over different timeframes;
2. Improving sustainable soil and land management (e.g., carefully channelling the expansion of crop and grazing land to mitigate the loss of carbon storage that results from land-use change);
3. Increasing energy use efficiency;
4. Promoting sustainable mechanisation (e.g., increasing the availability of suitable machinery in combination with proper agronomic management to reduce greenhouse gas emissions from various farm and processing operations); and
5. Developing simple and robust scientific tools to guide the decision-making of farmers on a seasonal and long-term basis.

Sustainable soil and land management

The negative impacts of climate change on land and soil resources can be reduced and optimise its potentially positive effects by implementing targeted and adapted sustainable soil and land management practices and selecting the most appropriate land-use systems for a given environment. Sustainable soil and land management encompasses options that allow different user groups to manage their resources, including water, crops, livestock and associated biodiversity, in ways that are best suited to the prevailing biophysical, socioeconomic and climatic conditions. Land resource planning is an essential entry point and process to choose the most suitable land-use systems for accommodating the often competing uses of land. It can also promote the adoption of locally adapted sustainable soil and land management practices and enable decision-makers and communities to put in place more resilient land-use systems that can support climate change mitigation and adaptation (FAO, 2021).

Water management

Water management is a critical component of climate-smart agriculture strategies. The agriculture sectors consume 70% of total global freshwater resources. Water demand will grow in the coming decades, primarily due to population growth and economic development. Changes in irrigation water demand are also expected. Climate change will also affect the design and operation of water infrastructure. Climate change adaptation options for water management will necessarily combine policies, institutions, investments, crop and water management practices and capacity development.

Conservation and sustainable use of genetic resources

There is a need to enhance conservation and sustainable use of genetic resources for food and agriculture and gather more and better information on these resources. Traditional and novel uses of genetic resources can increase the adaptability, resilience and yield of production systems and enhance their contribution to climate change mitigation. It is necessary to raise awareness about the important role genetic resources play in climate change adaptation and mitigation, and strengthen national capacities to sustainably manage their genetic resources.

Management of energy

The food sector currently accounts for around 30% of the world's total end-use energy consumption, and most of this energy is coming from fossil fuels. Thus, the energy used in food chains accounts for more than one-third of the total emissions from food chains. To address the challenges of climate change, it is required to scale up energy-smart food chains, relying on adequate access to modern energy services through

1. Improved energy efficiency,
2. Increased use and production of renewable energy,
3. Sustainable bioenergy, and
4. A water-energy-food nexus approach that connects the use and consumption of water, energy and food.

CLIMATE SMART LIVESTOCK PRODUCTION

Livestock provide multiple benefits including food, clothing, fuel, draught power, income and employment. They also contribute to nutrient cycling in the soil, and can represent a type of insurance for small-scale producers in developing countries that can safeguard food and income in the event of climate- and weather-associated crises. In many arid and semiarid regions, livestock production systems represent the only viable system of food production. Climate change presents a range of challenges for livestock producers. Livestock production will be directly affected by changes in temperature and water availability. Climate change will also affect livestock production through its impacts on the supply and quality of pasture and forage crops, the production and prices of feed grains, and modifications in the distribution ranges of livestock diseases and pests. Livestock is also responsible for 14.5% of anthropogenic greenhouse gas emissions. Livestock production systems account for up to half of the technical mitigation potential of the agriculture, forestry and other land use (AFOLU) sectors. This potential can be realised through management options that sustainably intensify livestock production, promote carbon sequestration in rangelands and reduce emissions from manure. A reduction in consumer demand for livestock products can also contribute to climate change mitigation (FAO, 2021).

For an animal to maintain homeothermy (no change in body temperature, other than normal circadian rhythms), the ambient environment and the animal must exchange heat at a rate that permits balancing the metabolic heat production and the energy gains/losses through the four transfer modes, i.e., radiation, convection, conduction and evaporation. Ruminant animals primarily adjust evaporative heat loss to maintain homeo-thermy during brief exposures to adverse weather, but will reduce feed intake to lower heat produc-tion during prolonged hot weather. Swine and poultry primarily adjust heat production to main-tain homeothermy. To alter the microclimate of an animal effectively through housing or environmental modification the following factors should be considered (WMO, 2012).

Site selection

The selection of a site for housing of animals is the most important factor to minimise the effects of local weather. Climatic factors vary with height above the ground at a specific location. Observations of the microclimates in a general location will reveal much variation in thermal conditions resulting from terrain features, differential exposure, wetlands, rivers, type and height of vegetation, human activities and other factors. Proper selection of a site to emphasise factors for enhanced heat dissipation will have long-term protection benefits.

The recommended ambient temperature and relative humidity for chickens vary with their age (table 9.7).

Table 9.7. Recommended temperature and relative humidity for chicken

Age of the bird (in days)	Recommended ambient temperature (°C)	Recommended relative humidity (%)
1	33	50
7	31.5	50
14	28.5	60
21	28.5	65
28	26	70
42	23	85

Source: Hulzebosch, 2006

At the time of hatching, ducklings require a high temperature of about 30° C. They are not yet able to regulate their body temperature and require supplemental heat such as that provided by a brooder. As they grow older they become better able to produce and conserve heat, and regulate their body temperature. After a duckling is fully covered with feathers, they are able to maintain proper body temperature even when the outside temperature is low.

Table 9.8. Recommended temperatures for ducks

Age of ducks in days	Recommended temperature in °C
1	30
7	27
14	23
21	19
28	15
35	13
42	13
49	13
Developing breeders	13
Laying breeders	13

Source: Scott and Dean, 1991

Windbreaks

Windbreaks have an important role to reduce the climatic impacts on animals in tropical and subtropical regions. Grazing animals are protected from strong winds, especially during cold weather. Structures or trees can markedly reduce wind speed and can be beneficial to the survival of exposed animals, especially the newborns. A windbreak acts as a barrier that lowers the wind speed near the ground surface and diverts and splits the air stream. The protection achieved is determined by the configuration, height, density and thickness of the trees in a belt. The use of leguminous trees or shrubs can be a practical means to counteract the effects of wind and heat stress, as well as to improve animal diet.

Shades

Shades may be provided for protecting farm animals in hot climates. In a tropical region, the solar irradiance is high even during the winter. It is reported that when dairy cows are given access to adequate shades, milk production is increased. Enclosed shelters are not recommended for tropical climates because of the decreased natural air velocity and sanitation. In temperate regions, partially enclosed shelters can reduce the thermal radiation received by animals during hot weather.

Genetic improvement for adaptation

Many local breeds are adapted to harsh living conditions and they are able to survive, grow and reproduce in conditions of poor nutrition, and even under infections of parasites and disease causing microorganisms. Any population must be genetically heterogeneous in order to be able to survive under the challenges of the changing environment. Thus, adaptation strategy should be identifying and strengthening local breeds that have adapted to local climatic stress and feed sources and improving local genetics through cross-breeding with heat and disease tolerant breeds. This is the basis for genetic improvement in livestock. If climate change is faster than natural selection, the risk of survival and adaptation of the new breeds is greater (Hoffmann, 2008).

CLIMATE SMART FISHERIES AND AQUACULTURE

In the fisheries and aquaculture sector, the major impacts of climate change include severe weather events, increased flooding in coastal and riparian zones, sea level rise, shifts in the distribution range of important species and ocean acidification. The fisheries and aquaculture sector also has the capacity for increasing carbon removal from the atmosphere through farming of seaweeds and improved coastal management mainly with the protection and management of mangrove forests and estuaries. Healthy ecosystems and sustainable use of fisheries and aquaculture resources are the foundation of climate change adaptation and mitigation for the fisheries and aquaculture sector. The ecosystem approach to fisheries and aquaculture provides the needed framework to holistically address climate change across marine and coastal ecosystems. Climate-smart agricultural strategies that capture fisheries can adapt to lower emissions include reducing fuel use in the global fishing fleet, reducing overcapacity and reducing the carbon imprint of the international trade in fish and fisheries products, which are the most highly traded food commodity.

There is a wide range of potential adaptation options for fisheries. To build resilience to the effects of climate change and derive sustainable benefits from fisheries and aquaculture river basins, watersheds and coastal zones are to be effectively managed in an integrated approach. In absence of careful planning, aquatic ecosystems, fisheries and aquaculture can potentially suffer as a result of adaptation measures applied by other sectors such as increased use of dams and hydropower in catchments. Mitigation solutions reducing the carbon footprint of fisheries and aquaculture will require innovative approaches. Mangrove conservation reduces carbon emissions from deforestation. Innovative but environmentally safe ways are to be explored to sequester carbon in aquatic ecosystems, and developing low-carbon aquaculture production systems (Mohanty et al., 2010). Use of herbivorous fishes may be used as a potential tool to help ecosystems recover from climate change impacts. This approach might be particularly suitable for recovery of coral reefs, which are acutely threatened by climate change (Mumby and Harborne, 2010). Farming of shellfish, such as oysters and mussels, is not only good business, but also helps clean coastal water. Culturing aquatic plants help to remove waste from polluted water. Fisheries and aquaculture need to be blended into National Climate Change Adaptation Strategies.

INTEGRATED PRODUCTION SYSTEMS

Integrated production systems use some of the products, by-products or services of one production component as inputs for another production component within a single farming operation. In integrated systems, the production components are mutually supportive and mutually dependent. Integrated production systems include agroforestry, crop-livestock, rice-fish, fish-livestock and food-energy systems, as well as less widespread systems, such as aquaponics (FAO, 2021). Aquaponics is a system of aquaculture in which the waste produced by farmed fish or other aquatic creatures supplies the nutrients for plants grown hydroponically, which in turn purify the water. By optimising efficiency in the use of resources, integrated production systems can achieve the synergistic objectives of climate-smart agriculture. The high efficiency in recycling resources (e.g., converting waste into biogas) creates a system with minimum environmental impact and lowers operating costs, as there is less need for inputs (e.g., fertiliser, feed and energy). The diversification of resources and incomes associated with integrated production offers farmers a greater number of risk management strategies and options to adapt to the impacts of climate change. Moreover, the emissions intensities of integrated systems are typically lower than those of specialised systems. However, successful integration rests on the flexibility to reduce trade-offs and competition between the various production components of the agricultural system, which demands substantial technical knowledge, labour, and sometimes upfront investments that will only pay off over a relatively long period of time. Sustainable production intensification from integrated agricultural systems requires a clearer understanding of the impacts of changes in climate and climate variability on these systems (FAO, 2021).

Developing sustainable food systems and value chains

Climate-smart agriculture approaches are required in the entire value chain from production to consumption. It is important to take a holistic, systems view of climate impacts and vulnerabilities to identify climate-smart interventions to adapt to and mitigate climate change. FAO's sustainable food value chain development (SFVCD) approach involves an analysis of three interconnected levels such as

1. The core value chain (i.e., the stages from production to disposal),
2. The extended value chain including support services, and
3. The enabling environment, both societal and natural elements.

By analysing the interlinked value-adding activities, including the production, aggregation, processing, distribution, consumption and disposal of products that originate from agriculture, forestry or fisheries, and the environments in which they are embedded, it is possible to identify interventions at all levels for sustainable value chains and food systems. According to FAO (2021) interventions for climate adaptation and mitigation to value chains at all levels include

1. Investments in infrastructure (e.g., storage, roads) and packaging;
2. Efforts to reduce food loss and waste;
3. Optimising energy and input usage;
4. Implementing policies; and
5. Generating knowledge to influence behaviour change in the value chain and food systems (e.g., reducing consumption and improving extension services).

ENABLING ENVIRONMENT FOR CLIMATE SMART AGRICULTURE

Scaling up climate-smart agriculture to trigger the desired transformation in agricultural production systems requires supportive policies, institutions and financing, that together create an enabling environment for climate-smart agriculture at local, national and international levels (FAO, 2021). Changes in production that might be expected based on certain climate indicators may not occur due to other factors related to human capital (e.g., level of education, age, ethnicity, gender of producers), economic conditions (e.g., relative prices, input and output market development, credit availability) and the policy environment (Asfaw et al., 2015; Arslan et al., 2015). The response of agricultural producers to climate change and variability will depend on the socioeconomic position of the household. Poor farmers are likely to take measures to ensure their survival, while wealthier farmers make decisions to maximise profits (Asfaw et al., 2016). For this reason, the impacts of climate change are expected to affect different segments of the rural population differently.

Role of policies for climate-smart agriculture

At the national level, making the transition to climate-smart agriculture requires not only strong political commitment but greater coherence, coordination and integration among the various sectors dealing with the drivers that are influencing climate change, agricultural development, and food and nutrition security. In this regard, the development of enabling policies and the promotion of an adequate regulatory and legislative environment is crucial. Public policies should work towards an integration of new climate-smart agriculture policies and support measures at the national and local levels. Representatives from the entire range of stakeholder groups involved in all sectors and at all levels need to participate fully in this coordination and integration process. It is necessary to recognise the value of indigenous knowledge and farming practices when developing coherent cross-sectoral policy frameworks. It is important to identify and enhance synergies between the different policy objectives and address and compensate for trade-offs where necessary. The public sector can play a key role in creating an enabling policy and legal environment for climate-smart agriculture, which can permit private sector and civil society stakeholders to make timely, well-informed and efficient decisions related to securing food production, adapting to climate change and reducing and removing greenhouse gases (FAO, 2021).

Role of institutions for climate-smart agriculture

The ability of agricultural producers to adapt to climate change requires that rural communities have access to up-to-date information, inputs and out markets, and microfinance and insurance.

Access to information: One of the key constraints to the widespread adoption of climate-smart agriculture concerns the flow of information and knowledge. Information on the range of climate-smart agriculture options, particularly those well suited to local conditions, is often scarce. For example, inadequate information can increase the risk of planting expensive seeds that may not survive or produce poor yields (Asfaw, Di Battista and Lipper, 2014; Asfaw et al., 2015). Information available to producers on climate-smart agriculture options that are well adapted to their particular setting is likely to be an important determinant for adoption. Information may come from a number of sources, including government extension programmes and non-governmental organisations. Increasing access to reliable information is crucial to facilitating adaptation as it can improve the decision-making capacities of farmers with regard to their farming systems. One key role of institutions is the production and dissemination of knowledge and information, which can range from the impacts of climate change on production and marketing conditions, to the development of regulations and standards. Climate change increases uncertainty. This increases the value of information and the importance of institutions that are engaged in conducting agricultural research, providing extension services, collecting and analysing agricultural production and marketing statistics and disseminating climate-related information (Campbell et al., 2011).

Access to microfinance and insurance: Adopting climate-smart agriculture practices requires that producers make increased investments that extend over a relatively long period of time. McCarthy et al. (2011) have documented the extended transition times needed to realise the benefits of climate-smart agriculture in the terms of productivity or increased resilience. During the transition, the returns to agriculture can be low or negative, and thus, some form of financing to support this transition is necessary. This transition cost is likely to be higher for poorer farmers, which makes the challenge of delivering the benefits of climate-smart agriculture to those most in need even greater (McCarthy et al., 2011). The capacity of farmers to make the required adjustments depends in large part on the existence of policies and institutions that can support their access to credit and insurance.

Access to agricultural input and output markets: The decisions farmers make about the type of technologies and practices they adopt are determined by their benefits and costs, which are affected by the access these farmers have to supplies of inputs and output market chains. Improved market access, which increases the returns on land and labour, is a critical factor for the adoption of new climate-smart agriculture practices. However, many smallholders in vulnerable areas continue to face complex challenges in adoption of climate-smart agriculture practices. Addressing these constraints requires innovative institutional arrangements and partnerships that improve the local availability and utilisation of climate-smart agriculture inputs and create effective market linkages that offer more stable and better prices to producers. Producer organisations have the potential to mitigate the effects of imperfect markets by establishing contractual links to input and output markets and promoting economic coordination in liberalised markets, which can leverage market functions for smallholder farmers (Shiferaw and Hailemariam, 2007). This can be expected to stimulate adoption of climate-smart agriculture practices, which will in turn drive the process of agricultural commercialisation in rural areas.

ADAPTATION AND MITIGATION STRATEGIES IN AGRICULTURE

Adaptation strategies

Adaptation in agriculture is the norm rather than the exception. Adaptation strategies will vary with agricultural systems, location, and scenarios of climate change considered. There are several adaptation measures that the agricultural sector can undertake to cope with climate change.

1. Changing planting dates
2. Planting different varieties or crop species
3. Development and promotion of alternative crops
4. Developing new drought and heat-resistant varieties
5. More use of intercropping
6. Using sustainable fertiliser and tillage practices (improving soil drainage, no-till, etc.)
7. Improved crop residue and weed management
8. More use of water harvesting techniques
9. Better pest and disease control for crops
10. Implementing new or improving existing irrigation systems (reducing water leakage, soil moisture conservation throgh mulching)
11. Improved livestock management (providing housing and shade, change to heat-tolerant breeds, change in stocking rate, altered grazing and rotation of pasture)
12. More use of agroforestry practices
13. Improved forest fire management (altered stand layout, landscape planning, dead timber salvaging, clearing undergrowth, insect control through prescribed burning)
14. Development of early-warning systems and protection measures for natural disasters (droughts, floods, tropical cyclones, etc.)

Key mitigation technologies in agriculture

The National Mission for Sustainable Agriculture (of India) has documented the following mitigation technologies in agriculture (NMSA, 2010).

Improved crop seeds, livestock and fish cultures

- Promoting use of biotechnology
- Research and promotion of C4 pathways in C3 plants

- Conserving indigenous genetic resources
- Public-private partnership (PPP) in Research and Development (R&D) , management and dissemination of improved varieties
- Conserving agricultural heritage

Water use efficiency

- Promoting water use efficiency in agriculture
- R&D in the areas of energy efficient water system
- Developing mechanism for integrated management of rainwater, surface and groundwater
- Policy instruments for PPP
- Strengthening local institutions in managing water allocation and utilisation

Pest management

- Efficient, safe and environmentally sound methods of pest management
- Research, commercial production and marketing of biopesticides
- Developing insect forecasting models
- Decision and information support system for pest and disease surveillance
- Institutional mechanism for quick response in case of disaster

Improved agronomic practices

- Improved agronomic practices to reduce farm losses
- Conservation and precision farming
- Knowledge management
- Soil conservation
- Policy instruments for optimum land use

Nutrient management

- Production and use of biofertilisers
- Developing nutritional strategies for managing heat stress in dairy animals
- Strengthening capacity of existing soil testing labs
- Quality standards and quality control systems for raising confidence among users

Agricultural insurance

- Developing various models for risk management
- Designing users' friendly Decision Support System (DSS) to assess risks and develop contingent plans
- Weather based crop insurance schemes
- Region-specific contingency plans based on vulnerability and risk scenario

Credit support

- Developing new forms of credit assessment and risk management system
- Promoting microfinance
- Developing mechanisms to enhance the flow of credit to critical infrastructure
- Up-scaling *kisan* credit card scheme
- Designing customised credit policies and programmes to mitigate risks

Markets

- To formulate market-aligned R&D programmes
- Improving supply chain efficiency
- Creation of new market infrastructure
- Supporting community partnerships in developing food and forage banks
- Strengthening access to quality and timely inputs by farmers

Access to information

- Minimising information asymmetry through information and communication technology-based system
- PPP to develop technology based solution to market related problems
- Building an Information and Communication Technologies (ICT) enabled knowledge management network
- To create, manage and develop national resource portal

Livelihood diversification

- Mitigating risks by supplementing income from off-farm activities
- Crop diversification
- Crop-livestock-fisheries farming system

REFERENCES

Anderson, K. E. and Carter, T. A. 2007. Hot Weather Management of Poultry. *Poultry Science and Technology Guide No. 30*. North Carolina State University, USA.

Arslan, A., McCarthy, N., Lipper, L., Asfaw, S., Cattaneo, A. and Kokwe, M. 2015. Climate smart agriculture: Assessing the adaptation implications in Zambia. *Journal of Agricultural Economics* **66**(3): 753-780.

Asfaw, S., Di Battista, F. and Lipper, L. 2014. Food security impact of agricultural technology adoption under climate change: micro-evidence from Niger. ESA Working Paper No. 14-12. FAO, Rome.

Asfaw, S., Maggio, G. and Lipper, L. 2016. Gender, climate shock and welfare: Evidence from Malawi. *Journal of Development Studies* **53**(7): 1-21.

Asfaw, S., McCarthy, N., Arslan, A., Cattaneo, A. and Lipper, L. 2015. What determines farmers' adaptive capacity? - Empirical evidence from Malawi. *Journal of Food Security* **8**(3): 643-664.

Bajagai, Y. S. 2011. Global climate change and its impacts on dairy cattle. *Nepalese Veterinary Journal* **30**: 2-16.

Battin, J., Wiley, M. W., Ruckelshaus, M. H., Palmer, R. N., Korb, E., Bartz, K. K. and Imaki, H. 2007. Projected impacts of climate change on salmon habitat restoration. *Proceedings of National Academy of Science, USA* **104**: 6720-6725.

Campbell, B., Mann, W., Melendez-Ortiz, R., Streck, C. and Tennigkeit, T. 2011. Addressing agriculture in climate change negotiations: a scoping report. Meridian Institute, Washington DC.

Cochrane, K., Young, C. D., Soto, D. and Bahri, T. 2009. Climate change implications for fisheries and aquaculture: overview of current scientific knowledge. FAO Fisheries and Aquaculture Technical paper No. 530, FAO, Rome.

EEA. 2001. Total E. U. CH_4 emissions: EEA Factsheets. European Environment Agency, Copenhagen, Denmark.

FAO. 2016. State of Food and Agriculture Report: Climate change agriculture and food security. Food and Agriculture Organization, Rome. http://www.fao.org/3/a-i6030e.pdf

FAO., 2021. Climate Smart Agriculture Source Book. http://www.fao.org/climate-smart-agriculture-sourcebook/about/en/

FAOSTAT. 2013. FAOSTAT database. Food and Agriculture Organisation of the United Nations. Rome. http://faostat.fao.org/

Hahn, G. L. 1999. Dynamic responses of cattle to thermal heat loads. *Journal of Animal Sciences* **77** (Supplement 2): 10-20.

Hartung, E. and G. J. Monteny. 2000. Methane (CH_4) and nitrous oxide (N_2O) emissions from animal husbandry. *Agrartechnische Forschung* **6**: E62-E69.

Hartung, J. 2003. Contribution of animal husbandry to climatic changes. In: Interactions between Climate and Animal Production. EAAP Technical Series No. 7. Wageningen Academic Publishers, Wageningen, the Netherlands.

Hoffmann, I. 2008. *Livestock Genetic diversity and Climate Change Adaptation.* Livestock and Global Change conference proceeding. May 2008, Tunisia.

Hulzebosch, J. 2006. Effective heating systems for poultry houses. *World Poultry* **22**(2): 18-19.

IPCC. 2007. Climate Change 2007: The Fourth Assessment Report of the Intergovernmental Panel on Climate Change. Cambridge University Press, U. K.

IPCC. 2014. Climate Change 2014: Synthesis Report. Contribution of Working Groups I, II and III to the Fifth Assessment Report of the Intergovernmental Panel on Climate Change. (eds. Pachauri, R. K. and Meyer, L. A.). IPCC, Geneva, Switzerland. 151 pp.

Kongshaug, G. 1998. Energy consumption and GHG emissions in fertiliser production. IFA technical conference, Marrakech. p 18.

Lenka, S., Lenka, N. K., Sejian, V. and Mohanty, M. 2015. Contribution of Agriculture Sector to Climate Change. In: Sejian, V., et al. (eds.). Climate Change Impact on Livestock: Adaptation and Mitigation. Springer India, New Delhi.

Lipper, L. and N., Zilberman. 2018. A short history of the evolution of the climate smart agriculture approach and its links to climate change and sustainable agriculture. In: Lipper, L., McCarthy, N., Zilberman, D., Asfaw, S. and Branca, G. (eds.). Climate Smart Agriculture: Building Resilience to Climate Change, Food and Agriculture Organisation, Rome.

MAFF. 2000. Climate change and agriculture in the United Kingdom. Ministry of Agriculture, Fisheries and Food, London.

Marcogliese, D. J. 2008. The impact of climate change on the parasites and infectious diseases of aquatic animals. *Scientific and Technical Review of International office of Epizootics* **27**(2): 467-484.

McCarthy, N., Lipper, L. and Branca, G. 2011. Climate smart agriculture: smallholder adoption and implications for climate change adaptation and mitigation. MICCA Working paper No. 4. FAO, Rome.

Mohanty, B. P., Mohanty, S., Sahoo, J. K and Sharma, A. P. 2010. Climate change: impacts on fisheries and aquaculture. In: Simard, S. (ed.) Climate Change and Variability. InTech Open, Croatia. pp. 119-138.

Monteny, G. J. 2003. Strategies for reducing the effects of animal husbandry on climate. *In*: Interactions between Climate and Animal Production. *EAAP Technical Series No. 7*. Wageningen Academic Publishers, Wageningen, the Netherlands.

Mumby, P. J. and Harborne, A. R. 2010. Marine reserves enhance the recovery of corals on Caribbean reefs. *PLoS One* **5**(1): e8657. https://doi.org/10.1371/journal.pone.0008657

NMSA. 2010. Strategies for Meeting the Challenges of Climate Change. National Mission for Sustainable Agriculture, Department of Agriculture and Cooperation Ministry of Agriculture New Delhi.

Patz, J. A., Campbell-Lendrum, D., Holloway, T. and Foley, J. A. 2005. Impact of regional climate change on human health. *Nature* **438**: 310-317.

Scott, M. L. and Dean, W. F. 1991. Nutrition and management of ducks. Ithaca Publishers, New York.

Shiferaw, B. and Hailemariam, T. 2007. Structure and functioning of chickpea markets in Ethiopia: Evidence based onanalyses of value chains linking smallholders and markets. IPMS Working Paper 6, ILRI, Nairobi, Kenya.

Signor, D., and Cerri, C. E. P. 2013. Nitrous oxide emissions in agricultural soils: a review. *Pesquisa Agropecuária Tropical* **43**(3): 322-338.

Thornton, P., Herrero, M., Freeman, A., Mwai, O., Rege, E., Jones, P. and McDermott, J. 2008. Vulnerability, climate change and livestock - research opportunities and challenges for poverty alleviation. International Livestock Research Institute, Kenya.

West, J. W. 2003. Effects of heat stress on production in dairy cattle. *Journal of Dairy Science* **86**: 2131-2144.

WMO. 2012. Guide to Agricultural Meteorological Practices, WMO No. 134. World Meteorological Organisation, Geneva, Switzerland.

10

Contingent Planning for Weather Hazards

Agriculture is an economic activity that is highly dependent upon weather and climate in order to produce the food and feed necessary to sustain human life. In spite of technological improvements that increase agricultural production, weather hazards have caused significant yield reductions in some years at different regions. Weather hazards include cyclones, droughts, floods, hailstorms, heatwaves and coldwaves. These hazards affect crops and agricultural productivity negatively if they are not checked. The interrelationship between agriculture and these abiotic factors is unavoidable. The cropping pattern, selection of crops, cultural operations, application of farm inputs, harvesting and even storage and transport is decided by the prevailing weather conditions.

DROUGHT

Drought is a relative term used with reference to deficiency of rainfall when compared to normal rainfall of a given location. However, drought cannot be defined in terms of amount of rainfall or number of days without rain. Since ancient times droughts have had far-reaching effects on humankind by causing the failure of crops, decreasing natural vegetation, and depleting water supplies. Drought is different from other hazards; in that it develops slowly. It is a hazard that requires many months to emerge and that may persist for many months or years thereafter. This type of hazard is known as a 'creeping' hazard. Generally, this occurs when a region receives consistently below average precipitation.

Causes of drought

Drought can be caused by a number of factors. The most important one though relates to the amount of water vapour in the atmosphere as it creates precipitation. Generally, rainfall depends on the amount of water vapour carried by regional atmosphere, combined with the upward forcing of the air mass containing that water vapour. If these combined factors do not support precipitation volumes sufficient to reach the surface, the result is a drought.

Human activities like over farming, excessive irrigation, deforestation, and soil erosion can also directly trigger droughts. Widespread cutting down of trees for fuel or for clearing forests for agriculture reduce the soil's ability to hold water that causes desertification leading to drought. Constructing a dam on a large river may help provide electricity and water to irrigate farmland near the reservoir. However, it may also cause drought downstream by severely reducing the flow of water.

Common causes of drought in India include inadequate monsoon rainfall, high temperature coupled with high wind speed causing higher evapotranspiration, inadequate water availability due to high water loss in storage, distribution and utilities, shift in agricultural practices (from low to moderate and high water demanding crops), high rate of increase in human and animal population in a specific location, and overexploitation of surface water and groundwater.

Impacts of drought

One of the sectors where the immediate impact of drought is felt is agriculture. The impact of a drought may be either direct or indirect and varies in nature as well as intensity. The extent and intensity of drought impact is determined by prevailing economic conditions, the structure of the agricultural sector, management of water resources, food grain reserves, etc. The direct impact of drought is generally classified under four categories, viz., physical, economic, social and environmental. The relative and absolute magnitudes of each impact will however, depend on specific regional characteristics. Droughts cause a loss of assets in form of crops, livestock, fisheries and productive capital as these are immediate consequences of water shortage (NDMA, 2010). Many of these losses are then passed on to consumers in the form of higher commodity pricing. The social impacts include increased chance of conflict over commodities, fertile land, and water resources. Other social impacts include abandonment of cultural traditions, loss of homelands, changes in lifestyle, and increased chance of health risks due to poverty and hygiene issues. The environmental impacts of drought include loss in species biodiversity, migration changes, reduced air quality, and increased soil erosion.

Drought preparedness and mitigation

Drought preparedness, coupled with appropriate mitigation actions and programmes, can reduce and, in some cases, eliminate many of the impacts associated with drought. Drought planning process has three components; monitoring and early warning, risk assessment, and mitigation. The monitoring and early warning component of a drought plan essentially provides the foundation on which timely decision can be made by decision makers at all levels, from farmers to national policy makers. Given drought's creeping characteristics, monitoring the hydrological system can detect drought's early onset and its potential impacts. There should be interdisciplinary approach to reduce and mitigate the impacts of drought. The meteorological and hydrological observations should be systematically collected and processed to support and strengthen the scientific networks for the enhancement of technical capacities in meteorology, hydrology and other related fields.

Conservation practices minimise the disruption of the soil's structure, composition and natural biodiversity, thereby reducing erosion and soil degradation, surface runoff, and water pollution. The established practices of soil and water conservation are crop rotation, contoured row cropping, terracing, minimum tillage practices, erosion control structures, water retention and detention structures, windbreaks and shelterbelts, litter management, and reclamation of salt-affected soil. Soil and water conservation can be approached through agronomic and engineering measures. Agronomic measures include contour farming, off-season tillage and deep tillage, mulching and providing vegetative barriers on the contour. These measures prevent soil erosion and increase soil moisture. Alternate land use system like agroforestry, i.e., integrating tree crops with annual herbaceous crops, may be encouraged. Carefully planned crop rotation can help to minimise erosion and allow farmers to plant less water-dependent crops in drier years.

Engineering measures differ with location, slope of the land, soil type, and amount and intensity of rainfall. These include soil and water conservation through regulated runoff, collection of surplus rainwater, checking evaporational and seepage loss of water and recycling of collected water for life saving irrigation. These can be achieved by making water harvesting structures, scooping the land, opening ridges and furrows, contouring across the slope, and compartmental bunding, etc. Contour bunds, trenches and stone walls can prevent soil erosion and obstruct the flow of runoff. The retained water increases soil moisture and recharges the groundwater. Check dams (check dams are temporary structures constructed with locally available materials) and other gully-plugging structures can be constructed to restrict the runoff and soil erosion. Percolation ponds can be constructed by excavating a depression to form a small reservoir, or by constructing an embankment in a natural ravine or gully to form an impoundment. These features store water for livestock and recharge the groundwater. Rainwater can be collected from roofs or other suitable catchments and stored for later use. The wastewater (sewage) can also be recycled by proper treatment and purification. Water-supply projects can also be implemented for drought mitigation, with a view to strengthen drought preparedness. Activities such as water-use planning, rainwater harvesting, runoff collection using surface and underground structures, improved management of channels and wells, exploration of additional water resources through drilling and dam construction, are implemented as a part of a drought-mitigation plan.

The physiological approaches include application of suitable chemicals in crops as antitranspirants, growth retardants, and radiation reflectants for reducing the transpiration rate, checking the shoot growth and encouraging the root growth, and reducing the radiation, respectively which ultimately may reduce the impact of drought on crops. Genetic approaches for reducing the adverse impacts of drought aimed at developing new plant types, known as 'ideotypes' which possess wider adaptability, shorter growth duration, drought resistance and high genetic yield potentiality. Plant breeders have developed many short duration varieties for various crops. These varieties are characterised by effective and extensive deep root system, drought tolerance, resistance to diseases, effective photosynthetic behaviour, dwarf plant types with lesser number of erect leaves and high yield potentiality. Thus, the drought can be evaded or escaped by substituting the traditional long duration varieties with these ideotypes.

FLOODS

A flood is an overflow of water that submerges land which is usually dry. Flooding may occur as an overflow of water from waterbodies, such as a river or lake, in which the water overtops, resulting in some of that water escaping its usual boundaries. Floods are recurrent phenomena since time immemorial. Almost every year some parts of the world or the other are affected by the floods of varying magnitude. Even in the same country, while some parts are suffering from devastating floods, other parts experience severe drought.

Causes of flood

The years in which actual rainfall exceeds the normal value by twice the standard deviation or more are defined as the years of flood. Most often floods are a secondary hazard resulting from other meteorological processes, such as prolonged rainfall, localised and intense thunderstorms, or onshore winds. Inadequate capacity of the rivers to contain within their banks the high flows brought down from the upper catchment areas following heavy rainfall, leads to flooding. In coastal areas, floods are caused by cyclones and typhoons. Other causes include backing up of waters in tributaries at their outfalls into the main river often with synchronisation of floods in them, and landslides blocking stream courses resulting in the backwater overflowing river banks. However, the root cause of flood is excessive rainfall which occurs mainly in the monsoon months. Because of the varying rainfall distribution, many a times, areas which are not traditionally prone to floods also experience severe inundation. Areas with poor drainage facilities get flooded by accumulation of water from heavy rainfall.

Flood may also occur due to an accumulation of rainwater on saturated grounds. Excess irrigation water applied to command areas and increase in groundwater levels due to seepage from canals and irrigated fields also are factors that accentuate the problem of waterlogging. The problem is exacerbated by factors such as silting of the riverbeds, reduction in the carrying capacity of river channels, erosion of beds and banks leading to changes in river courses, obstructions to flow due to landslides, synchronisation of floods in the main and tributary rivers and retardation due to tidal effects (NDMA, 2008).

Effects of floods

The primary effects of floods include loss of life, damage of crops and livestock, loss of properties, damage to buildings and other structures including bridges, roads and transport infrastructure, sewerage systems, roadways, canals, and disruption of power supply. Flood waters typically inundate farmland, making the land unworkable and preventing crops from being planted or harvested, which can lead to shortage of food, both for humans and farm animals. Entire harvests for a region may be lost in extreme flood circumstances. Some tree species may not survive prolonged flooding of their root systems. Floods are followed by epidemics. Supply of drinking water and restoration of sanitation are post flood challenges. Secondary effects of flooding include coastal erosion and soil erosion. Economic hardship due to a temporary decline in tourism, rebuilding costs, or food shortage leading to price increases is a common after-effect of severe flooding. The flood problem in India is mostly confined to the states located in the Indo-Gangetic plains, northeast India and occasionally in the rivers of central India.

Frequent or smaller floods can also bring many benefits, such as recharging groundwater, making soil more fertile and increasing nutrients in some soils. Flood waters provide much needed water resources in arid and semiarid regions where precipitation is very unevenly distributed throughout the year. Flooding can spread nutrients to lakes and water reservoirs, which can lead to increased biomass and improved fisheries for a few years.

Flood preparedness and mitigation

Flood forecasting has been recognised the most important reliable and cost-effective non-structural measure for flood mitigation. Forecasting of floods allows for precautions to be taken and people to be warned so that they can be prepared in advance for flooding conditions. A long time-series of historical data that relates streamflows to the measured past rainfall events can have fairly accurate flood forecasts. Coupling this historical information with real-time knowledge about volumetric capacity in catchment areas, such as spare capacity in reservoirs, groundwater levels, and the degree of saturation of area aquifers is also needed in order to make the most accurate flood forecasts. Radar estimates of rainfall and general weather forecasting techniques are also important components of good flood forecasting. The intensity and height of a flood can be predicted with plenty of lead time, if good quality data is available. In India, the Central Flood Board was set up in 1954. The India Meteorological Department (IMD) and Central Water Commission (CWC) are working in close coordination for issue of flood forecasts. IMD has 10 flood meteorological offices and CWC has 20 flood forecasting divisions located in different flood prone river basins in the country.

Various mitigation measures can be taken to reduce the adverse impact of floods. In order to achieve control over the flow and height of the water carried by the river, the channel, floodplain or watershed must undergo some physical alterations.

1. Dams, retention basins or reservoirs on mainstreams or tributaries can be constructed to store excessive water and release it gradually after the threat is over.
2. Levees or floodwalls can be constructed to confine flood waters to a floodway, thereby reducing flood damage.
3. Channel improvement works can be taken up to increase the efficiency of drainage system. These include straightening to remove undesirable bends, deepening and widening to increase size of waterways; clearing to remove trees and other obstructions; lining with concrete to increase efficiency.
4. Watershed treatment can also be employed to help the soil on slopes to become more absorbent of rainfall until flood height is receded. Watershed treatment involves crop rotation, construction of terrace, contour strip cropping; selective planting and reforestation.
5. Any type of construction or land filling should not be allowed on the lateral boundaries of the designated floodways. This is done to ensure that the flow of water is not obstructed.

Other mitigation measures include change in crop management practices, flood insurance, temporary evacuation of persons, livestock and property from the path of flood waters, etc. The crops and varieties which can withstand waterlogging conditions for a fairly long period should be preferred in the flood prone areas.

TROPICAL CYCLONE

Major storms in the tropics are known as tropical cyclone. Depending on its location and strength, a tropical cyclone is referred to by names such as hurricane, typhoon, tropical storm, cyclonic storm, tropical depression, and simply cyclone. All forms of storm fall under the definition of tropical cyclone which is a synoptic scale cyclonic circulation with organised convection and surface wind circulation originating over tropical waters. Tropical cyclones are characterised by large low pressure centres and numerous thunderstorms that produce strong winds and heavy rain. Heavy rains, however, can produce significant inland flooding, and storm surges can produce extensive coastal flooding up to 40 km from the coastline.

Storm surge is the abnormal rise of sea level which occurs when the cyclone moves from ocean to continent. It is a disastrous and destructive feature of cyclone, which could cause massive destruction in the coastal region and misery to the coastal human population and livestock.

Cyclones and their impact in India

Cyclones are known to cause severe damage to infrastructure through high-speed winds. Very strong winds accompanying a cyclonic storm damage trees, installations, residences, communication systems, etc., which results in loss of life and property. Torrential rainfall (more than 30 cm/hour) associated with cyclones is another major cause of damages. Unabated rain gives rise to unprecedented floods. Heavy rainfall from a cyclone is usually spread over wide area and cause large scale soil erosion and weakening of embankments. A severe tropical cyclone causes storm surge; as a result of which seawater inundates low lying areas of coastal regions drowning human beings and livestock. This also erodes beaches and embankments, destroys vegetation and leads to reduction of soil fertility.

Recurring frequency of cyclones leads to drastic transformation in the pattern of coastal land use by turning fertile areas into saline wastelands due to marine incursions. In addition, standing crops are destroyed; salt deposits from saline water render farms unproductive and even barren. An entire cropping season is lost and areas affected by severe storm surge are sometimes not cultivable during subsequent seasons also. Very often harvested food crop stocks are also destroyed. Horticultural crops suffer heavy losses. Livestock including farm and dairy cattle, poultry, sheep and goats also suffer.

Although tropical cyclones are known for destruction they cause, they may be important factors in the precipitation regimes of places they affect, as they may bring much-needed precipitation to dry regions, relieving drought conditions. Tropical cyclones also help maintain the global heat balance by carrying warm, moist air from the tropics to the temperate latitudes and polar regions. This helps to maintain equilibrium in earth's troposphere and maintains relatively stable and warm temperature worldwide.

Cyclone preparedness and mitigation

Although it is not possible to completely avoid natural disasters, their effects can be minimised by taking some known long and short-term structural and non-structural mitigation measures such as developing proper early warning systems, creating awareness at all levels in the concerned communities, coastal afforestation, construction of shelters, embankments, dykes, coastal roads, bridges, canals, etc. through better preparedness, mitigation measures and improved response mechanisms (NDMA, 2008).

The India Meteorological Department has introduced a 4-stage warning system as described below to issue cyclone warnings to the disaster managers.

1. **Pre-cyclone watch:** The pre-cyclone watch is issued when a depression forms over the Bay of Bengal irrespective of its distance from the coast and is likely to affect Indian coast in future. It is issued at least 72 hours in advance of the commencement of adverse weather. It is issued at least once a day.
2. **Cyclone alert:** The cyclone alert is issued at least 48 hours before the commencement of the bad weather when the cyclone is located beyond 500 km from the coast. It is issued every three hours.
3. **Cyclone warning:** The cyclone warning is issued at least 24 hours before the commencement of the bad weather when the cyclone is located within 500 km from the coast. Information about time and place of landfall are indicated in the bulletin.
4. **Post landfall outlook:** The post landfall outlook is issued 12 hours before the cyclone landfall, when the cyclone is located within 200 km from the coast. More accurate and specific information about time and place of landfall and associated bad weather is indicated in the bulletin.

The sector that is most affected by tropical cyclones is agriculture and food security. The sharpest falls in agricultural production has been recorded during those years that have experienced the occurrence of severe cyclones in India. Various options have been conceptualised for protecting the crops (NDMA, 2008).

1. The crop calendar for major crops should be adjusted to normal cyclone periods.
2. The cultivation of high yielding varieties and hybrids of rice in the coastal areas, which can withstand surge waters, floods and cyclonic winds without damaging the grain and straw quality, should be encouraged.
3. Short height crops should be grown to avoid lodging in cyclone prone areas.

4. Emergency input supply with emphasis on fertiliser and insecticide should be ensured.
5. Establishment of community-based threshing yards and drier facilities should be encouraged
6. The farmers should be advised to insure their crops and livestock.
7. Proper application of sufficient fertilisers after inundation of storm surge to reduce salinity.
8. Alternative storage facilities and processing units (boiler and dry pans) should be established.

Strategies for protecting standing crops

Standing crops will be protected through improvement of drainage systems by executing the following activities (NDMA, 2008).

1. Strengthening of bunds all along canals and tanks.
2. De-silting of tanks, canals and drains.
3. Widening and deepening of drains and canals.
4. Weeding of tanks and canals from water hyacinth and other plants which obstruct normal flow.
5. Removal of encroachments from canals and tanks.
6. Diversion of drains and canals.
7. Alignment of new canals and drains keeping in view that water will be flowing out freely during disaster periods.
8. Drains which are directly joining the sea should be maintained regularly.
9. Tidal banks (with flap gates) should be constructed along irrigated lands where these are near the coast.
10. Main canals may be closed during cyclone periods, giving preference to drainage to adjacent fields/sectors.

Strategy for protecting livestock

The following steps may be taken to protect the domestic animals and birds from cyclone (NDMA, 2008).

1. The animals and birds should be shifted to a higher and safer place as soon as warnings for impeding cyclones are received.
2. Cowsheds to house the animals should be constructed in a way that they can withstand strong cyclone winds.

3. Animal sheds should be constructed with lightweight timber, coconut and palm tree leaves that will not harm the animals even if they collapse during cyclonic winds
4. Sufficient nutritious food may be preserved which can last for 10-15 days, and dry grass compressed to bales or blocks with machines under high pressure, can be stored in small areas.
5. Animals will be vaccinated against communicable diseases during emergency periods. The birds can be protected from diseases by getting them vaccinated beforehand.
6. The carcasses of animals and birds should be buried in a pit over which lime is sprinkled. The surroundings must be kept clean by sprinkling with lime or insecticides.
7. Clean and unpolluted water should be provided to the animals and the water may be treated with chlorine or bleaching powder before giving it to the animals. Poultry birds should be provided with clean water by mixing antibacterial material.
8. Increase in the overall fodder resource through raising and preserving to meet the requirements during the crisis period. By-products of agriculture/ industrial outputs are to be used as fodder. The cultivation of hybrid nappier grass should be encouraged.
9. Some non-conventional resources are to be developed for use as fodder for animals. Molasses is to be used as an alternate fodder during cyclone periods and periods of non-availability of fodder to protect the animals from disaster impacts and also from scarcity of food/fodder. To avoid deficiency in proteins and minerals, animals are be fed urea-molasses and multi-vitamin food.

Contingent planning should be made sufficiently ahead. The farmers may be provided with the required inputs like seeds/seedlings and fertilisers timely to go for a contingent crop to compensate the crop loss due to cyclonic storms and storm surges. Absolute security against cyclone hazard cannot be achieved, but an effective cyclone warning response coupled with the above mentioned short-term and long-term management options can definitely reduce loss (NDMA, 2008).

REAL TIME CONTINGENCY PLANNING

Indian agriculture is mostly dependant on monsoon. The weather aberrations such as delayed onset and early withdrawal of monsoon, intermittent and prolonged dry spells, and prolonged monsoon necessitate changes in the agricultural planning and strategy. These frequent weather aberrations round the year in one or the other part of the country impact agricultural production.

Thus, the contingency measures on real-time basis are to be implemented to minimise the losses in agriculture and allied sectors and to improve the efficiency of the production systems. Real Time Contingency Planning (RTCP) is considered as any contingency measure, either technology related (land, soil, water, crop) or institutional and policy based, which is implemented based on real time weather pattern (including extreme events) in any crop growing season (Srinivasarao et al., 2013).

The real-time contingency measures expected to fulfil the following objectives (Srinivasarao et al., 2013).

1. To establish a crop with optimum plant population during the delayed onset of monsoon
2. To ensure better performance of crops during seasonal drought (early/mid and terminal drought) and extreme events, enhance performance, improve productivity and income
3. To minimise damage to horticultural crops/produce
4. To minimise physical damage to livestock, poultry and fisheries sector and ensure better performance) to ensure food security at village level
5. To enhance the adaptive capacity and livelihoods of the farmers

RTCP measures for various weather aberrations in arable crops

The following measures have been suggested in arable crops as real time contingency plan during various weather aberrations (Venkateswarlu et al., 2011; Umate et al., 2011; Srinivasarao et al., 2016).

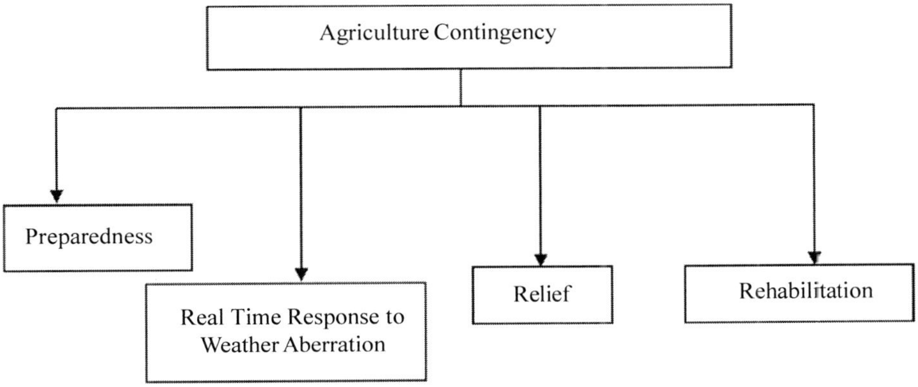

Fig. 10.1. Initial preparedness for drought - for implementation of real time contingency planning
Source: Srinivasarao et al., 2015

Delayed onset of monsoon

Beyond the sowing window, choice of alternate crops or cultivars depends on the farming situation, soil, rainfall and cropping pattern in the location and extent of delay in the onset of monsoon.

Early-season drought

1. Resowing within a week to 10 days with subsequent rains for better plant stand when germination is less than 30%.
2. Thinning in small-seeded crops.
3. Interculture to break soil crust and removal of weeds, and creation of soil mulch for conserving soil moisture.
4. Avoidance of topdressing of fertilisers till favourable soil moisture is available.
5. Opening conservation furrows at 10 to 15 m intervals.
6. Ridge and furrow across the slope for effective moisture conservation in wide spaced crops.
7. Pot watering along with gap filling when the crop stand is less than 75% in crops like cotton.
8. Foliar spray of 2% urea during prolonged dry spells wherever ground/surface water is available.

Mid-season drought

1. Plant protection.
2. Supplemental/protective irrigation, if available.
3. Repeated interculture to remove weeds and create soil mulch to conserve soil moisture.
4. Avoidance topdressing of fertilisers until receipt of rains.
5. Opening conservation furrows for moisture conservation.
6. Foliar spray of 2% KNO_3 or 2% urea solution or 1% water soluble N-P-K fertilisers.
7. Opening of alternate furrows.
8. Surface mulching with crop residues.

Terminal drought

1. Life-saving or supplemental irrigation, if available may be provided.
2. Crops may be harvested at physiological maturity with some realisable yield or may be used as fodder.

3. Preparation for winter sowing in double-cropped areas.
4. Prefer contingency crops like horsegram and cowpea or dual-purpose forage crops on receipt of showers under receding soil moisture conditions.

Unseasonal heavy rainfall events

1. Resowing of crops.
2. Provision of surface drainage.
3. Application of hormones / nutrient sprays to prevent flower drop or promote quick flowering / fruiting and plant-protection measures against pest/disease outbreaks with need-based prophylactic/curative interventions.
4. At crop maturity stage, prevention of seed germination and harvesting of produce.

Floods

1. In sand-deposited crop fields/fallows, ameliorative measures include early removal or ploughing in of sand (depending on the extent of deposit) for facilitating *rabi* or next *kharif* crop.
2. In silt-deposited Indo-Gangetic Plains, early *rabi* crop plan is suggested in current cropped areas and current fallow lands.
3. Draining out of stagnant water and strengthening of field bunds.
4. In flood prone land areas, measures include alternate crop plans for receding situations.
5. Community nursery raising.
6. Retransplanting in damaged fields and transplanting new areas or direct seeding depending on seed availability.
7. Prevention of premature germination of submerged crop at maturity or of harvested produce by spray of salt solution.

Heatwave

1. Light frequent irrigation.
2. Foliar spray with thiourea or KNO_3 at recommended dose.

Coldwave

1. Light frequent irrigation during evening hours.
2. Basin mulching.
3. Smoking during night.
4. Application of supplemental fertiliser dose.

Frost

1. Change in planting time to avoid sensitive stages coinciding with frost period.
2. Thatching for young plants.
3. Mulching/ground cover to prevent loss of heat.

Hailstorm

1. Use of anti-hail guns and anti-hail nets.
2. Trenches may be opened to drain out excess water from the field.
3. If crop is at harvesting stage, field crops may be harvested immediately and dried to avoid fungal growth.

Cyclones

1. Providing field drainage, staking and propping of plantation crops.
2. Cleaning and drying of harvested field crops.

RTCP measures for various weather aberrations in fruit crops

The following measures have been suggested in fruit crops as real time contingency plan during various weather aberrations (Venkateswarlu et al., 2011; Umate et al., 2011; Rao et al., 2014).

Heatwave

1. For fruit crops like mango, pear, guava, citrus and litchi irrigation, if available may be provided.
2. These crops may be provided shade with wild bush.

Coldwave

1. For mango, litchi, pear and guava light irrigation may be provided frequently.
2. Mulching will be very much helpful.
3. Thatching and creating smoke screens and lighting of fire where irrigation facilities are not available.

Frost

1. At vegetative stage, damaged branches and twigs should be pruned and Bordeaux paste 1% is applied to avoid fungal infection.
2. At reproductive stage, hormonal spray NAA 20 ppm + 1% urea may be applied to prevent flower drop.

3. In mango, pear, litchi and guava light irrigation may be given frequently.
4. Mulching is very useful to these crops.
5. Thatching and creating smoke screens and lighting of fire.

Hailstorm

1. In mango spraying of potassium nitrate @ 1% to arrest flower and fruit drop, if cloudy conditions persist after the hail occurrence, orchards may be sprayed with fipronil (5% EC) @ 1.5 ml/l to control hoppers and thrips and hexaconazole @ 1 ml/l water to control powdery mildew disease.
2. Anti-hail nets may be beneficial at flowering and fruit bearing stages of mango.
3. If banana fruit bunch is beyond three quarters maturity at the time of damage, fruits are to be harvested. At other stages, severely damaged leaves are pruned up to base of petiole.
4. Booster dose of urea and potash may be applied to the partially damaged banana orchard. Potassium sulphate @ 5 g/l and urea @ 10 g/l water solution may be spayed on plants with newly emerged bunches.
5. Mancozeb @ 2.5 g/l or carbendazim @ 1.0 g/l water is spayed on leaves, pseudo-stems and developing fruits to avoid secondary infection in banana.

Floods

1. Excess water should be drained out from the orchards. Proper drainage is to be ensured since all fruit crops are sensitive to continuous submergence.
2. The field bunds should be strengthened.
3. If possible, the short duration fruit crops may be transplanted in new areas.
4. Urea 2% solution may be sprayed for 2-3 times at 7-10 days interval in mango crop.
5. In banana, deep trenches between plant rows are to be opened to improve drainage.

Cyclone

1. If the damage is severe to the mango plants, replanting is preferred.
2. Propping and earthing up should be done to damaged and fallen mango trees.
3. Manuring should be done in the basin.
4. Broken and damaged branches in fruit trees should be pruned and applied with 1% Bordeaux paste.

5. In banana, proper propping and staking with bamboo poles will avoid lodging.
6. Wind damaged plant should be pruned using disinfected secateurs and cut ends must be smeared with 1% Bordeaux paste.
7. At flowering stage of mango, hormones and multi-nutrients may be sprayed to promote fruit setting. Pollination techniques may be supplemented to improve fruit setting.
8. Windbreaker trees may be planted around the orchard to give protection to the fruit trees from high speed winds.

RTCP measures for various weather aberrations in livestock sector

The following measures have been suggested in livestock sector as real time contingency plan during various weather aberrations (Venkateswarlu et al., 2011; Umate et al., 2011).

Drought

1. Biomass of dried-up crops can be used as fodder. Azolla and silage should be given to milch animals.
2. Hay is enriched with 2% urea molasses solution or 1% common salt solution and fed to livestock.
3. Unproductive livestock may be culled out.
4. Adequate supply of drinking water must be ensured and wallowing of animals in waterbodies/ resources should be restricted.
5. Deworming of animals should be carried out to all animals entering into relief camps. Sick animals should be identified and ring vaccination must be performed at 8 km radius in case of any disease outbreak.
6. Feed supplementation should be arranged only for productive birds with household grain and shell grit (calcium) for laying birds. Weak birds should be culled.
7. Cool hygienic drinking water mixed with vitamins should be given to birds.

Floods and cyclone

1. In case of early forewarning, the crops that can be useful as feed or fodder in future may be harvested and properly stored from inundation by flood waters.
2. Sufficient dry fodder is to be transported to the flood affected villages to meet the fodder requirements.
3. Urea may be applied in the inundated areas to enhance the biomass production.

4. Proper drainage facility surrounding the livestock shed should be ensured to prevent waterlogging.

Heatwave

1. The animals should be allowed preferably early in the morning or late in the evening for grazing.
2. In severe cases, vitamin C and electrolytes should be added in drinking water.
3. The birds should not be allowed for scavenging.
4. Weak birds may be culled.
5. Sufficient cool hygienic water should be provided to the birds.
6. Livestock shelters should be provided with cooling equipments. Supply of electricity should be ensured by generator or solar energy or biogas.

Coldwave

1. Animals should be allowed for grazing between 10 AM to 3 PM.
2. Heaters may be arranged in shelters.
3. In case of high productive animals 25-50 ml of edible oil may be added in concentrated and fed to the animals.
4. Lime powder should be sprinkled in the animal shed during coldwaves to neutralize ammonia accumulation.

REFERENCES

NDMA. 2008. National Disaster Management Guidelines: Management of Flood. National Disaster Management Authority, Government of India, New Delhi, 134 pp.

NDMA. 2008. National Disaster Management Guidelines: Management of Cyclones. National Disaster Management Authority, Government of India, New Delhi, 156 pp.

NDMA. 2010. National Disaster Management Guidelines: Management of Drought. National Disaster Management Authority, Government of India, New Delhi, 78 pp.

Rao, V. U. M., Bapuji Rao, B., Sikka, A. K., Subba Rao, A. V. M., Singh, R. and Maheswari, M. 2014. Hailstorm threat to Indian agriculture: a historical perspective and future strategies. All India Coordinated Research Project on Agrometeorology, ICAR-Central Research Institute for Dryland Agriculture, Hyderabad. 36 pp.

Srinivasarao, Ch., Lal, R., Prasad, J. V. N. S., Gopinath, K. A., Singh, R., Jakkula, V. S., Sahrawat, K. L., Venkateswarlu, B., Sikka, A. K. and Virmani, S. M. 2015. Potential and challenges of rainfed farming in India. *Advances in Agronomy* **133**: 113-181.

Srinivasarao, Ch., Ravindra Chary, G., Mishra, P. K., Nagarjuna Kumar, R., Maruthi Sankar, G. R., Venkateswarlu, B. and Sikka, A. K. 2013. Real time contingency planning:

Initial experiences from AICRPDA. All India Coordinated Research Project on Dryland Agriculture, ICAR-Central Research Institute for Dryland Agriculture, Hyderabad. 63 pp.

Srinivasarao, Ch., Ravindra Chary, G., Rani, N. and Baviskar, V. S. 2016. Real time implementation of agricultural contingency plans to cope with weather aberrations in Indian agriculture. *Mausam* **67**(1): 183-194.

Srinivasarao, Ch., Venkateswarlu, B., Lal, R., Singh, A. K. and Kundu, S. 2013. Sustainable management of soils of dryland ecosystems of India for enhancing agronomic productivity and sequestering carbon. *Advances in Agronomy* **121**: 253-329.

Umate, M. G., More, G. R., Prasad, Y. G., Ravindra Chary, G., Mandal, D. K., Ramana D. B. V., Sarkar, D. and Venkateswarlu, B. 2011. District level contingency plans for weather aberrations in Marathwada region, Maharashtra. Marathwada Krishi Vidyapeeth, Parbhani, Maharastra and ICAR-Central Research Institute for Dryland Agriculture, Hyderabad. 272 pp.

Venkateswarlu, B., Singh, A. K., Prasad, Y. G., Ravindra Chary, G., Srinivasarao, Ch., Ramana, D. B. V. and Rao, V. U. M. 2011. District level contingency plans for weather aberrations in India. ICAR-Central Research Institute for Dryland Agriculture, Hyderabad. 136 pp.

11

Cropping Systems for Sustainable Agriculture

A system is a set of components that are interrelated and interact among themselves. A system may be natural, social or artificial. Natural systems consist of all the materials both physical and biological and are independent of human beings. Our role in relation to natural systems is to try to understand these systems and make use of them as per our needs. Social systems involve relationships between individuals, groups, communities, and not between things. Artificial systems do not exist in nature and they are of human creation to serve our purposes. All the agricultural systems come under artificial system, developed from either or both of natural and social systems.

In agriculture, the ideal management practices were developed for individual crops through regular research. Farmers grow different crops based on climatic and soil requirements. Besides they have their own choices and requirements, and before selecting a crop for cultivation they usually analyse its market demand, economic viability, availability of resources, etc. Very often there is an established and well defined relationship among the crops cultivated by a farmer, either in some form of spatial mixture or in some temporal sequence. Therefore, at present emphasis is being given to develop system based production technology for the crops grown in a year or beyond. This will give a scope to study the relationship among the crops, their environment and the production inputs required.

A cropping system refers to a set of crop systems, making up the cropping activities of a farm system. It comprises all components required for the production of a particular crop and interrelationships between them and environment (TAC-CGIAR, 1978). A cropping system usually refers to a combination of crops in space and time. The term cropping system essentially represents a philosophy of maximum crop production per unit area of land within a calendar year or relevant time unit with minimum natural resource degradation (Gangwar et al., 2012). Combinations in space occur when crops are interplanted and combinations in time occur when crops occupy different growing periods. Cropping system is an important component of a farming system and it represents the cropping pattern used on a farm and their interactions with farm resources, other farm enterprises and available technology which determine their makeup. In the recent past, cropping systems approach has gained importance in agriculture and related enterprises.

Cropping systems approach enables to address issues pertaining to maximising system productivity on annual basis. It also ensures utilisation of resources with higher efficiency by considering various interactions and direct, residual and cumulative effect in soil-plant-environment system.

SCOPE OF CROPPING SYSTEM

1. Optimising the use of locally available resources by combining the different components of the farm system, so that they complement each other and have the greatest possible synergetic effects.
2. Reducing the use of off-farm, external and non-renewable inputs with the greatest potential to damage the environment or harm the health of farmers and consumers, and a more targeted use of the remaining inputs used with a view to minimising variable costs.
3. Relying mainly on resources within the agroecosystem by replacing external inputs with nutrient cycling, better conservation, and an expanded use of local resources.
4. Improving the match between cropping patterns and the productive potential and environmental constraints of climate and landscape to ensure long-term sustainability of current production levels.
5. Working to value and conserve biological diversity, both in the wild and in domesticated landscapes, and making optimal use of the biological and genetic potential of plant and animal species.
6. Taking full advantage of local knowledge and practices, including innovative approaches not yet fully understood by scientists although widely adopted by farmers.

CLASSIFICATIONS OF CROPPING SYSTEM

Depending on the resources and technology available, different types of cropping systems are adopted in an area.

Monocropping

Monocropping refers to growing only one crop on a particular land year after year. Many farmers plant only one crop in the same place year after year due to climatic factors or resource constraints. In low rainfall areas groundnut or cotton or millets are grown in uplands year after year. Similarly, rice is grown, as it is not possible to grow any other crops, in canal irrigated areas, and under water logged conditions.

Advantages of monocropping

1. Monocropping increases farming efficiency of a farmer. When the farmer grows only one type of crop he can specialise in that crop and it is a more profitable way to farm than switching crops around each year.
2. Since the crop is the same over the whole farm the crop management decisions are simplified.
3. Only the machinery and tools necessary to maintain the crop needs to be purchased.
4. The seasonal work is predictable allowing scheduling of casual labours and machinery.
5. It reduces the overall costs.
6. The production technology becomes simple and much skill may not be required.

Limitations of monocropping

1. If the market falls then income is severely affected due to great dependency on one crop.
2. Monocropping is more susceptible to pests and diseases. One pest or disease can wipe out an entire crop and there is no income for that year. To save the affected crop more chemicals are to be used.
3. Different plants have different nutritional needs. Using the same crop repeatedly will tend to pull soil nutrients unevenly. This will increase the application of higher doses of chemical fertilisers.
4. The effects of monocropping on the environment are severe when pesticides and fertilisers make their way into groundwater or air, creating pollution.
5. Monocropping is usually associated with high levels of mechanisation which leaves the soil compact and the soil structure is distorted. In the long-run soil health is deteriorated.
6. In monocropping the land remains bare between harvest of a crop and planting of the next crop for a substantial period which increases the erosion.

Multiple cropping

Growing two or more crops on the same piece of land in one calendar year is known as multiple cropping. It is the intensification of cropping in time and space dimensions, i.e., more number of crops within a year and more number of crops on the same piece of land at any given period. Multiple cropping makes effective use of inputs such as soil, water, fertiliser, etc. Thus output per unit area increases. It can be practised in annual food crops, fodders, vegetables, fruit plants and perennial crops. It includes intercropping, mixed cropping, sequence cropping, crop rotation, relay cropping and paira cropping, etc. For successful multiple cropping, availability of irrigation facility and other input resources including labour should be ensured at right time.

Principles of multiple cropping

1. The crops should be agroclimatically suitable and technically feasible.
2. The most profitable crops capable to utilise growth resources and land in best manner should be chosen.
3. The crop combinations and input rations should be compatible with farmer's skill, enterprise preference, capital, assured irrigation facility.
4. The essential inputs like improved seeds, fertilisers, pesticides and labour should be available locally.
5. Marketing, storage and transportation facilities of the produce are to be ensured before choosing the crops.

Advantages of multiple cropping

1. Land is better utilised through multiple cropping.
2. In multiple cropping, incidence of crop failure owing to biotic agents is minimised. The loss of any crop due to biotic or abiotic stress may be compensated to some extent by another crop.
3. Overall production and yield per unit of area is increased.
4. Cost of production is reduced in multiple cropping as compared to individual crops grown for the same level of production.
5. Pest and disease infestation is checked.
6. Different type of products can be produced at a time in multiple cropping.
7. Multiple cropping provides a balance diet for a farm family
8. Multiple cropping helps in maintaining soil fertility provided suitable crops such as legumes are included in the cropping system. It improves or maintains the soil health.
9. Weed menace is reduced in multiple cropping. It narrows the space available for weeds to grow and sometimes may hamper their growth through exudation of allelochemicals.

10. An important aspect of multiple cropping is the utilisation of nutrients more efficiently as the crops growing on the same piece of land simultaneously would have different nutritional requirements.
11. Multiple cropping is also important from marketing point of view. As more than one product is available simultaneously if the selling price of one product is less in the market, the other will be there to compensate.
12. It generates more employment opportunity by having crops round the year.

Limitations of multiple cropping

1. Multiple cropping may reduce the farming efficiency of a farmer. When a farmer grows different type of crops, he cannot specialise in production technology of all the crops. He has to acquire the latest technologies for the crops grown which may not be possible at all the time.
2. Sometimes the crop management decisions become difficult and often the advice of the professionals may be required.
3. The survival of pests becomes easier in a multiple cropping. Pests can easily shift from one crop to another crop. Sometimes the insect pests and diseases get more favourable environment to flourish, thus diminishing and deteriorating crop yields.
4. Intercultural practices are sometimes difficult to be carried out in the crops grown together.
5. Weed management may also be difficult.
6. Initial investment may be high with procurement of different types of inputs including equipments and machineries for a number of crops at a time.

Intercropping

Practice of intercropping is as old as agriculture. But scientific approach to it is relatively new. Intercropping refers to growing of two or more dissimilar crops simultaneously on the same piece of land. The recommended optimum plant population of the base crop is suitably combined with appropriate additional plant density of the associate crop. Various workers have classified and defined intercropping in many ways. However, the objective remains the same, i.e., the intensification of cropping both in time and space dimensions and to raise productivity per unit area by increasing the pressure of plant population. It has better utilisation of growth resources than sole cropping. Intercropping results in more efficient use of solar energy and harnessing benefits of positive interactions in crop associations. These advantages are, in general more pronounced in wide-spread crops and stress environments (Gangwar et al., 2012). Intercropping can be practised as annual plants with annual plants intercrop, annual plants with perennial plants intercrop and perennial plants with perennial plants intercrop.

Intensive monocropping is less complicated for big-scale farmers with the fullest utilisation of machines, while smallholder farmers do not have steady access to markets and only produce enough food for their family members under subsistence farming. Intercropping ensures their livelihood. Therefore, intercropping is mainly practised on small farms. Moreover, intercropping is known to produce stable yields from diversified crops with less use of inputs for nutrient supply and plant protection, focusing on sufficient food under healthy environmental conditions (Maitra et al., 2021).

Types of intercropping

Intercropping may be of various types based on the method of sowing and objective of cropping.

Mixed intercropping: It is an intercropping where two or more crops are grown simultaneously intermingled in the same piece of land with no distinct row arrangement. This type of intercropping can be suitable for grass–legume intercropping in pastures (Gulwa et al., 2017). The mixed intercropping is commonly observed to fulfil the requirement of food and forage where the land resource is a limiting factor (Undie et al., 2012). Sometimes it is also referred to as mixed cropping (VonCossel et al., 2019).

Row intercropping: It is an intercropping where at least one crop is planted in regular rows, and crop or other crops may be grown simultaneously in row or randomly with the first crop. The row intercropping is a usual practice targeting maximum and judicious use of resources and optimisation of productivity (Varma et al., 2017).

Mixed row intercropping: It is growing two or more crops simultaneously in the same piece of land intermingled within a distinct row arrangement.

Strip intercropping: It is growing two or more crops simultaneously in different strips on sloppy lands wide enough to permit independent cultivation, but narrow enough for the crop to interact with each other. Strip intercropping is known to enhance greater radiation use efficiency in marginal and poor lands (Yang et al., 2015). A combination of soil conserving and depleting crops are taken in alternate strips running perpendicular to the slope of the land or the direction of prevailing winds. An important objective of strip cropping is the reduction of soil erosion and harvesting of yield output from poor lands.

Relay intercropping: Growing two or more crops simultaneously during at least a part of the life cycle of each. A second crop is planted after the first crop has reached its reproductive stage but before it is ready for harvest. This practice is beneficial in terms of resolving time conflict and limitation of soil moisture for plantation of various crops (Balde et al., 2011). For example, delayed sowing is an important reason for yield decline in field pea or chickpea that can be avoided through relay cropping of these crops in standing rice crop. Relay cropping can also fetch certain other benefits such as usage of residual nutrients and moisture from the previous crop and reduced planting costs.

Parallel intercropping: In this practice two crops are selected which have different growth habits and have a zero competition between each other and both of them express their full yield potential. Greengram or blackgram with maize, and greengram or soybean with cotton are the examples of parallel intercropping.

Companion cropping: In companion cropping the yield of one crop is not affected by other. In other words, the yield of both the crops is equal to their pure crops. The standard plant population of both crops is maintained. Rapeseed, wheat, potato, etc. can be taken as companion crops with sugarcane.

Additive intercropping: Here the plant population of the main or base crop remains unchanged as a pure crop stand but one or more minor crops are grown simultaneously with it by utilising the inter-row space. Wheat, rapeseed or potato can be taken as intercrops by utilising the inter-space of two rows of sugarcane planted with its normal spacing.

Substitutive intercropping: In the substitutive or replacement series of intercropping, the crops grown together are known as component crops or intercrops. Here, one component crop is introduced by the replacement of the other crop and no crop is sown with its fullest population as seeded in respective sole cropping. In this system, a definite proportion of a crop is sacrificed and the component crop is introduced in that place. Sometimes to obtain yield advantage, plant population may be increased compared to their density adopted in the pure stands (Baker and Blamey, 1985). In such an intercropping system, competition among species is relatively less than additive series.

Alley intercropping/alley cropping: Food crops are grown in alleys formed by hedge or shrubs or trees. *Glyricidia*+groundnut or any suitable annual crop is an example of alley cropping. It is an agroforestry system.

Trap crops: These crops are grown in the main cropped field in definite rows to trap insect pests.

Filler cropping: Growing of short duration crops in between the newly established perennial crops for few years to fill the space and to utilise the resources. Coconut+pineapple, mango+greengram, guava+turmeric are the examples of filler cropping. It is also an agroforestry system.

Principles of intercropping

1. Before putting any intercrop with the main crop it is very essential to know the prerequisites of the companion crops such as soils and water requirement compatibility, competition for space, sunshine and air, compatibility for pests and diseases, duration and yielding potential and time of sowing and harvesting. The selected intercrop should be compatible with the main crop in its water, nutrients and soil requirement.
2. The crops grown in association should have complementary effects rather competitive effects. The subsidiary crop should be of shorter duration and of faster growing habits to utilise the early slow growing period of main crop and they must be harvested when main crop starts growing.
3. The proper spatial arrangement is required to maximise cooperation and minimise the competition between or among the crops selected for intercropping. An intercropping is successful only when a suitable companion crop is selected to grow with the main crop.
4. Overcrowding should always be avoided by optimising the population of intercrops. A standard plant population of main crop should be maintained and plant population of subsidiary crops can be increased or decreased as per demand of the situation.
5. The crops of different maturity dates should be preferred to reduce the competition for resources during peak demand period.
6. In an intercropping system the component crops should have different root depths so that they can draw water and nutrients from different layers of soil. This will minimise competition for nutrients, water and root respiration.
7. The intercrop should not be an alternate host for pest and diseases of the main crop. Component crops of similar pests and disease pathogens and parasite infestations should not be chosen. Crops selected should be of different families to avoid pests and diseases.
8. Component crops should have similar agronomic practices.
9. Erect growing crops should be intercropped with cover crops like pulses. These check the weeds and reduce soil erosion. The losses of water due to evaporation are also reduced.
10. The planting method and management should be simple, less time consuming, economical and profitable so that it may have wider adoptability.
11. Fast growing crops should be taken with slow growing crops.
12. Crops should have least allelopathic effect.

Advantages of intercropping

1. Additional income is realised from the companion crop or crops. Thus, intercropping gives higher income per unit area than sole cropping.
2. If the base crop is damaged due to any unfavourable climatic conditions or spread of insect pests and diseases, the companion crop may give some income. So it acts as an insurance against failure of crop in abnormal year.
3. Legumes grown as companion crops always benefit the principal crop through N-fixation. Legumes also leave a huge quantity of nitrogen rich organic matter in soil in the form of roots and leaf litter.
4. Quick growing companion crops always suppress the growth of weeds thriving in the inter-spaces of the main crop.
5. Intercropping practices encourage better utilisation of growth resources like nutrient, water, light and space.
6. Incidence of insect pests and diseases is low.
7. Employment opportunity is created throughout the year. It offers best utilisation of labour, machine and power throughout year.
8. Soil erosion due to runoff is checked as the soil is covered with vegetation most part of the year.
9. Intercrops maintain soil fertility as the nutrient uptake is made from both layers.

Limitations of intercropping

1. The fertiliser management is difficult in intercropping practice because the nutrient requirement of the crops is different.
2. Harvesting of different crops become difficult as the seeding and harvesting time of the crops vary.
3. If the selection of crops in an intercropping system is not compatible, the growth of one crop may be suppressed by another crop. There may be competitive effects among component crops.
4. Incidence of insect pests and diseases is increased because sometimes one crop may serve as alternate host for pests and diseases.
5. The grower has to acquire the latest technologies for the crops grown which may not be possible at all the time.
6. Intercultural practices are sometimes difficult in an intercropping system since the time and requirements of agronomic practices for different crops in a system are different. Mechanisation is also difficult.

7. Application of irrigation water, nutrients, pesticides or herbicides is difficult in an intercropping system and for proper management of the system some skill is required.
8. It is labour intensive.
9. There may be allelopathic effect.

Mixed cropping

Growing of two or more crops simultaneously on the same piece of land, without any definite row arrangement is termed as mixed cropping. It is mainly practised in areas where climatic hazards such as flood, drought, frost, etc. are frequent and common. Mixed cropping is also practised with a view to achieve multiple requirements of food, feed and fibre. Mixed cropping is especially important for fodder crops where it can provide enormous quantities of feedstuff for supporting sustainable livestock production. In mixed cropping, the time of sowing of all the crops is almost the same. However, they may mature at a time (wheat+gram, wheat+barley or wheat+mustard) or they may mature at different times (pigeon pea+jowar, sesame+greengram, wheat+mustard or bajra+groundnut).

Types of mixed cropping

Based on method of sowing and objective of cropping, mixed cropping can be classified into several groups.

Mixed cropping: The seeds of different crops are mixed together and then sown either in lines or broadcasted. The system is not scientific. Agricultural operations and harvesting of the crops become difficult.

Guard crops: The main crops is grown in the centre, surrounded by hardy or thorny crops such as safflower around pea or wheat, mesta around sugarcane, jowar around maize, etc. with a view to provide protection to the main crop.

Augmenting crops: When sub-crops are sown to supplement the yield of the main crop, the sub-crops are termed as augmenting crops. Mustard is sown on the borders of wheat fields to harvest mustard leaves for greens during the initial stages.

Multistoried cropping or multitier cropping: Growing plants of different height in the same field at the same time is termed as multistoried cropping. It is mostly practised in orchards and plantation crops for maximum use of solar energy even under high planting density. Examples of multitier cropping are *Eucalyptus*+papaya+berseem, sugarcane+potato+onion, sugarcane+*toria*+ potato, and coconut+pineapple+turmeric/ginger.

Principles of mixed cropping

1. All crops in a mixed cropping system do not fail under adverse climate conditions.
2. An epidemic attack of any insect, pest or disease kills only one crop without affecting the rest of the crops.
3. The different types of crops are grown with an objective of fulfilling the daily need for cereals, pulses and oilseeds of a farm family.
4. The sub-crops should not affect the growth of the main crop.
5. The sub-crops should mature earlier or later than of the main crop.
6. One of the crops should preferably be a legume.
7. The crops in a mixed cropping system should have different growth habits and nutrient requirements.
8. The crops should have different rooting depths and ramification.
9. The crops should not be very exacting in climatic requirements.

Advantages of mixed cropping

1. Mixed cropping ensures efficient utilisation of land, labour, equipment and other resources.
2. It utilises available space and nutrients to maximum extent possible.
3. It provides safe guard against hazards of diseases and pests.
4. It also reduces the effects of adverse weather conditions for the farmer as planting and harvest is at different times of the season.
5. Mixed cropping secures daily requirements like cereals, pulses, oilseeds, etc. to a farm family.
6. It improves yield. According to some scientific basis there is 10-25% increase in the yield in mixed cropping than monocropping.
7. It gives balanced cattle feed with both legume and grass fodders.
8. Costs of input decrease as compared to individual crop growing cost.

Limitations of mixed cropping

1. Some crops are very specific in the type of soil they need for maximum profits.
2. Crops for mixed cropping if not chosen properly have chances of competition between the crops for nutrients.
3. It also may reduce the fertility of the soil as more than one crop is grown at a time in the same piece of land.

Besides these, all the limitations for intercropping are also applicable for mixed cropping.

Table 11.1. Difference between intercropping and mixed cropping

Intercropping	Mixed cropping
1. The main object is to utilise the space left between two rows of main crop.	1. The main object is to get at least one crop under favourable conditions.
2. More emphasis is given to the main crop.	2. All crops are cared equally.
3. There is no competition between both crops.	3. There is competition between/among the crops grown.
4. Intercrops are of short duration and are harvested much earlier than main.	4. Usually the crops are almost of the same duration.
5. Sowing time may be same or different.	5. It is same for all crops.
6. Crops are sown in different rows.	6. Either sown in rows or mixed without considering the population of the crops.

Sequential cropping

Sequential cropping refers to growing crops in sequence within a crop year, one crop being sown after the harvest of the other. Here crop intensification is only in the time dimension without any spatial intensification like intercropping. One has to manage only one crop at a time in the same field. Intensification in sequential multiple cropping through introduction of non-conventional crops/short duration crop cultivars and intensive input management, is a common way of increasing land use efficiency especially in irrigated ecosystems (Gangwar et al., 2012). Irrigation facility, availability of short duration, photo-insensitive and thermo-insensitive high yielding varieties and other production inputs should be ensured for realising the maximum benefits from the sequence cropping.

Types of sequential cropping

Double cropping: Cultivation of two crops in succession on a piece of land in a year is colled double cropping. Growing rice–wheat, rice–groundnut or pigeon pea–maize in a year in sequence are the examples of double cropping.

Triple cropping: It refers to cultivation of three crops in succession on a piece of land in a year. Growing rice-maize-greengram, rice–potato–okra or pigeon pea-maize-cowpea in a year in sequence are the examples of triple cropping.

Quadruple cropping: It refers to cultivation of four crops in succession on a piece of land in a year. Growing rice–fodder maize–greengram–greens, rice–potato–greens–okra or greengram–fodder maize–fodder cowpea–okra in a year in sequence are the examples of quadruple cropping.

Ratoon cropping or ratooning: It refers to raising a crop with regrowth coming out of roots or stalks after the harvest of the crop. Here, more than one harvest is done from one sowing/planting. Thus, ratooning consists of allowing stubbles of the original crop after harvesting and to raise another crop. In sugarcane, 3-4 ratoon crops can be taken, but here a cropping year is not the time limit as it is 10-18 months duration crop. Seasonal vegetables like okra, chilli, etc. can be taken as ratoon crops from the stubbles in the following season within a cropping year.

Gangwar and Singh (2011) have documented the efficient alternative cropping systems for various agroclimatic zones of the country. The existing mean productivity of 9.2 t/ha in various locations can be increased to 16.9 t/ha (table 11.2) through high productive systems identified at various locations.

Table 11.2. Identified high productive cropping systems (sequential cropping) for selected locations

Locations	Prevailing cropping sequence		High productive alternative system	
	System	System rice equivalent yield (t/ha)	System	System rice equivalent yield (t/ha)
Jammu, J & K	Rice–wheat	11.3	Rice–marigold–french bean	30.1
			Rice–potato–onion	29.5
Ludhiana, Punjab	Rice–wheat	13.2	Maize–potato–onion	27.9
			Groundnut–potato–bajra (fodder)	23.3
Modipuram, U. P.	Rice–wheat	12.9	Maize–potato–sunflower	24.2
			Rice–wheat–greengram	15.9
Sabour, Bihar	Rice–wheat	11.0	Rice–potato–onion	29.0
			Rice–wheat–maize	15.7
Bhubaneswar, Odhisha	Rice–rice	6.7	Rice–maize–cowpea	17.4
			Rice–maize–greengram	14.8
Coimbatore, Tamil Nadu	Cotton–sorghum –finger millet	4.1	Beet root–greengram–maize+cowpea	7.1
			Chillies+onion–Sunhemp– okra+coriander	6.6
Thanjavur, Tamil Nadu	Rice–rice–sesame	13.7	Rice–rice–brinjal	18.3
			Direct seeded rice –rice–maize + blackgram	17.4
S.K. Nagar, Gujarat	Groundnut–wheat	4.1	Groundnut–wheat –sesame	7.0
			Groundnut-onion-greengram	5.0
Bengaluru, Karnataka	Hybrid cotton –sunflower	7.0	Maize–groundnut	12.2
			Maize–sunhemp –sunflower	11.3
Hyderabad, Telangana	Rice–rice	7.9	Maize–onion	12.3
			Maize–tomato	12.1
Mean		9.2		16.9

Source: Gangwar and Singh, 2011

Advantages of sequential cropping

1. It increases the overall productivity of the system.
2. It also influences the soil conditions favourably.
3. It ensures efficient utilisation of available resources like nutrients, water and solar energy.
4. Positive residual influence of previous crop with respect to plant nutrients on succeeding crops is realised.

Disadvantages of sequential cropping

1. Pests and disease from a crop may be carried over to the succeeding crop.
2. Weed population and species may sometimes be shifted from one crop to second crop.
3. The growth of second crop may be adversely affected by allelopathic effect of first crop. Sunflower has allelopathic effect on succeeding crops like mustard and soybean by inhibiting germination of their seeds.

Table 11.3. Difference between intercropping and sequential cropping

Intercropping	Sequential cropping
1. Growing of two or more crops simultaneously on the same piece of land in a cropping season.	1. Growing of two or more crops in sequence on the same piece of land in a farming year.
2. Intercropping is intensive cropping mainly in space dimension.	2. Sequential cropping is intensive cropping mainly in time dimension.

Crop rotation

Crop rotation refers to growing of different crops alternatively on the same piece of land in a definite sequence or process of growing different crops in succession on a piece of land in a specific period of time with an objective to get maximum profit from least investment without impairing the soil fertility. This may also be defined as the repetitive cultivation of an ordered succession of crops (or crops and fallow) on the same land and one cycle may take one or more years to complete.

Principles of crop rotation

1. Crops best suited to the climatic conditions and soil type should be selected.
2. The crops with tap roots should be followed by those with fibrous root system. This helps in proper and uniform use of nutrients from the soil and root do not compete with each other for uptake of nutrients.

3. The leguminous crops should be grown after non-leguminous crops. Legumes fix atmospheric nitrogen in the soil and add more organic matter to the soil in the form of huge quantity of root and foliage.
4. More exhaustive crops should be followed by less exhaustive crops. Exhaustive crops like sugarcane, maize, etc should be followed by low inputs demanding crops like oilseeds and pulses.
5. The crop of the same family should not be grown in succession because they act like alternate hosts for pests and diseases.
6. The crop rotation should provide maximum employment to the farm family, farm machineries and equipments available.
7. Selection of the crops should be based on local demand, storage, transport and marketing facility.
8. Grower's preference and his socioeconomic conditions and resource mobilising power should also be considered.
9. The selection of the crops should be based on farmer's knowledge and skill.
10. The latest production technology of the crops should readily be available.
11. On slopes which are prone to soil erosion, alternately erosion-promoting crops like millets or other row crops and erosion-resisting crops like legumes should be adopted.
12. Under dryland farming or partially irrigated areas, the crops like pulses, sunflower which can tolerate drought spells to some extent should be selected. Similarly in low lying and flood prone areas the crops like rice and jute which can tolerate water stagnation should be selected.

Benefits of crop rotation

1. Soil fertility is restored by fixing atmospheric nitrogen by integrating legumes in a rotation.
2. Proper selection of crops in a rotation encourages soil microbial activity and improves physico-chemical properties of the soil.
3. The succeeding crop utilises the residual nutrients.
4. The soil is also protected from erosion, salinity and acidity.
5. It controls pests and diseases if dissimilar crops are taken in succession.
6. It also restricts the weeds growth.
7. Farmers get better price for their produce because of its higher demand in the locality or in the market.
8. The family labour, power, equipment and machineries are well employed.
9. Proper utilisation of all the resources and inputs could be made.

Limitations of crop rotation

1. Specialisation in many crops is not possible always.
2. Requirement of equipments and machineries varies from crop to crop.
3. Allelopathic effect of preceding crop affects the succeeding crop.
4. Similar types of crops serve as alternate hosts for pests and disease.

PLANT INTERACTIONS

In intensive cropping, when crops are grown in association (intercropping or sequential cropping) interaction between different component crop species occurs, which is essentially a response of one species to the environment as modified by presence of another species. This phenomenon is commonly referred as interference or interaction.

Nature of interactions

The nature of interactions between two components can be described on the basis of observable net effect of one component on another in a system. The basic principle deciding the nature of interactions depends upon the ability of different components to capture and use the most limiting essential growth resources effectively. The capture of the limiting resource (e.g., light, water or nutrients) depends on the number, surface area, distribution and effectiveness of the individual elements within the canopy or root system of the species or mixture of species involved.

Complementary interaction: In a system if the component crops help each other, by creating favourable conditions for their growth in such a way that the system provides a greater yield than the yield of their corresponding sole crops, then the interaction between the components is said to be complementary in nature. The complementary interaction is of two types, viz., spatial and temporal. When complementarity is achieved simply by varying the proportion of the desired species, it is known as spatial complementary interaction. In this, the geometry of the system is altered by changing the spacing and density of the components. Therefore, under complementary association the proportion of each species can be modified to suit one's requirements without any sacrifice. When two species complement each other, the capture of the major limiting growth resources by the mixture always exceeds that by the corresponding sole stands. The two components also complement each other through more effective use of growth resources by utilising them at specific times due to difference in growth pattern and resource requirement. There are substantial opportunities for temporal complementarities if species make their major demands on available resources at different times, thereby reducing the possibility of competition.

Supplementary interaction: If the two components interact in such a way that the yield of one component exceeds the yield corresponding to its sole crop without affecting the yield of other component, the interaction is known to be supplementary in nature. Therefore, there will be an additional yield from one of the crops; thus the overall yield of the system will be more than that of sole crops of both the components. In this interaction one component is maximising its resource utilisation without limiting it for the other component.

Competitive interaction: In this system the components interact in such a way that the increase in the yield of one component leads to decrease in the yield of other component due to competitive interaction. One species may have greater ability to use the limiting factor and will gain at the expense of the other. The intensity of competition is the greatest when requirements are similar, and the growth and development proceed synchronously for the component crops. Thus, there is extensive overlap between species in their resource requirements, resulting in a severe competition. To reduce the competition in a system, the components should be selected in such a manner that they can utilise the finite resources (water, nutrients, solar energy etc.) at different times and show different growth and development phases from each other.

Further subdivision of the interactions as suggested by Vandermeer (1989) is given in the following table.

Table 11. 4. Analysis of interactions between two populations 'A' and 'B'

Type of interaction	Effect of the interaction on the population		Nature of the interaction
	A	B	
Mutualism	+	+	Interaction favourable to the two populations
Facilitation	+	0	Interaction favourable for A but not obligatory; B not affected
Commensalism	+	0	Interaction obligatory for A; B not affected
Neutralism	0	0	None of the populations affects the other in crop land
Parasitism/predation	+	–	Interaction obligatory for A; B is inhibited
Amensalism	–	0	A inhibited; B not affected
Competition andinterference	–	–	Each population is inhibited by the others use of (above or below ground) growth resources

0: No significant interaction; +: Advantage for the population in question (growth, survival, reproduction etc.); –: Disadvantage for the population in question

Factors affecting interactions

The interactive relations among the component crops of any system are affected by many factors such as choice of species, population of species, site factors, management practices, etc. Many of these factors can be manipulated for better production from the system.

Species of crops: The growth pattern of crop plants varies from species to species and thus, affects the interactions with the associated components. Some crops may perform better in association with a particular crop species, whereas the yield of other crops may reduce with the same crop species.

Plant population: Yield per plant linearly correlated with the available space. In close planting, leaf surface per plant and unit area is reduced. Closely planted crops have a uniform root spread. Widely spaced crops have a circular root distribution which is beneficial to crops. High plant density leads to seedling mortality. Build-up and spread of pest and diseases are more in high plant density without proper management.

Site factors: Variations in climatic (rainfall, temperature, humidity, etc.), edaphic (physical, biological and chemical properties) and physiographic features of an area affect the plant growth which leads to variation in the interactive relationship of component species in system.

Management practices: The interactive relationship between or among the different crop species also varies with management practices adopted for the system. Through various management practices the interactions among the species in a system can be manipulated.

Interactions in intercropping system

In an intercropping system, always there is an interaction between different component crops. This interaction may be positive (beneficial) or negative (harmful). These positive or negative interactions can be direct or indirect or both. Both positive and negative interactions may further be divided into above or below ground interaction.

The main effects of positive interactions in intercropping systems are increased productivity, improved soil fertility, nutrient cycling, soil conservation, water conservation, weed control and microclimate improvements. Similarly, the effects of negative interactions include light competition, nutrient competition, water competition, pests and diseases and allelopathy. However, these interaction processes are interdependent and the manifestation of their effects is influenced by the environmental conditions.

Positive interactions

Increased productivity: The biomass productivity of an intercropping system is generally greater than that of a monocropping system. The basis for the potentially higher productivity is due to the capture of more growth resources (e.g., light, nutrient and water) or due to improved soil fertility through biological nitrogen fixation by the legumes. The presence of two or more species of different growth habits leads to a net effect of complex interactions resulting into microclimate amelioration, thus affecting transpiration rate, photosynthesis and energy balance of the associated crops. All these factors may translate into increased productivity from the system.

Improved soil fertility: The potential for micro-site enrichment by some crop species is an extremely important aspect of intercropping when a legume is integrated. In alley cropping use of fast growing, nitrogen-fixing trees (e.g., *Leucaena leucocephala* and *Gliricidia sepium*) can substantially increase the soil fertility. A major feature of this system is to produce a large quantity of biomass which enriches the soil. Presence of rhizosphere microflora and mycorhiza associated with one of the crops may lead to mobilisation and availability of nutrients which may benefit the associated crops also. Legumes in an intercropping system supply part of N fixed by symbiosis to non-legumes like in case of maize+cowpea and rice+groundnut.

Nutrient cycling: Soil nutrients are considered among the least resilient components of sustainability. A fundamental principle of sustainability therefore is to return to the soil the nutrients removed from it through harvests, runoff, erosion, leaching, etc. In intercropping systems, these nutrient loss pathways are minimised. The legume components provide additional nutrient inputs through biological nitrogen fixation. If component crops are of different root depths, they can capture nutrients from different layers of soil. The nutrients captured from deeper layers by the deep-rooted crops are transferred to the soil through litter (leaves, crop residues, root debris, etc.) decomposition.

Soil conservation: Intercropping systems usually have more plant population and cover the soil surface for greater period in a year as compared to monocropping. This is highly effective in controlling soil erosion. The thick vegetation provides a semipermeable barrier to surface movement of water and reduces the impact of raindrops on the soil and minimises splash.

Water conservation: Thick vegetation increases the infiltration of rainwater while simultaneously reduces evaporation from the soil.

Weed control: Another potentially positive interaction in intercropping systems is related to weeds. Higher plant population causes shade on the soil surface and shade suppresses light-demanding weeds. Weed biomass yield is positively correlated with available solar energy.

Annidation: Complementary interactions which occur both in space and time are referred to annidation. Certain crops require less light intensity and high relative humidity. Such an altered microclimate is provided when such crops are grown in between tall growing components in an intercropping system. This is annidation in space. Growing turmeric, ginger, pineapple or black pepper in coconut plantation uses this principle. The component crops utilising resources from different soil layers in pigeon pea and rice or groundnut intercropping system is an example of annidation in space. When two crops of widely varying duration are planted, their peak demands for light and nutrients are likely to occur at different periods, thus reducing competition. This is annidation in time. When early maturing crop is harvested it becomes favourable for late maturing crop in intercropping systems like sorghum+ pigeon pea, groundnut+ pigeon pea, maize+greengram, etc.

Negative interactions

As all the crops in an intercropping system utilise the same reserves of growth resources, negative interactions are likely to occur in every crop plant association. The major yield decreasing effects arise from competition for light, water, and nutrients, as well as from interactions through allelopathy.

Competition for light: In intercropping, shading by the taller crop plants reduces light intensity at the short height crop and this is the most limiting factor in many situations, particularly those with relatively fertile soils and adequate water availability. The relative importance of light will decrease in semiarid conditions as well as on sites with low fertility soils. Since crops differ in their responses to poor nutrition, competition for light may either be reduced or amplified by a shortage of nutrients. The growth duration of the component crops play a major role to minimise competition for light. The difference in maturity of component crops should be at least 30 days to get the maximum benefit from an intercropping system. Proper choice of crops and genotypes, adjustment of population density, judicious proportion of each component in the mixture and correct planting pattern are the ways of increase the light use efficiency. The best way to orient crop rows in the northern hemisphere is north to south. This gives the most sun exposure and allows for ample air circulation. When crops are planted east to west, the rows tend to shade each other.

Competition for nutrients: If the peak demand period for the nutrients coincides for all the component crops in an intercropping system there is a competition for nutrients. The effect of nutrient competition is more severe for the crops when the root systems of different crops confined to the same soil horizon and availability of plant nutrients in the rhizosphere is limited. Competition for nutrients also depends upon the root spacing of dominated and aggressive crops.

Competition for water: Competition for water is likely to occur in intercropping systems at some period of time except in areas with well distributed rainfall or continuous supply of irrigation water. The competition depends on the severity of the drought and the drought tolerance of the crop plants. It plays a major role in the productivity of the systems despite the use of drought-tolerant and drought-adapted plants, especially in dry areas. Competition for water also depends on the method of irrigation, earliness of water demand, root extension (lateral and vertical growth) of the component crops.

Microclimatic modification for pests/diseases: Bacterial and fungal diseases may increase in shaded, more humid environments in an intercropping system. Reduced temperature and humidity fluctuations under shade can also have a favourable effect on spread of pests and diseases. One crop can be a host of insect pests and diseases for another crop of the system.

Allelopathic interaction: Allelopathy refers to the inhibition of growth of one plant by chemical compounds (allelochemicals) that are released into the soil from neighbouring plants. Some crops may be unsuitable as intercrops because they secrete toxins into soil which will adversely affect the associated crops. Allele chemicals are reported to be present in practically all plant tissues, including leaves, flowers, fruits, stems, roots, rhizomes and seeds. The effects of these chemicals are dependent mainly upon the concentration as well as the combination in which one or more of these substances is released into the environment. The organic compounds released this way are often phytotoxins. Such toxic chemicals may be injurious to microbes and even to the seedlings of those plants releasing them. The effects of the chemicals may result in complete inhibition of growth or retarded growth. Allelopathic compounds may be released into the environment by volatilisation, leaching from living or dead tissues, exudation from roots and decay of plant tissues. The mechanism of action of allelochemicals is diverse and includes

1. Inhibition of cell division and elongation,
2. Inhibition of action of gibberellins or indole acetic acid that induces growth,
3. Reduction of mineral uptake,
4. Retardation of photosynthesis,
5. Inhibition or stimulation of respiration,
6. Inhibition or stimulation of stomatal opening, and
7. Inhibition of protein synthesis and changes in lipid and organic acid metabolism.

Interactions in sequential cropping system

In sequential cropping, the preceding crop has considerable influence on the succeeding crop mainly by changes in soil conditions, presence of allelopathic chemicals, shift in weed flora and carry over effects of fertilisers, pests and diseases.

Soil conditions: A few problems are encountered for land preparation in sequential cropping system. The time available for seedbed preparation is less in high intensity cropping system. Due to the effect of the previous crop, the field may not be in proper condition to carry out field operations. Field preparation becomes difficult after transplanted rice because the soil structure is destroyed due to puddling. The establishment of a pulse crop after rice is also difficult. Delay in sowing or planting is the most common problem in sequential cropping systems because the field preparation and planting of the succeeding crop is delayed. Short duration varieties of crops are selected to fit well in sequential cropping systems. Photo- and thermo-insensitive varieties are essential for successful sequence cropping system.

Allelopathic interaction: Crops like sorghum and sunflower leave toxic chemicals in the soil which do not allow germination of subsequent crops. Roots of cucumber, leaves of *Eucalyptus*, decomposing residues of sunflower are known to produce allele chemicals affecting the growth of succeeding crops.

Weed flora: Weed number and species differ in the succeeding crop due to the effect of the previous crop. It is reported that wheat crop that follows rice suffers from high density of *Phalaris minor*. Weed management practices are followed for the entire system as a unit instead of considering weed problem of individual crops. Herbicides applied to the previous crop may be toxic to the succeeding crop. Higher dose of atrazine applied to sorghum crop affects germination of succeeding pulse crop. Herbicide and its dose for the preceding crop should, therefore, be recommended with the consideration of succeeding crop.

Soil fertility: The previous leguminous crop leaves considerable amount of nitrogen in the soil for the succeeding crop. The legumes also add a huge quantity of organic matter rich in nitrogen in the soil in the form of root residues and leaf litter. P applied to the previous crop is also available for the succeeding crop.

Pests and diseases: The infestation of pests and diseases are more in sequence cropping system due to continuous cropping. The pests and diseases in crop stubbles and other residues of the previous crop may infect the subsequent crop. However, carry over effects of insecticides are not observed.

EFFICIENT CROPPING SYSTEMS

Widespread occurrence of second-generation problems, such as over-mining of soil nutrients, decline in factor productivity, reduction in profitability, lowering of groundwater tables and build-up of pests including weeds, diseases and insects has been reported during post-green revolution era in most of the intensively cultivated, high-productivity, cereal based production systems, which are threatening their sustainability (Gangwar and Prasad, 2005). Studies carried out under AICRP on Cropping Systems have resulted in identification of appropriate duration of varieties for some popular crop sequences for different regions of the country. Several alternative cropping systems with high land-use efficiency and monetary returns have been identified for different regions and agroecological/farming situations of the country. These multiple cropping systems include sequential as well as intercropping systems, including several variants such as mixed cropping, relay cropping, alley cropping, parallel multiple cropping, multistoried cropping, etc.

Cropping systems for better nutrient use

Inclusion of legumes in the cropping system has been known since times immemorial. Legumes play a vital role in increasing indigenous nitrogen production in the field. These help in solubilising insoluble P in soil, improving the soil physical environment, increasing soil microbial activity and restoring organic matter. The carryover of N derived from legume grown, either in crop sequence or in intercropping system for succeeding crops, is also important. Thus, the scope for exploiting direct and residual fertility due to legumes has a great potential.

Sorghum, pearl millet, maize and castor are usually grown in drylands and marginal and sub-marginal lands. Sorghum yield increased when sown after cowpea, greengram, and groundnut. Legumes like groundnut or cowpea provide an equivalent to 60 kg N/ha on the subsequent crop of pearl millet. Incorporation of whole plant of summer greengram or blackgram into soil (after picking pods) before transplanting rice is very much beneficial to rice–wheat system and up to 25% N can be saved in rice. Legumes with indeterminate growth are more efficient in N_2 fixation than determinate types. Fodder legumes in general are more potent in increasing the productivity of succeeding cereals. The carryover of N for succeeding crops may be 60-120 kg in berseem, 75 kg in cluster bean, 35-60 kg in fodder cowpea, 68 kg in chickpea, 55 kg in blackgram, 54-58 kg in groundnut, 50-51 kg in soybean, 50 kg in *Lathyrus*, and 36-42 kg per ha in pigeon pea (Ghosh et al., 2007).

The major consideration in N management in intercropping systems is to quantify the 'direct transfer' of N from legume component to the non-legume component.

However, their crop components have different requirement for nutrients. Cereals have less P but high N requirement, while legumes possess effective mechanism for symbiotic N fixation but have a high P requirement. Because of this, fertiliser use is considered complicated. N economy through intercropped legumes is to be correctly assessed in different intercropping systems. In legume and cereal intercropping system, the maize-pigeon pea is highly suitable with a minimum competition for nutrients, while in legume and legume intercropping system, pigeon pea–groundnut system is the most efficient one in terms of resource use efficiency. Pigeon pea starts flowering after the maize has been harvested and its period of greatest nutrient demand occurs when the maize has already completed its growth cycle. However, in sorghum–pigeon pea system, grand growth period of both the crops falls exactly at the same time as a result of which tall cereal adversely affects the growth of the associated pigeon pea. Further, higher amounts of N applied to sorghum–pigeon pea system would result in better growth and yield of sorghum and consequently the growth and yield of pigeon pea will be adversely affected. Therefore, moderate quantity of N (25-50 kg N/ha) should be applied to sorghum–pigeon pea system under dry land condition to achieve higher yield advantage (Stern, 1993).

Cropping systems for weed management

Intermittent crop substitution/diversification in sequential cropping systems has been found to reduce some obnoxious weeds to a considerable extent, thereby reducing herbicide need to a great extent in areas where such weeds have assumed alarming proportions due to continuous adoption of a certain cropping system. Johnson grass (*Sorghum halepense*) becomes predominant weed in continuous maize cultivation but can be controlled by rotating with cotton. Similarly, a change from rice–wheat and rice–potato system to any other system, not involving rice in rainy reason, tends to reduce population of *Phalaris minor* in wheat considerably. The crops that quickly form a shade canopy and are allelopathic in nature have an adverse impact on weeds sensitive to shade and can be adopted as weed control measures. Berseem (*Trifolium alexadnrinum*) may be taken as a break crop successfully for reducing weed problem in continuous rice–wheat system without any monetary loss. Weed management through cover crops, intercrops, relay cropping are best strategies for cultural control of weeds.

Cropping systems for disease and pest management

Besides minimising herbicide use for weed control, insecticide/fungicide use can also be minimised to a considerable extent through cropping system approach. Sorghum ear head fly damage is extremely rare where pigeon pea is planted in

alternate rows. Incidence of root rot of cotton caused by *Rhizoctonia solani* fungus is appreciably reduced by intercropping of moth bean (*Phaseolus aconitifolius*). The moth bean intercropping causes moderating effect on soil temperature increase which is unfavourable for the parasitic activity of fungus. It is also reported that coriander, garlic or fennel (*Foeniculum vulgare*) intercropped with autumn planted sugarcane prevents top borer attack in sugarcane and there is no need to apply insecticides for the control of top borer in sugarcane intercropping with these spices crops.

Cropping systems for higher water-use efficiency

Irrigation is a commonly used platform for crop intensification because it offers a point where the potentiality of other inputs is fully exploited. Making this sustainable intensification, more crop and profit per drop of water approach should be followed which is possible only through carefully designed cropping systems (Gangwar et al., 2012). Irrigation water is a costly and scarce resource and its availability for agriculture is expected to further go down due to increased demand for domestic and industrial uses. Though water use efficiency can be increased by genetic and environmental manipulation of the crops, it can also be increased by decreasing the evapotranspiration and other losses of water, such as conveyance, application, distribution and deep percolation. In cropping systems perspective, water use efficiency can be increased by identification of appropriate crop combinations in various systems. More remunerative and less water consuming crop rotations have been standardised at different locations in the country. Rice–mustard–sesame, rice–mustard–greengram and rice–potato–greengram rotations have been identified more water efficient systems at Memari in West Bengal (Sharma and Rajput, 1990). At Kharagpur, rice–wheat, rice–mustard and rice–potato are more viable sequences under lesser water input. At Chiplima (Odisha) rice–mustard, rice–wheat–greengram and rice–potato–sesame rotations were identified as most water efficient systems. In Tarai conditions of Uttar Pradesh, under higher level irrigation rice–lentil and rice–wheat cropping systems were better while under lowest levels of *kharif* and *rabi* irrigations soybean–wheat proved the best sequence (Singh and Lal, 1994). Water productivity can be considerably increased through relay intercropping systems (Bastia et al., 2008; Ved Prakash et al., 2007).

Cropping systems in context of climate change

Whenever, a cropping system becomes unsustainable due to any reason, it is replaced by a new emerging system. A farmer is interested in cropping systems that have a balance mix of higher biological productivity and economic returns with lesser risk and which possesses more stability over a period of time under fluctuating environmental situations. Indian agriculture is highly prone to the risks due to climate change, especially to drought, because two third of the agricultural land in India is rainfed, and even the irrigated system is dependent on monsoon. Flood is also a major problem in many parts of the country. Increase in temperature can reduce crop duration, increase crop respiration rates, alter photosynthesis process, affect the survival and distributions of pest populations and thus developing new equilibrium between crops and pests, hasten nutrient mineralisation in soils, decrease fertiliser use efficiencies, and increase evapotranspiration.

Development of new crop varieties with higher yield potential and resistant to multiple stresses (drought, flood, salinity) will be the key to maintain yield stability. On-farm water conservation techniques, micro-irrigation systems for better water use efficiency and selection of appropriate crop has to be promoted. Adjustment of planting dates to minimise the effect of high temperature induced spikelet sterility can be used to reduce yield instability so that the flowering period does not coincide with the hottest period. The conventional puddled transplanted rice and intensively tilled wheat cropping system should be changed to other cropping systems such as maize–wheat, pulse–wheat, maize–pulse, oilseeds-wheat and direct-seeded rice–wheat. These systems have less demand for water and nutrient and use resources more efficiently.

EFFICIENCY ASSESSMENT

Several indices have been developed and are being used for evaluating system productivity, input use efficiency and economic efficiency of multiple cropping systems.

Assessment of land use

The main objective of cropping systems is to use available resources efficiently. There are several indices to compare the efficiency of different cropping systems in terms of land use.

Multiple cropping index (MCI)

This measures the sum of area planted to different crops and harvested in single year divided by the total cultivated area times 100.

$$MCI = \frac{\sum_{i=1}^{n} ai}{A}$$

where, n = total number of crops,

ai = area occupied by i^{th} crop planted and harvested within one year and

A = total cultivated land area available.

This index gives the percent utilisation of the land area and is variant of the cropping intensity which does not account for the cultivars grown in mixed or intercropping systems.

Land utilisation index (LUI)

Land utilisation index has been defined as the number of days during which the crops occupied the land during a year divided by 365. This index can be expressed as a fraction or as a percentage. It gives only an idea how the land has been put to use. If the index is 1, the land has not been left fallow and if more than 1, it tells the specifications of intercropping and relay cropping etc.

Cultivated land utilisation index (CLUI)

This index is calculated by summing the products of land area planted to each crop, multiplied by the actual duration of that crop and divided by the total cultivated land area available during 365 days.

$$\text{CLUI} = \frac{\sum_{i=1}^{n} a_i d_i}{A \times 365}$$

where, n = total number of crops,

ai = area occupied by i^{th} crop,

di = days that i^{th} crop occupied ai and

A = total cultivated land area available during the 365 days period.

Cropping intensity (CI)

Cropping intensity refers to raising of a number of crops from the same field during one agricultural year. It is the ratio of effective crop area harvested to the physical area.

$$\text{Cropping intensity} = \frac{\text{Gross cropped}}{\text{Net sown area}} \times 100$$

Thus, higher cropping intensity means that a higher portion of the net area is being cropped more than once during one agricultural year. This also implies higher productivity per unit of arable land during one agricultural year. The cropping intensity shows great spatial variation in India, with higher levels in northern plains. Lower levels are found in dry, rainfed regions of Rajasthan, Gujarat, Maharashtra and Karnataka.

Cropping intensity index (CII)

Cropping intensity index is a measure of farmer's actual land use in area and time relationships for each crop or group of crops compared to the total available land area and time, including the land available for production.

$$CII=\frac{\sum_{i=1}^{Nc} ai\,ti}{A0T\sum_{j=1}^{M} AjTj}$$

where,

Nc = total number of crops grown by a farmer during the time period T;

ai = area occupied by the ith crop;

ti = duration of the ith crop;

A0 = farmer's total cultivated land area available for use during the entire time period T;

M = total number of fields temporarily available to the farmer for cropping during time period T;

Aj = land area of jth field;

Tj = time period that j^{th} field occupies.

Specific crop intensity index (SCII)

SCII is a derivative of cropping intensity index and determines the amount of area-time devoted to each crop or group of crops compared to the total time available to the farmer for crop production under study.

$$SCII=\frac{\sum_{k=1}^{Nk} ak\,tk}{A0T+\sum_{j=1}^{M} AjTj}$$

where,

Nk = total number of crops within a specific designation such as vegetable crops or field crops grown by the farmer during the time period T;

ak = area occupied by the kth crop;

tk = duration of the kth crop;

A0 = farmer's total cultivated land area available for use during the entire time period T;

M = total number of fields temporarily available to the farmer for cropping during time period T.

Using this formula vegetable intensity index, field crop yield index, rice or wheat crop yield index etc. can be calculated.

Relative cropping intensity index (RCII)

It is the modification of CII and determines the amount of area and time allotted to one crop or group of crops related to area-time actually used in the production of all crops.

$$\text{RCII}= \frac{\sum_{k=1}^{Nk} \text{ak tk}}{\text{A0T} + \sum_{i=1}^{N} \text{aiti}}$$

These indices (SCII and RCII) can be used for classifying farmers, based on the crops grown in how much area and for what time period it occupies the field. These indices help to measure shifts of various crops among farm of different sizes and determining whether the consistent types of cropping pattern occur within various farm size strata. These indices also help to know how intensity cultivated land area has been utilised. But none of these indices takes productivity into account and cannot be used for comparing different cropping systems and evaluating their efficiency in utilisation of the resources other than the land.

Rotational intensity (RI)

This is calculated by counting the number of crops in a rotation and multiplied by 100 and then divided by duration of the rotation.

$$\text{RI} = \frac{\text{Number of crops grown in a rotation}}{\text{Duration of rotation}} \times 100$$

Assessment of yield advantage

Crop equivalent yield (CEY)

Since several crops are grown in intercropping or sequential cropping systems, it is not logical to compare the total yield of different crops in one system with the other. Thus yields from all the crops grown mixed or intercropped or sequentially cropped are converted to the yield of main crop of the system based on price of the produce. The crop equivalent yield is estimated as follows

$$\text{CEY} = \sum_{i=1}^{n} (\text{Yi}, \text{ei})$$

where, Yi = yield of the ith component and

ei = equivalent factor of ith component which can be calculated as $\frac{\text{Pi}}{\text{Pm}}$, where

Pi is the price of a unit weight of ith crop and Pm is the price of a unit weight of main crop.

This type of comparison is valid only while considering the gross returns. However, it does not indicate the net gain obtained from a cropping system. This also does not explain the land use pattern of the cropping systems.

Relative yield total (RYT)

It is assumed that in inter-genotypic competition the biomass yield of each component is strictly proportional to the share of environmental resources it can acquire. Based on this assumption in a 1:1 ratio, mixtures are competing for same supplies of environmental resources, the proportional increase of per plant yield of one component will tend to equal the proportional decrease of per plant yield of the other. This implies that the mean of plant relative yield (ratio of the per plant yields in a mixture) to that in monoculture will have a value close to unity. However, the use of RYT is restricted to replacement series of intercropping only.

$$\text{RYT} = \frac{1}{2}\left(\frac{\text{Yij}}{\text{Yii}} + \frac{\text{Yji}}{\text{Yjj}}\right)$$

where, Yij and Yji are yields of i and j species in a mixture respectively and Yii and Jjj are their respective monoculture yield in 1:1 ratio.

Land equivalent ratio (LER)

The concept of land equivalent ratio is the most widely accepted index for evaluating the effectiveness of all forms of mixed cropping. LER is the relative land area under sole crop required to produce the same yield as obtained under a mixed or an intercropping system at the same management level. It is similar to the concept RYT. However, unlike RYT, LER can be used for any intercropping treatments and not just for replacement series. It is calculated as sum total of the ratios of yield of each component crop in an intercropping or mixed cropping system to its corresponding yield when grown as a sole crop.

$$\text{LER} = \text{La} + \text{Lb} = \frac{\text{Ya}}{\text{Sa}} + \frac{\text{Yb}}{\text{Sb}}$$

where, La and Lb are the LERs for individual crops in the mixture, Ya and Yb are the individual crop yields in an intercropping situation, Sa and Sb are the yields of species a and b as sole crops.

LER provides a standardising basis so that crops can be added to form combined yield. It can also be used to assess the competitive abilities of the component species of any intercropping situation. The LER has the great merit that it gives an accurate assessment of the greater biological efficiency of the intercropping situation. When LER = 1, there is no advantage to intercropping over sole cropping, when LER < 1, more land is needed to produce a given yield by each component as an intercrop. Intercropping is beneficial only when LER > 1.

Relative crowding coefficient (RCC)

Relative crowding coefficient is a measure of the relative dominance of one component crop over the other in an intercropping or mixed cropping system. The coefficient (K) is determined separately for each component crop.

For crop 'a' in association with crop 'b', the relative crowding coefficient is

$$Kab = \frac{Yab}{Yaa - Yab} \times \frac{Zba}{Zab}$$

where, Y_{ab} is the yield of species 'a' in association with 'b', Y_{aa} is the pure stand yield of species 'a', Z_{ab} and Z_{ba} represent the sown proportion of 'a' (in association with 'b') and 'b' (in association with 'a'), respectively.

Similarly, for crop 'b' in association with crop 'a', the relative crowding coefficient is

$$Kba = \frac{Yba}{Ybb - Yba} \times \frac{Zab}{Zba}$$

where, Y_{ba} is the yield of species 'b' in association with 'a', Y_{bb} is the pure stand yield of species 'b'.

The product of the two coefficients, K_{ab} and K_{ba} designated as K, denotes yield advantage if K > 1, no advantage if K = 1 and disadvantage if K < 1. With the index more than a unity will mean that the given crop is more competitive than the associate crop and vice-versa. Major limitation of RCC is that the comparison of the coefficients for each crop cannot give a quantitative measure of the intercrop competition, but can only indicate that a given crop is more or less competitive.

Aggressivity

Aggressivity gives a simple measure of how much the relative yield increase in species 'a' is greater than that for species 'b' in an intercropping system and can be expressed as Aab.

$$Aab = \frac{Yab}{YaaZab} - \frac{Yba}{YbbZba}$$

where, Yaa and Ybb are pure stand yields of species 'a' and 'b' respectively; Yab and Yba are mixture yields of species 'a' in combination of species 'b' and species 'b' in combination of species 'a' respectively; Zab and Zba are sown proportions of species 'a' in mixture with 'b' and species 'b' in mixture with 'a' respectively.

Aggressivity reveals the simple difference between the extent to which crop 'a' and 'b' vary from their respective expected yields. However, since this is a simple difference, the interpretation of intercropping situations may become difficult if the values are identical in different situations.

Competitive ratio (CR)

Considering the drawback of aggressivity values, the use of competitive ratio has been proposed. Here instead of taking the difference of two terms in aggressivity, the ratio of these terms are taken and designated as competitive ratio (CR).

$$\text{CRa} = \frac{\text{Yab}}{\text{YaaZab}} \div \frac{\text{Yba}}{\text{YbbZba}}$$

This can also be written as

$$\text{CRa} = \left(\frac{\text{Yab}}{\text{Yaa}}\right) \div \left(\frac{\text{Yba}}{\text{Ybb}}\right) \times \left(\frac{\text{Zba}}{\text{Zab}}\right) \text{or} \left(\frac{\text{La}}{\text{Lb}}\right) \times \left(\frac{\text{Zba}}{\text{Zab}}\right)$$

In the above equations, all their abbreviations are the same as in case of aggressivity. However, La and Lb are the LER for crops 'a' and 'b', respectively.

Competition index (CI)

Competition index is the product of two equivalent factors of component crops. Equivalent factor for component 'a' means the number of plants of component 'a' which is equally competitive to one plant of component 'b'. If the equivalence factor is < 1 for a component, the component is more competitive than the other. If the index is < 1, it will be an advantage.

$$\text{Equivalence factor for 'a'} = \frac{\text{yield of 'b' / plant}}{\text{yield of 'a' / plant}}$$

$$\text{Equivalence factor for 'b'} = \frac{\text{yield of 'a' / plant}}{\text{yield of 'b' / plant}}$$

CI = Equivalence factor for 'a' X Equivalence factor for 'b'

Competition coefficient (CC)

Competition coefficient is the ratio of the relative crowding coefficient (RCC) of any species in the mixture.

$$CC = \frac{\text{RCC of a given species}}{\text{Total RCC of all crops in mixture}}$$

It is used to find out the relative crowding from which maximum yield can be obtained without any adverse effect on any of the species.

Area time equivalency ratio (ATER)

It is a modification of LER and takes in account the duration of crop and permits an evaluation of crops on yield per day basis. ATER is the ratio of number of hectare-days required in monoculture to the number of hectare-days used in intercropping to produce identical quantities of each of the component crops.

$$ATER = \frac{La \times Da + Lb \times Db}{T}$$

where, La and Lb are relative yields or partial LERs of component crops 'a' and 'b', Da and Db are duration of crops 'a' and 'b' and T is the total duration of the intercropping system.

ATER > 1 indicates a more efficient use of area-time, e.g., hectare-days by the intercrop.

Relative productivity efficiency

It refers to the capacity of the diversified system for production in relation to the existing system and expressed in percentage.

Relative productivity efficiency

$$= \frac{\text{Total yield of diversified cropping system - total yield of existing cropping system}}{\text{Total yield of existing cropping system}} \times 100$$

Positive figure show the superiority of the new system over the existing one in percentage and considered desirable. Any positive figure of more than 20% is considered worth recommending for extension use.

Sustainable yield index (SYI)

This index is used to assess the sustainability of a cropping system or management practice which can be considered for wider adoption. It identifies the practices systems giving the maximum sustainable yield.

$$SYI = \frac{Y - Sd}{Y\,max}$$

where, Y is the estimated average yield of different practices or systems over years, Sd is the estimated standard deviation and Ymax is the observed maximum yield among all the systems over years.

Less fluctuation in yield over the years reflects positive side of stability. The stability index ranges from 0 to 1. While calculating SYI, the negative value of (Y – Sd) is taken as zero since yield is always a positive quantity. Sd is used to quantify the risk associated with the average performance of a system. When Sd = 0 and Y = Ymax, SYI is 1. This is an ideal system or practice which gives consistently maximum yield in all the years. But invariably in biological systems Sd is always greater than zero due to variations in yield over years because of climatic factors. The value of SYI nearing to unity shows higher stability reflecting that the system is more sustainable.

Assessment of economic advantage

Economic viability of a cropping system can be assessed by converting inputs and products into monetary values. However, the major weakness of this monetisation is that costs and prices of inputs and products rarely remain static and fluctuates from season to season and place to place. Hence, a profitable cropping system in one year may not be so in the next year. For evaluation of cropping systems several indices are being used.

Gross return

This is total income from the farm, by virtue of sales of entire farm produce. The main drawback in this is that the cost involved in production is not considered.

Cost of cultivation

This is a supplementary index to indicate the amount of capital resources needed to adopt a particular cropping system. The resource poor farmer will lower cultivation cost and cropping system with low profit over the system yielding high profit with high investment in many cases.

Net return

It is the income obtained after deducting cost of cultivation from the gross return. The monetary difference of input and output value is more meaningful than the gross return. But variable cost like rent of land, interest on capital, land revenue etc. are generally excluded while computing net return.

Benefit cost ratio

This index provides an estimate of the benefit a farmer derives for the expenditure incurred in adopting a particular cropping system. This measure shows the return per each rupee investment. Wider the ratio more is the benefit.

$$\text{B: C ratio} = \frac{\text{Gross reurn}}{\text{Cost of cultivation}} \quad \text{(From economist point of view)}$$

$$\text{B: C ratio} = \frac{\text{Net reurn}}{\text{Cost of cultivation}} \text{ (From agronomist point of view)}$$

Relative economic efficiency (REE)

It refers to the capacity of the diversified cropping system for realising net profit in relation to the existing system and expressed in percentage. Positive figure shows that the new system is more profitable over the existing one and considered desirable.

Relative economic efficiency

$$= \frac{\text{Net return of diversified cropping system - net return of existing cropping system}}{\text{Net return of existing cropping system}} \times 100$$

Income equivalent ratio (IER)

The ratio of the area needed under sole cropping to produce the same gross income as one hectare of intercropping at the same management level. IER is the conversion of LER into economic terms.

$$\text{IER} = \frac{\text{Iab}}{\text{Iaa}} + \frac{\text{Iba}}{\text{Ibb}}$$

where, Iab and Iba are the gross income of component 'a' and 'b' in an intercropping situation, Iaa and Ibb are the gross income of component 'a' and 'b' as sole crops.

Relative net return index (RNRI)

RNRI is the ratio between the monetary value of base and intercrop plus differential cost of cultivation of intercropping and sole cropping of base crop and monetary value of base crop as sole crop.

$$\text{RNRI} = \frac{\text{Pi Yi} + \text{Pj Yj} \pm \text{Dij}}{\text{Pi Yii}}$$

where, Yi and Yj are the yields of ith base crop/ha and j^{th} intercrop/ha, respectively in ij^{th} crop combination; Pi and Pj are the prices of ith base crop and j^{th} intercrop respectively; Yii is the yield of ith sole crop/ha and Dij is the differential cost of cultivation of ij^{th} crop combination in comparison to i^{th} sole crop.

This index appears to be more appropriate for comparisons in intercropping situations since it takes into account the cost of cultivation and thus, is said to be based on net returns, one of the important requirements for comparisons.

Sustainable value index (SVI)

This index is used to assess the sustainability of a cropping system on monetary terms. In cropping systems where more than one crop is involved, the economic assessment is ideal one than the biological assessment. To work out the SVI, the monetary values of the economic produce are used instead of yield values. It is worked out by the same way as SYI.

$$SVI = \frac{Y - Sd}{Y\ max}$$

where, Y is the estimated average net profit of a system over years, Sd is the estimated standard deviation and Ymax is the observed maximum profit among all the systems over years.

REFERENCES

Baker, C. M. and Blamey, F. P. C. 1985. Nitrogen fertiliser effects on yield and nitrogen uptake of sorghum and soybean, grown in sole cropping and intercropping systems. *Field Crops Research* **12**: 233-240.

Balde, A. B., Scope, L. E., Affholder, F., Corbeels, M., DaSilva, F. A. M., Xavier, J. H. V. and Wery, J. 2011. Agronomic performance of no-tillage relay intercropping with maize under smallholder conditions in Central Brazil. *Field Crops Research* **124**: 240-251.

Bastia, D. K., Garnayak, L. M. and Barik, T. 2008. Diversification of rice (*Oryza sativa*)-based cropping system for higher productivity, resource use efficiency and economics. *Indian Journal of Agronomy* **53**(1): 22-26.

Gangwar, B. and Prasad, K. 2005. Cropping system management for mitigation of second generation problems in agriculture. *Indian Journal of Agricultural Sciences* **75**(2): 65-78.

Gangwar, B. and Singh. 2011. Efficient Alternative Cropping Systems. Project Directorate for Farming Systems Research, Modipuram, Meerut, India. pp. 339.

Gangwar, B., Ravisankar, N. and Prasad, K. 2012. Agronomic research on cropping systems in India. *Indian Journal of Agronomy (Special Issue)* **57**: 171-181.

Ghosh, P. K., Bandyopadhyay, K. K., Wanjari, R. H., Manna, M. C., Misra, A. K., Mohanty, M. and Subba Rao, A. 2007. Legume effect for enhancing productivity and nutrient use-efficiency in major cropping systems - an Indian perspective: a review. *Journal of Sustainable Agriculture* **30**(1) 2007: 59-86.

Gulwa, U., Mgujulwa, N., Beyene, S. T. 2017. Effect of grass–legume intercropping on dry matter yield and nutritive value of pastures in the Eastern Cape Province, South Africa. *Universal Journal of Agricultural Research* **5**(6): 355-362.

http://www.jnkvv.org/PDF/13042020134922Unit%20II.pdf

https://imp.center/agri/farming-system-concept-and-scope/

Maitra, S., Hossain, A., Brestic, M., Skalicky, M., et al. 2021. Intercropping - a low input agricultural strategy for food and environmental security. *Agronomy* **11**(2): 1-28.

Sharma B. R. and Rajput, A. L. 1990. Sustained rice productivity in eastern region of India through efficient water management techniques. *Indian Journal of Agronomy* **35**: 85-90.

Singh, M. and Lal, B. 1994. Grain yield and economic returns under varying crop sequences and levels of rainy season and winter season irrigations. *Indian Journal of Agronomy* **39**: 211-215.

Stern, W. R. 1993. Nitrogen fixation and transfer in intercrop systems. *Field Crops Research* **34**(3-4): 335-356.

TAC-CGIAR. 1978. Quinquennial Review Report, International Institute of Tropical Agriculture, 1978. Technical Advisory Committee of the Consultative Group for International Agricultural Research.

Undie, U. L., Uwah, D. F. and Attoe, E. E. 2012. Effect of intercropping and crop arrangement on yield and productivity of late season maize/soybean mixtures in the humid environment of South Southern Nigeria. *Journal of Agricultural Research* **4**(4): 37-50.

Vandermeer, J. 1989. The ecology of intercropping. Cambridge University Press. Cambridge, U.K.

Varma, D., Meena, R. S. and Kumar, S. 2017. Response of mungbean to fertility and lime levels under soil acidity in an alley cropping system in Vindhyan region. *International Journal of Chemical Studies* **5**(4): 1558-1560.

Vedprakash, N. K., Bhattacharriyya, R. Kumar, M. and Srivastava, A. K. 2007. Productivity, economics, energetics and soil properties of vegetables based relay intercropping systems. *Indian Journal of Agronomy* **52**(4): 300-304.

Von Cossel, M., Wagner, M., Lask, J., Magenau, E., Bauerle, A., Von Cossel, V., Warrach-Sagi, K. and Winkler, B. 2019. Prospects of bioenergy cropping systems for a more social-ecologically sound bioeconomy. *Agronomy* **9** (10): 1-32.

Yang, F., Wang, X. C., Liao, D. P., Lu, F. Z., Gao, R. C., Liu, W. G., et al. 2015. Yield response to different planting geometries in maize-soybean relay strip intercropping systems. *Agronomy Journal* **107**(1): 296-304.

12

Integrated Farming Systems for Sustainable Agriculture

Integrated farming system is a multidisciplinary whole farm approach, which enables the farmers to identify opportunities and threats and act accordingly. It is a dynamic approach which can be applied to any farming system around the world. It is very effective in solving the problems of small and marginal farmers. The approach aims at increasing income and employment from small-holding by integrating various farm enterprises and recycling crop residues and by-products within the farm itself. It involves attention to detail and continuous improvement in all areas of a farming business through informed management processes. Integrated farming combines the best of modern tools and technologies with traditional practices available at a given location and situation. Preserving and enhancing soil fertility, maintaining and improving a diverse environment and the adherence to ethical and social criteria are indispensable basic elements of integrated farming systems.

'Farming' is a process of harnessing solar energy in the form of economic plant and animal products. 'System' implies a set of interrelated practices and processes organised into functional entity, i.e., an arrangement of components or parts that interact according to some process and transforms inputs into outputs (Fresco and Westphal, 1988).

The International Organisation of Biological Control describes integrated farming as a farming system where high quality food, feed, fibre and renewable energy are produced by using resources such as soil, water, air and nature as well as regulating factors to farm sustainably and with as little polluting inputs as possible. Integrated farming system (IFS) is a commonly and broadly used word to explain a more integrated approach to farming as compared to monoculture approaches. In this system an interrelated set of enterprises are maintained and by-products or wastes from one production system becomes an input for another production system, which reduces cost and improves production and/or income. Thus, IFS works as a system of systems (Chan, 2006; Soni et al., 2014). FAO (1977) stated that 'there is no waste', and 'waste is only a misplaced resource which can become a valuable material for another product' in IFS. For example, paddy straw, by-product from rice crop can be used as a valuable input for mushroom cultivation or dry fodder for dairy animals. Similarly spent of mushroom cultivation (used straw) can be used as a raw material in compost or vermicompost pits and by-products from dairy unit like dung can be used as fish feed or raw material for vermicompost unit. The farming system is essentially cyclic, organic resources - livestock - land - crops. Therefore, management decisions related to one component may affect the others. Combining ecological sustainability and economic viability, the integrated livestock farming system maintains and improves agricultural productivity while also reducing negative environmental impacts.

Different scientists have defined a farming system differently. However, many definitions, in general, convey the same meaning that it is the strategy to achieve profitable and sustained agricultural production to meet the diversified needs of farming community through efficient use of farm resources without degrading the natural resource base and environmental quality. Farming system is a decision making unit comprising the farm household, cropping and livestock system that transform land, capital and labour into useful products that can be consumed or sold (Fresco and Westphal, 1988). Lal and Miller (1990) defined farming system as a resource management strategy to achieve economic and sustained agricultural production to meet diverse requirements of farm livelihood while preserving resource base and maintaining a high level of environment quality. On the other hand, a farming system is the result of complex interactions among a number of interdependent components, where an individual farmer allocates certain quantities and qualities of four factors of production, viz., land, labour, capital and equipments to which he has access (Mahapatra, 1994).

Farming system represents an appropriate combination of farm enterprises (cropping systems, horticulture, livestock, fishery, forestry, and apiculture) and the means available to the farmer to raise them for profitability. It interacts adequately with environment without dislocating the ecological and socioeconomic balance on one hand and attempts to meet the national goals on the other (Jayanthi et al., 2002). Thus, integrated farming is a system which tries to imitate the nature's principle, where not only crops but, varied types of plants, animals, birds, fish and other aquatic flora and fauna are utilised for production. These are combined in such a way and proportion that each element helps the other; the waste of one is recycled as resource for the other through 'low external input and sustainable agriculture' (LEISA) techniques (Das, 2013).

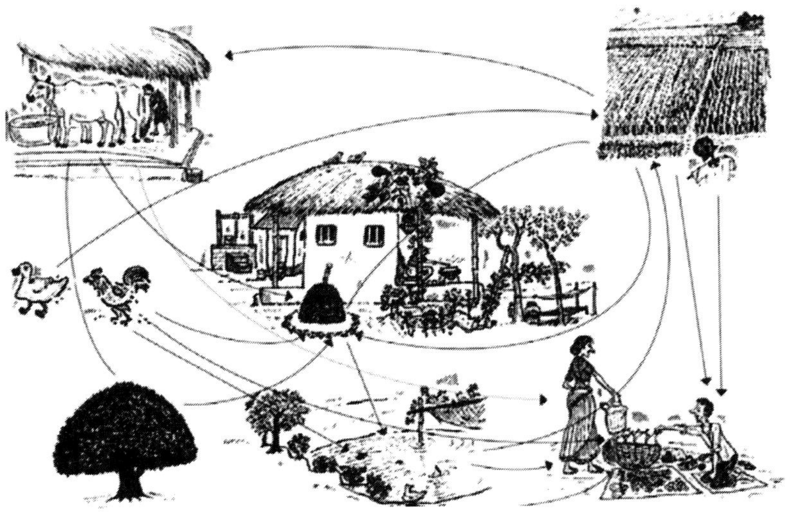

Fig. 12.1. Integration of several components on the farm- a schematic representation

Source: http://www.agriculturesnetwork.org/magazines/india/family-farming/integrated-farming

Okigbo (1995) defines IFS as a mixed farming system that consists of at least two separate but logically interdependent parts of a crop and livestock enterprises. Enterprises in the integrated farming system are mutually supportive and depend on each other (Csavas, 1992). Mixed farming system consists of components such as crops and livestock that coexist independently from each other. In this farming, integrating crops and livestock serves primarily to minimise the risk and not to recycle resources. But in IFS, crops and livestock interact to create a synergy, with recycling allowing the maximum use of available resources (Sasikala et al., 2015). Crop residues can be used for animal feed, while livestock and livestock by-product production and processing can enhance agricultural productivity by intensifying nutrients that improve soil fertility, reducing the use of chemical fertilisers. A high integration of crops and livestock is often considered as a step forward, but small farmers need to have sufficient access to knowledge, assets and inputs to manage this system in a way that is economically and environmentally sustainable over the long term (FAO, 2001). Thus, IFS is an integrated set of elements/components and activities that farmers perform in their farms under their resources and circumstances to maximise the productivity and net farm income on a sustainable basis (Singh and Ratan, 2009). The integration is made in such a way that the product, i.e., output of one enterprise/ component should be the input for the other enterprises with high degree of complementary effects. The rationale of IFS is to minimise the wastes from the various subsystems on the farm and thus it improves employment opportunities, nutritional security and income of the rural people (Panke et al., 2010).

CHARACTERISTICS OF FARMING SYSTEMS

The term 'farming system', in its broadest sense, is any research that views the farm in holistic manner and considers interactions (between the components and of components with the environment) in the system (CGIAR, 1978). This type of research is most appropriately carried out by interdisciplinary teams of scientists who in association with extensionists, continually interact with the farmers in the identification of problems and in devising ways to solve them. It aims at generating and transferring technologies to increase the resource productivity for an identified group of farmers.

The farming system research activities are to be farmer-oriented and system-oriented with problem-solving and interdisciplinary approach. The farming system research also compliments mainstream disciplinary research, tests the technology in on-farm trials and provides feedback from the farmers. The strategy of farming system research should emphasise that the research agenda should be determined by explicitly farmers' needs through an understanding of the existing farming systems rather than its perception by the researchers. The farming systems research and extension should be dealt in holistic manner on farmers' participatory mode with problem-solving approach, keeping genders activity, interdisciplinary and interactive approach.

Farming system research is thus, an approach to agricultural research and development that view the whole farm as a system and focus on the interdependencies between the components under the control of members of the household and how these components interact with each other in respect of physical, biological and socioeconomic factors not under the household's control (Shaner et al., 1982). Many authors define characteristics of farming system research in different ways. The farming system core characteristics may be summarised as below (Behera, 2013; Kumar, 2013).

1. **It is problem-solving:** As an applied problem-solving approach, it emphasises on developing and transferring appropriate technologies to overcome the production constraints through the diagnosis of biophysical, socioeconomic and institutional constraints that influence the technological solutions.
2. **It is holistic:** The whole farm is viewed as a system encompassing interacting subsystems; and no potential enterprise is considered in isolation.
3. **It acknowledges the location specificity of technological solutions:** Recognising the location specific nature of agricultural production problems, it emphasises on testing and adaptation of technological solutions based on agroecological and socioeconomic specificities.
4. **It defines specific client groups:** Emphasis is made on the identification of specific and relatively homogeneous groups of farmers, with similar

problems and circumstances for whom the technology is to be developed, as the specific client group. On the basis of common environmental parameters, production patterns and management practices, relatively homogeneous recommendation domains need to be identified.

5. **It is farmer participatory:** It revolves round the principle that successful agricultural research and development efforts should start and end with the farmers (Rhoades and Booth, 1982). Farmer participation is ensured at different stages of technology generation and transfer processes such as problem diagnosis, design and implementation of on-farm trials, and providing feedback through monitoring and evaluation.

6. **It gives weightage to ITK system:** The indigenous technical knowledge (ITK) which is time-tested at the farmers' level for sustainability through a dynamic process of integrating new innovations into the system as they arise, has to be properly understood by the scientists and utilised in their research activities.

7. **It is concerned with 'bottom-up' research strategy:** It begins with an understanding of the existing farming systems and the identification of key production constraints.

8. **It is interdisciplinary:** It lays greater emphasis on interdisciplinary cooperation among the scientists from different areas of specialisation to solve the agricultural problems that are of concern to farmer.

9. **It emphasises extensive on farm activities:** It involves problem analysis through diagnostic surveys, on-farm testing of the developed technologies, and providing feedback through evaluation to influence the research agenda of the research stations. It provides a structural framework for the farmers to express their preferences and apply their evaluation criteria for selecting technologies suiting to their circumstances.

10. **It is gender sensitive:** While explicitly acknowledging the gender-differentiated roles of farm family in agriculture, it emphasises the critical review of farming systems in terms of activities, analysis, access and control over the resources and benefits, and their implications in developing relevant research agenda.

11. **It is iterative:** Instead of trying to know everything about a system at a time, it requires step-by-step analysis of only key functional relationships.

12. **It is dynamic:** It involves recurrent analysis of the farming systems, permitting continuous learning and adaptations.

13. **It recognises interdependencies among multiple clients:** The generation, dissemination and adoption of relevant technologies to improve

the productivity and sustainability of agriculture require productive and interactive linkages among the policy planners, scientists, developmental agencies and farmers. The approach attaches more importance for this critical factor.

14. **It focuses on actual adoption:** It is to be judged by the extent to which it influences the production of socially desirable technologies that diffuse quickly amongst specified groups of farmer clients.
15. **It focuses on sustainability:** It seeks to harness the strengths of the existing farming practices, and to ensure that the productivity gains are environmentally acceptable. Towards preserving the natural resource base and strengthening the agricultural production base, it attempts to develop technologies that are environment friendly and economically viable.
16. **It complements experiment station research:** It only complements, but does not substitute, the on-station research. It has to draw upon the scientific knowledge and technologies generated at research station. It has to be kept in mind that approach is not being promoted as panacea for all the maladies of local agricultural production systems.

SCOPE OF IFS

Farming enterprises include crops, livestock, poultry, fish, tree crops, plantation crops, etc. A combination of one or more enterprises with cropping, when carefully chosen, planned and executed, gives greater dividends than a single enterprise, especially for small and marginal farmers. Farm as a unit is to be considered and planned for effective integration of the enterprises to be combined with crop production activity (CARDI, 2010). Integration of farm enterprises depends on the following factors.

1. Soil and climatic features of the selected area
2. Availability of resources, land, labour and capital
3. Present level of utilisation of resources
4. Economics of proposed integrated farming system
5. Managerial skill of the farmer

There are a number of situations and conditions that can be alleviated by an integrated farming system. The following situations are ideal for the introduction of IFS (CARDI, 2010).

1. The farmer wishes to improve the soil quality
2. The farm household is struggling to buy food or below the poverty line
3. Water is stored on-farm in ponds or river-charged overflow areas

4. Fertilisers are expensive or the recommended blend is unavailable
5. Soil salinity has increased as a result of inorganic fertiliser use
6. The farmer is seeking to maximise profits on existing holding
7. The farm is being eroded by wind or water
8. The farmer is looking to reduce chemical control methods
9. The farmer wants to reduce pollution or waste disposal costs

In agriculture, though crop production is the main activity, the income obtained from it may hardly be sufficient to sustain the farm family throughout the year. Assured regular cash flow is possible when the crop is combined with other enterprises. Judicious combination of enterprises, keeping in view of the environmental conditions of a locality will pay greater dividends. At the same time, it will also promote effective recycling of residues/wastes.

A grower can use IFS to

1. Improve productivity,
2. Regulate nutrient and material flows,
3. Increase on-farm biodiversity,
4. Limit disease, and
5. Reduce the smell of some livestock operations.

GOALS OF FARMING SYSTEMS RESEARCH

Farming system research aims to meet the following interrelated goals (Plucknett et al., 1987).

1. To understand the physical and socioeconomic environment within which agricultural production takes place.
2. To gain an understanding of the farmer in terms of his or her skills, constraints, preferences and aspirations.
3. To comprehend and evaluate existing important farming system, in particular the practice and performance of these systems.
4. To improve the identification of problem and opportunities for change in existing farming system and thereby better focus research on specific key aspects.
5. To enhance the capacity of research organisations to conduct research on priority problems.
6. To conduct research on new or improved practices or principles and to evaluate these for possible testing on-farms.

7. To evaluate new or improved systems or system components on-farms in major production areas under normal farm conditions.
8. To assist the extension, monitor the adoption and assess the benefits of improved farming systems.

OBJECTIVES OF IFS

The overall objective of integrated farming systems is to evolve technically feasible and economically viable farming system models by integrating cropping with allied enterprises for irrigated, rainfed, hilly and coastal areas with a view to generate income and employment from the farm. The major objectives of integrated farming systems can be listed as below (CARDI, 2010; Behera, 2013).

1. Maximisation of yield of all component enterprises to provide steady and stable income at higher levels.
2. Rejuvenation/amelioration of system's productivity and achieve agroecological equilibrium.
3. Control the build-up of insect pests, diseases and weed population through natural cropping system management and keep them at low level of intensity.
4. Reduction in the use of chemical fertilisers and other harmful agrochemicals and pesticides to provide pollution free, healthy produce and environment to the society at large.
5. Utilisation and conservation of available resources and effective recycling of farm residues within system and to maintain sustainable production system without damaging resources/environment.

ADVANTAGES OF IFS

Farming systems approach introduces a change in farming techniques for higher production from the farm as a whole with the integration of all the enterprises. The farm products other than the economic products, for which crops are grown, can be better utilised for productive purposes in the farming systems approach. A judicious mix of cropping system with associated enterprises like dairy, poultry, fishery, apiary, sericulture, piggery, etc. suited to the given agroclimatic conditions and socioeconomic status of farmers would bring prosperity to farmer (Panda, 2004).

Integrated farming system has the advantages of increasing economic yield per unit area per unit time, profitability, sustainability and provides balanced nutritious food for the farmers, pollution free environment and provide opportunity for effective recycling of one product as input to other component, money round the year and solve the energy, fodder, fuel and timber crisis, avoids degradation of forests and enhance the employment generation, increase input use efficiency and finally improve the livelihood of the farming community (Suba Reddy and Ramakrishna, 2005). Thus the IFS is part of the strategy to ensure sustainable use of the natural resources for the benefit of present and future generations. The major benefits of integrated farming system can be summarised as below.

1. **Productivity:** IFS improves space utilisation and increase productivity per unit area. Thus it provides an opportunity to increase economic yield per unit area per unit time by virtue of intensification of crop and allied enterprises.
2. **Profitability:** In IFS waste materials of one component are used as inputs of another component at the least cost. Thus cost of production is reduced and interference of middleman in procuring most of the inputs is eliminated. Therefore, return per rupee invested or benefit to cost ratio in the farming is increased.
3. **Sustainability:** In integrated farming systems, organic supplementation through effective utilisation of by-products of linked components as a measure is possible and this provides opportunity to promote soil health and to sustain the potentiality of the soil for much longer periods.
4. **Balanced food:** In IFS, components of different nature are linked enabling production of different sources of nutrition, namely, protein, carbohydrates, fats, minerals, vitamins, etc. from the same unit. It provides opportunity to mitigate malnutrition problem of the farm family.
5. **Pollution abatement:** In crop based activity, some of the organics are left as waste materials which in turn pollute the environment on decomposition. Application of huge quantities of fertilisers, pesticides, herbicides, insecticides, etc. pollutes soil, water and air. Much of the wastes could be converted/ recycled to some other forms of economic/ecological/social value, under the integrated farming system. Integrated farming will also help in environmental protection through effective recycling of waste from animal activities like piggery, poultry and pigeon rearing.
6. **Recycling:** Effective recycling of waste materials (crop residues and livestock wastes) is possible in IFS. Therefore, there is less reliance to outside inputs like fertilisers, agrochemicals, feeds, energy, etc.
7. **Income round the year:** IFS provides diversified products. It provides flow of money to the farmer round the year due to integration of enterprises like crops, birds, animals, mushroom, honey, fish, etc. Thus it provides opportunities as crop insurance cover as money round the year is obtained from different farm produces. There is higher net return to land and labour resources of the farming family.

8. **Adoption of new technology:** Resources farmer (big farmer) fully utilise technology. On the other hand money flow in IFS round the year gives an inducement to the small/ marginal farmers to go for the adoption technologies.IFS promotes technology infusion through research and development (R&D) integrated with indigenous/traditional knowledge.
9. **Saving energy:** IFS provides an opportunity to identify an alternative source to the fossil energy. Effective recycling technique of the organic wastes available in the system can be utilised to generate biogas.
10. **Meeting fodder crisis:** Every piece of land area is effectively utilised. Plantation of perennial legume fodder trees on field borders and farm boundary will greatly relieve the problem of non-availability of quality fodder to the farm animals. This practice also improves soil fertility through atmospheric nitrogen fixation. Cultivation of fodder crops as intercropping and as border cropping will also result in the availability of adequate nutritious fodder for animal components like milch cow, goat/sheep, pig and rabbit.
11. **Solving fuel and timber crisis:** The production level of fuel and industrial wood can be enhanced by integrating suitable perennial tree species in farming system (agroforestry) without affecting the crop productivity. This will also greatly reduce pressure on forest resource, thus it preserves the natural ecosystem.
12. **Employment generation:** Integrating crop with livestock enterprises would increase the labour requirement significantly and would help in reducing the problems of unemployment to a great extent. IFS provides enough scope to employ family labour round the year.
13. **Agroindustries:** Integrated farming also provides opportunities for agri-oriented industries, tourism and related tourism based activities. When produce of any component of the IFS is increased to commercial level there will be surplus for value addition, leading to development of allied agro-industries.
14. **Increasing input efficiency:** IFS provides good scope to use inputs in different components with greater efficiency and benefit cost ratio.
15. **Improving/maintaining soil health:** IFS improves soil fertility and soil physical structure through practice of appropriate crop rotation, cover crop, green manuring, composting farm organic wastes and recycling crop residues. Soil erosion can also be checked by agroforestry and proper cultivation of each part of land by integrated farming.
16. **Reduction of pests:** Adoption of appropriate crop rotation in the farming system can reduce the infestation of weeds, insect pests and diseases.

COMPONENTS OF IFS

Farmers make decisions about what to grow, what animals to keep, the level and type of inputs and the methods they will use. Their decisions are based upon a range of social, economic and environmental factors. The farmers' attitudes and level of knowledge are also important. However, the selection of enterprises must be based on cardinal principle of minimising the competition and maximising the complementarities between the enterprises (Annadurai et al., 1994). The productivity of farming system depends upon natural resource climate, available science and technology, policies, institutions and public goods, besides latest information of market technology and human resources. The following factors influence the choice of modules in an IFS.

1. Climatic conditions
2. Soil type
3. Farmers' preferences
4. Size of the farm
5. Knowledge, skill and technology
6. Storage, transport and marketing
7. Resource mobilising power
8. Credit facilities available
9. Socioeconomic conditions
10. Customs, sentiments and believes

The integrated farming system is in vogue from time immemorial, but needs systematic approach starting from diagnosis of the constraints, selection of enterprises to impact analysis in farmers' participatory mode. The focus of present-day agricultural development programmes should be reoriented and changed from commodity based to system-based approach which can gradually be facilitated by capacity building through vocational training and establishment of demonstration units. IFS is an integrated resource management strategy for obtaining economic and sustained crop and livestock production and preserving the resource base with high environmental quality. The entire philosophy of farming system revolves around better utilisation of time, money, resources and family labour. Integrated farming system is focused around a few selected, inter-dependent, interrelated and often interlinking production systems based on few crops, animal and related subsidiary enterprises. It involves the utilisation of primary and/or secondary produce of one system as raw material for other system, thus making the entire system mutually interdependent as one whole unit. In this context there is a necessity to develop models on integrated farming system for different categories of farmers under various agroecological zones. In the process of development of models, the following points should be taken into consideration.

- Sustainability of agriculture depends on the adoption of suitable farming system with inclusion of crop, animal and other allied enterprises.
- Appropriate utilisation of family labour, scientific management of available resources and recycling of agricultural waste by way of integrating different enterprises will make farming a profitable business.
- The farmers should be empowered with technological knowledge and marketing intelligence.

The integrated farming systems include the following components/modules.

1. Crops and cropping systems
2. Horticulture (vegetables, fruits, floriculture, kitchen/nutritional garden)
3. Livestock (cow, buffalo, sheep, goat, pig, etc.)
4. Fishery
5. Poultry (chicken, duck, quail, guinea fowl, pigeon, turkey, emu)
6. Agroforestry
7. Boundary plantation
8. Beekeeping
9. Mushroom cultivation
10. Biogas
11. Composting and vermicomposting
12. Vocational activities like leaf-plate making, bamboo/cane work, chalk/candle making, tailoring, embroidery, food processing, doll making, carpentry, hand paper, packaging, paper packet, spice making, *papad* making, , marketing agricultural produce, etc.

The crop activities in the IFS consist of grain crops (rice, wheat, maize, sorghum, pulses, soybeans), oilseeds (groundnut, sesame, rapeseed and mustard, linseed, niger, sunflower, castor), vegetable crops, plantation crops (coconut, arecanut), short duration fruit crops (papaya, banana, citrus, pineapple), root crops (cassava, sweet potato), sugarcane, tree crops (moringa, mulberry, teak, acacia, sissoo) and fodder crops. The selection of crops is dependent on preferences based on family consumption, market, soil type, and rainfall and type of animals raised.

Integrated farming system comprising crop and livestock has been sustainable over the centuries. In this system animals are raised on agricultural wastes and animal power is used for agricultural operations and animal excreta are used as manure and fuel. In coastal areas, the rice + fish system can stabilise the productivity and profitability in rice production areas. Cereal based crops in combination with livestock (poultry, dairy, sheep and goat) systems can improve the profitability of the farmers along with the sustenance of natural resources. The livestock activities in IFS consist of cows, goat, sheep, poultry, ducks, pigs and small ruminants. The selection of livestock is also dependent on preference based on family consumption, potential market and availability of resources. Livestock and livelihoods are intimately related and the ownership of livestock is more egalitarian than that of land (Swaminathan, 2010). The livestock component acts as a stabilising factor in the system, thus needs strengthening of the crop and livestock linkage to enhance the economic viability and sustainability of the farming systems (Baishya et al., 2007).

Crop-livestock integration involves the natural resources (crops, animals, land and water) in which these subsystems and their synergistic interactions have a significant positive and the greater total effect than the sum of their individual efforts. These systems increase crop yields and soil fertility and improve the livelihoods with promotion of stable households and increased economic output. The livestock farming systems can be grouped into small holder livestock production with little and low land and commercial livestock production dominating with poultry and dairy. The livestock production in IFS should conserve the natural resource base and raise the productivity through efficient utilisation, optimised allocation and rational management of available of resources.

Integrated farming systems are often less risky and, if managed efficiently, benefits from synergisms among enterprises, diversity in produce, and environmental soundness (Mahapatra, 2010). On this basis, IFS models have been suggested by several workers for the development of small and marginal farms across the country (Behera and Mahapatra, 1999; Singh et al., 2006). Dominant farming systems of the different agroclimatic regions are presented in the following table.

Table 12.1. Dominant farming systems in different agroclimatic regions of the country

Agroclimatic region	States	Most dominate farming systems in the region
Western Himalayas	J & K, Himachal Pradesh, Uttarakhand	1. Crops + dairy 2. Agrihorti system
Eastern Himalayas	Assam, Meghalaya	1. Crops + fish + cattle + piggery 2. Monocropping of rice and maize + piggery + backyard poultry + 2-3 cows
Trans Gangetic Plain	Punjab, Haryana	1. Crops + dairy
Upper Gangetic Plain	Uttar Pradesh	1. Crops + dairy
Middle Gangetic Plain	Uttar Pradesh, Bihar	1. Crops + dairy 2. Crops + fishery
Lower Gangetic Plain	West Bengal	1. Crops + dairy 2. Crops + dairy + poultry + duckery
Easter Plateau & Hills	Chhattisgarh, Jharkhand	1. Crops + dairy 2. Crops + backyard poultry + fish
Central Plateau & Hills	Madhya Pradesh	1. Crops + dairy
Western Plateau & Hills	Maharashtra	1. Crops + dairy 2. Crops + goatery 3. Crops + horticulture
Southern Plateau & Hills	Andhra Pradesh, Telengana, Tamilnadu, Karnataka	1. Crops + dairy 2. Crops + horticulture 3. Crops + horticulture + dairy
East Coast Plain & Hills	Odisha	1. Crops + dairy 2. Crops + dairy + fish
West Coast Plain & Hills	Maharashtra, Goa, Kerala	1. Crops + dairy 2. Coconut based homestead farming 3. Rice based farming systems
Western Dry Region	Rajasthan	1. Crops + dairy
Gujarat Plain & Hills	Gujarat	1. Crops + dairy
Islands	Andaman & Nicobar	1. Plantation crops + piggery 2. Crops + cattle + fish

Source: Singh, 2010

Low-cost interventions for IFS modules

A number of 'no cost' or 'low cost' interventions may be made in the modules of IFS for increasing overall productivity and profitability of the marginal and small farmers.

- Development nutritional garden and growing vegetables throughout the year with HYV and hybrids
- Growing climbing vegetable crops on trellis
- Climbers like bottle gourd and ash gourd trailing to the thatches (roof tops) of residential house and cattle shed
- Pruning of fruit trees like mango, guava and cashew trees, pruned materials used as fuel for cooking, as a substitution to the dry cow dung
- Country bean trailing to the fruit crops during *kharif*
- Use of interspace of fruit plant rows in orchards for fodder or pineapple cultivation
- Yam cultivation in fruit orchards
- *Azolla* culture in small pits for feeding dairy animals and poultry birds
- Feeding animals with the leaves of maize plants after the green cobs are harvested
- After harvest of cabbage and cauliflower the leaves are fed to animals
- Use of pulse by-product as cattle feed
- Artificial insemination of cows
- Introduction of coloured dual purpose poultry breed like Chhabro
- Paddy chaffs used as poultry house bedding material
- Use of chopped inflorescence and pseudo-stem of banana plants as poultry feed
- Use of leaves of winter vegetables like cauliflower, cabbage, radish, etc. as poultry feed
- Use of cow dung as fish feed
- Use of weed biomass for compost and vermicompost preparation
- Composting and vermicomposting of all types of organic farm wastes and by-products
- Value addition like preparation and marketing of ghee, *badi*, *papad*, pickle, chips, etc. out of farm produce

CROP DIVERSIFICATION FOR SUSTAINABILITY

In agriculture, crop production is the main activity. But in many cases, it is not sustainable mainly due to unscientific crop production practices, indiscriminate use of natural resources, and non-judicious use of available inputs with the grower. Thus, crop diversification with a more scientific approach is required to make the crop production sustainable. It intends to give a wider choice in the production of a variety of crops in a given area so as to expand production related activities on various crops and also to lessen risk.

Crop diversification can be considered as an attempt to increase the diversity of crops through, e.g., crop rotation, multiple cropping or intercropping, compared to specialised farming with the aim to improve the productivity, stability and delivery of ecosystem services (Garbach et al., 2017). It can be one measure to develop more sustainable production systems, develop value-chains for minor crops (Meynard et al., 2018) and contribute to socioeconomic benefits (Feliciano, 2019). Crop diversification practices can include higher crop diversity (Renard and Tilman, 2019), more diverse crop rotations (Reckling et al., 2016), mixed cropping (Bedoussac et al., 2015), cultivation of grain legumes in otherwise cereal dominated systems (Watson et al., 2017) and regionally adapted varieties or variety mixtures (Vijaya et al., 2019). Crop diversification and/or additional diversification measures like variation of seeding time or changing cropping patterns have the potential to lead to higher and more stable yields, increase profitability and lead to greater resilience of agroecosystems in the long term (Liu et al., 2019; Renard and Tilman, 2019). These practices have the potential to make cropping systems more diverse in space, time and genetics. Consequences of diversification are temporal shifts and ranges of phenological stages (relevant for biodiversity and adaptation to climate change), more frequent or continuous soil cover and more diverse management strategies, i.e., 'tillage', 'sowing dates', 'fertilisation', 'irrigation', 'harvesting' and also reducing labour peaks and economic risk (Hufnagel et al., 2020).

In Indian context, crop diversification is a demand driven, need based situation specific and national goal seeking continuous and dynamic concept and involves spatial, temporal, value addition and resource complementary approaches (Nishan, 2014). Diversified farms are usually more economically and ecologically resilient. While monoculture farming has advantages in terms of efficiency and ease of management, the loss of the crop in any one year could put a farm out of business and/or seriously disrupt the stability of a community dependent on that crop. By growing a variety of crops, farmers spread economic risk and are less susceptible to the radical price fluctuations associated with changes in supply and demand (Rana and Rana, 2011).

With the advent of modern agricultural technology, especially during the period of the 'green revolution' in the late sixties and early seventies, there is a continuous surge for diversified agriculture in terms of crops, primarily on economic considerations. The crop pattern changes, however, are the outcome of the interactive effect of many factors which can be broadly categorised into the following five groups (Hazra, 2001).

1. Resource related factors covering irrigation, rainfall and soil fertility.
2. Technology related factors covering not only seed, fertiliser, and water technologies but also those related to marketing, storage and processing.
3. Household related factors covering food and fodder self-sufficiency requirement as well as investment capacity.
4. Price related factors covering output and input prices as well as trade policies and other economic policies that affect these prices either directly or indirectly.
5. Institutional and infrastructure related factors covering farm size and tenancy arrangements, research, extension and marketing systems and government regulatory policies.

All these five factors are interrelated. The adoption of crop technologies is influenced not only by resource related factors but also by institutional and infrastructure factors. Similarly, government policies, both supportive and regulatory in nature, affect both the input and output prices. Likewise, special government programmes also affect area allocation and crop composition. Although the factors that influence the area allocation decision of farmers are all important, they obviously differ in terms of the relative importance both across farm groups and resource regions. While factors such as food and fodder self-sufficiency, farm size, and investment constraints are important in influencing the area allocation pattern among smaller farms, larger farmers with an ability to circumvent resources constraints usually go more by economic considerations based on relative crop prices than by other non-economic considerations. Economic factors play a relatively stronger role in influencing the crop pattern in areas with a better irrigation and infrastructure potential. However, the relative importance of these factors change over time. From a much generalised perspective, Indian agriculture is increasingly getting influenced more and more by economic factors. This need not be surprising because irrigation expansion, infrastructure development, penetration of rural markets, development and spread of short duration crop varieties and drought resistant crop technologies have all contributed to minimising the role of non-economic factors in crop choice of even small farmers (Hazra, 2001).

Approaches to crop diversification

There are several approaches to crop diversification for achieving the sustainability in agriculture.

1. The overall productivity of a farm may be increased by crop intensification with addition of more number of crops (intercropping, mixed cropping, sequential cropping, multitier cropping, etc.) to the existing crops or cropping systems.
2. The crops which are less suitable or productive under an agroecological region may be substituted with more suitable alternate crops. More remunerative crops but with high risk to a drought prone area should be substituted with low risk crops like short duration pulses and drought resistant oilseed crops. Enough elasticity should be kept in rotation so that if pest or diseases destroys a crop, another crop can be substituted.
3. Selection of crops should be based on land type and situation. Erosion resisting crops like legumes and fodder grasses that cover the land fast should be taken on sloppy lands which are prone to soil erosion.
4. Fertile and well-drained land should be utilised for important food-crop rotation, and less fertile land may be diverted for restorative crops or soil improving crops like legumes and crops with high foliage. Similarly, tolerant crops should be selected for problem soils like acidic, saline or alkali soils.
5. Conventional crops sometimes fail to provide stability of production over a period of time due to abiotic stresses. Tree-based cropping systems like alley cropping, agrisilvicultural or silvipastoral or silvihortipatoral systems are more sustainable in dry areas.
6. Marginal lands are permanently disadvantaged in terms of steep slopes, poor drainage, poor soil depth and harsh environments. Such lands can be developed into pastures or can put to use for tree farming. Ley farming, i.e., growing of grass or legumes in rotation with grain crops may also practised in these lands. Grasses improve soil structure while the legumes enrich soil nitrogen status. Grasses such as *Cenchrus* sp., *Chloris* sp., *Panicum* sp. can also be grown in mixture with perennial pasture legumes.
7. Degraded lands may be used with suitable tree plantation. Legume tree fodder trees should be preferred.
8. Low water requiring crops should replace the high water-demanding crops in dryland areas and water scarce regions.
9. Two or three crops in sequence are usually taken in a year under irrigated conditions. However, a dry crop should be included in the rotation to avoid

damage to the soil due to continuous irrigation. Rice–rice system should be discouraged.

10. The cropping intensity of a region varies with the amount of annual rainfall received. Intercropping or mixed cropping with the companion crops having varying peak moisture-demanding period may be encouraged in the areas where moisture may be enough to produce one crop but not enough to produce two in a sequence.
11. Root crops are well adapted to the humid lowland than non-rice cereals and legumes since high rainfall and humidity often adversely affect reproduction, ripening, drying and storage and increase pest and disease problems of cereals and grain legumes. In areas where drainage is not a problem, vegetables may be sustainable.
12. Low yielding varieties can be substituted with high yielding varieties to sustain the production. Always new high yielding varieties or hybrids which are resistant or tolerant to biotic and abiotic stresses should be given priority.
13. All the principles of crop rotation should be strictly followed to have sustainable production over the years.
14. Selection of the crops should always be based on soil, climate and market demand.

Advantages of an ideal crop diversification

An ideal crop diversification brings sustainability in agriculture with the following major advantages.

1. Inclusion of crops of different feeding zone and nutrient requirement could maintain the better balance of nutrient in soil. Growing crops of different root depths avoids continuous depletion of nutrients form same depth.
2. Crops requiring high irrigation when followed by crops requiring light irrigation, the overall water use efficiency and water productivity of the system is increased.
3. It improves soil structure, percolation and reduces chances of creation of hard-pan in subsoil and also reduces soil erosion.
4. There is an overall increase in yield of crops mainly due to maintenance of physical and chemical properties of soil. Soil fertility is restored by maintaining more organic matter, encouraging microbial activity and biological cycles and protecting the soil from erosion, salinity and acidity.
5. It helps in controlling insects, pests and soil borne diseases. It also controls weeds.

6. Diversification of crops reduces risk of financial loss due to unfavourable conditions.
7. The family needs of feed, food, fuel, fibre, etc. are fulfilled.
8. It facilitates even distribution of labour.
9. Farmers get a better price for his produce due to higher demand in local market. So there is regular flow of income over years.

Constraints in crop diversification

The major problems and constraints in crop diversification in India are primarily due to the following reasons with varied degrees of influence (Hazra, 2001).

1. Over 117 m ha (63%) of the cropped area in the country is completely dependent on rainfall.
2. Suboptimal and overuse of resources like land and water resources, causing a negative impact on the environment and sustainability of agriculture.
3. Inadequate supply of seeds and plants of improved cultivars.
4. Fragmentation of land holding less favouring modernisation and mechanisation of agriculture.
5. Poor basic infrastructure like rural roads, power, transport, communications, etc.
6. Inadequate post-harvest technologies and inadequate infrastructure for post-harvest handling of perishable horticultural produce.
7. Very weak agro-based industry.
8. Weak research - extension - farmer linkages.
9. Inadequately trained human resources together with persistent and large-scale illiteracy amongst farmers.
10. Host of diseases and pests affecting most crop plants.
11. Poor database for horticultural crops.
12. Decreased investments in the agricultural sector over the years.

Alternate cropping systems

Several alternate cropping systems have been identified and recommended by All India Coordinated Research Project on Cropping Systems for various agroclimatic regions of India. The yield increase over existing systems with alternate efficient cropping systems was found to be 40 to more than 300% in various agroclimatic zones.

Table 12.2. Alternate cropping systems under various agroclimatic zones

Agroclimatic zones	Location	Existing system	Alternative systems	% increase over existing system
Western Himalayas	Dhansaur	Rice–wheat	Rice–potato–onion	230.7
		Maize–wheat	Maize–potato–onion	338.5
	Pantnagar	Rice–wheat	Rice–potato–vegetable cowpea	123.3
Eastern Himalayas	Karimganj	Rice–potato	Rice–rajmah	42.5
Lower Gangetic Plains	Kakdwip	Rice–greengram	Rice–okra	106.9
Middle Gangetic Plains	Patna	Rice–wheat	Rice–wheat–greengram	59.8
Upper Gangetic Plains	Saini	Rice–wheat	Sesame–pea	97.8
Eastern Plateau and Hills	Kawardha	Soybean–gram	Soybean–tomato–cowpea	244
	Dhenkanal	Rice–greengram	Rice–tomato	116.9
	Gondia	Rice–wheat	Rice–wheat–cowpea	75.3
Central Plateau and Hills	Udaipur	Maize–wheat	Maize–wheat–okra	359.6
Western Plateau and Hills	Aurangabad	Soybean + pigeonpea	Soybean–gram–fodder maize	208.4
	Ahmednagar	Soybean–wheat	Soybean–onion	238.9
Southern Plateau and Hills	Warrangal	Rice–rice	Rice–maize	39.9
	Bangalore	Rice–finger miller	Rice–brinjal	250.7
	Paiyur	Rice–rice	Rice–tomato	42.2
East Coast Plains and Hills	Kendrapara	Rice–greengram	Rice–bittergourd	88.9
West Coast Plains and Ghats	Thiruvalla	Rice–rice	Rice–rice–okra	243.5
	Roha	Rice–cowpea	Rice-maize	105.7
Gujarat Plains and Hills	Thasara	Tobacco	Tobacco–fodder bajra	169.6
		Bajra–wheat	Bajra–lucerne (seed)	129.5
	Deesa	Pearlmillet–mustard	Greengram–fennel	55.5
		Castor	Fennel–fodder pearlmillet	41.1

Source: Gangwar et al., 2012

FRUIT CROPS IN IFS FOR SUSTAINABILITY

Fruit cultivation and tree based farming plays very important role in imparting sustainability to the integrated farming system. The advantages of fruit crops in integrated farming system may be summarised as follows.

1. Trees trap moisture from deeper layers of soil during drought due to their deep root system. They tolerate adverse weather conditions better than field crops. If annual crops fail due to adverse weather, tree component imparts stability to farmer's economy.
2. Root system of fruit trees penetrate into very deep in soil layers and thus draw plant nutrients from the zones that is not usually available to the herbaceous crops and would have been lost through leaching.
3. This provides higher system productivity and net return.
4. Trees fulfill diverse needs, viz., food, fuel, fodder, fruit, fibre and timber.
5. Fruit trees utilise the off-season rainfall.
6. Canopy of fruit trees slashes the raindrops and thus soil erosion is reduced in high rainfall areas.
7. Perennial fruit crops in the IFS provide employment opportunity round the year.
8. Pruning of perennial fruit trees provide firewood in rural areas.
9. Fruit trees in the IFS gives aesthetic value,
10. Biodiversity is maintained when perennial fruit crops are integrated in the system.
11. Perennial trees provide shade to household and the microclimate is moderated.

Fruit crops suitable for IFS

Perennial fruit crops can be used as hedges, individual trees or live fences in IFS, depending on the needs of the farmer, the nature of the fruit species, and the other components of the system. A number of fruit crops, viz., mango, jackfruit, custard apple, guava, sapota, litchi, ber, tamarind, lime, pomegranate, anola, etc. can be grown under integrated farming systems. Most of these fruit crops take 5-8 years for full canopy coverage. In all plantations, field crops and fodder crops can be grown as intercrop during initial years. Up to 3-4 years there is no restriction for choice of field crops and crops can be chosen depending upon climatic and edaphic conditions. Once the productivity of the intercrops is decreased these should be replaced with shade tolerant crops like arrowroot, turmeric, ginger, mango-ginger (*Curcuma amada*), *Aloe vera*, etc. These shade tolerant crops can be grown up to 8-9 years of fruit tree plantation and thereafter pineapple can be taken as intercrop in plantation.

Short duration fruit crops like papaya, banana, pineapple and perennial fruit crops like pomegranate, lime, custard apple, guava, ber, etc. can be planted on pond dykes, if fishery is a component in the IFS. Similarly, these fruit crops can also be planted on household boundaries.

Table 12.3. Fruit crops and their popular variety suitable for IFS

Crop	Variety	Spacing (m × m)	Bearing age (years)	Economic life (years)	Yield kg/plant
Mango	Amrapalli, Mallika, Kesar, Neelum, Sindhu, Bombay Green, Totapuri	10 × 10 (tall) 5 m × 5 (dwarf)	2-3	25-30	10 kg in 4th yr to 70 kg at full bearing
Jackfruit	Khajara, Gulabi, Champa, Hazar, NJT 1, NJC 1, Singapore	12 × 12	5-6	15-20	4-8 fruits in the beginning, 60-70 fruit at full bearing
Bael	Narendra bael 1, Narendra bael 2	10 × 10	3-4	25-30	10 kg in 4th yr 70 kg or more at full bearing
Custard apple	Bangalore selection, Saharanpur local, Mehboobnagar local	5 × 5	2-3	12-15	2- 3 kg at 3 years and 30 kg at full bearing
Guava	Allahabad Safed, Lucknow 49, Banaras Round	5-8 × 5-8	2	12-15	10 kg in 4th yr to 70 kg at full bearing
Aonla	Chakaiya, Banarasi, Krishna, Kanchan, NA 6, NA 7	7-10 × 7-10	3-4	15-20	5 kg in 4th yr 50 kg or more at full bearing
Pineapple	Queen, Kew, Jiant kew, Mauritious	0.7 × 0.6	10-12 months	03-04	800-1200 g fruit/ plant
Papaya	Pusa Delicious, Pusa Majesty, Pusa Dwarf, Pusa Nanha, Coorg Honey Dew, Taiwan	1.2-1.8 × 1.2-1.8	06-08 months	02-03	15-25 kg / plant up to 3-4 years
Banana	Dwarf Cavendish, Robusta, Grand Naine	1.8 × 1.8 (dwarf) 2.0-2.5 × 2.0-2.5 (tall)	12-14 months	02-03	15 -20 kg bunch weight/plant

Management of fruit trees in IFS

Certain criteria must be followed for the selection of fruit trees for tree based farming systems.

1. The crop should have a commercial potential at the local and regional level.

2. Its growth habit must be compatible with other crops or trees in the system, for example a crown that allows light to reach the associated crops.
3. Fruits should have the ability for prolonged storage and post-harvest processing and transport.
4. The individual fruit tree should have the potential for high production of fruits or biomass.
5. Fruit trees should not act as a host to pests and diseases that are a threat to the associated crops in the system.
6. These should have deep rather than surface root system.
7. Fruits should have the desirable characteristics for industrial uses, for example, pectin content for jelly making.
8. Fruit crops should possess the easier vegetative propagation methods.
9. Fruits should have good flavour and high quality.
10. There should have easy fruit harvesting methods.
11. Fruit crops should have short period of establishment.
12. There should be low incidence of pests and diseases.
13. Fruit trees should have multiple uses.

Optimising the use of light, water and nutrients during the formation and development of the fruit increases the quantity and quality of the fruit produced. Therefore, treatments to improve the capture of solar energy and application of water and plant nutrients will maximise the number of fruits that reach maturity. Abortion of fruits and flowers occurs when the inflorescence does not receive sufficient resources.

The main objective of fruit tree management in an integrated farming system is to maximise the value of fruit production without reducing the agricultural production significantly. The trees are to be managed in such a way that the agricultural crops do not suffer from inadequate levels of solar radiation, plant nutrients and soil moisture due to competition from the fruit crop component. Agricultural production is expected throughout the life of the fruit trees in IFS.

Pruning is the removal of plant parts to improve the form and growth of fruit trees. Branches are removed with minimum damage to cambium or growing tissue so that the wound will close in the shortest period of time and with the least possibility of wound infection. The major objectives of pruning are for training the plant, maintaining plant health, improving the quality of flowers, fruit, foliage and stems, and controlling growth. Pruning also promotes regular flowering and fruiting, and maintains a low crown that facilitates fruit harvest and the control of disease. Old, sick, damaged and diseased branches and sprouts should be pruned regularly. The season for pruning depends on the seasons for tree growth and development and on the climatic conditions of the site. It is better not to prune during the dry season when the crops need protection from the summer sun.

LIVESTOCK BASED IFS FOR SUSTAINABILITY

Livestock based integrated farming system provides an opportunity of increasing economic yield per unit area per unit time to small and marginal farmers. In this system, waste materials are effectively recycled by linking appropriate components. Thus, it minimises the environment pollution. Recycling of product, by-products and waste material in integrated farming system are the factors responsible for the sustainability of farming system. Due to integration of different livestock components with crops, production of eggs, meat and milk provide nutritional security and income generation round the year to the farmers (Kochewad et al., 2017).

Fish-livestock farming systems

Fish-livestock farming systems are recognised as highly assured technology where predetermined quantum of livestock waste obtained by rearing the livestock in the pond area is applied in pond to raise the fish crop without any other additional supply of nutrients. Both production and processing of livestock generate by-products that can be used for aquaculture. Direct use of livestock production wastes is the most widespread and conventionally recognised type of integrated farming. Production wastes include manure, urine and spilled feed; and they may be used as fresh inputs or be processed in some way before use. Based on the type of livestock used for integration there are many combinations in livestock-fish systems.

Cow-fish system: Manuring of fishpond by using cow dung is one of the common practices all over the world. A healthy cow excretes over 4000-5000 kg dung, 3500-4000 litre urine on an annual basis. Manuring with cow dung, which is rich in nutrients results in increase of natural food organism and bacteria in fishpond. A unit of 5-6 cows can provide adequate manure for 1 ha of pond. Cowshed should be built close to fishpond to simplify handling of cow manure. A cow requires about 7000-8000 kg of green fodder annually. Grass carp utilises the left over grasses, which are about 2500 kg. Fish also utilises the fine feed which consists of grains wasted by cows. In place of raw cow dung, biogas slurry could be used with equally good production. Twenty to thirty thousand kg of biogas slurry are recycled in 1 ha water area to get over 4000 kg of fish without feed or any fertiliser application. In addition to 9000 litres milk, about 3000-4000 kg fish/ha/year can be harvested from dairy-fish integration.

Pig-fish system: The waste produced by 30-40 pigs is equivalent to one tonne of ammonium sulphate. Exotic breeds like White Yorkshire, Landrace and Hampshire are reared in pigsty near the fishpond. Depending on the size of the fishponds and their manure requirements, such a system can either be built on the bund dividing two fishponds or on the dry-side of the bund. Pigsties, however, may also be constructed in a nearby place where the urine and dung of pigs are first allowed to the digestion chambers of biogas plants for the production of methane for household use. The liquid manure is then discharged into the fishponds through small ditches running through pond bunds. Alternately, the pig manure may be heaped in localised places of fishponds or may be applied in fishponds by dissolving in water. Pig dung contains more than 70% digestible feed for fish. The undigested solids present in the pig dung also serve as direct food source to tilapia and common carp. Forty pigs are enough to fertilise a fishpond of one hectare area. Fish like grass carp, silver carp and common carp are suitable for integration with pigs.

Chicken-fish system: Chicken raising for meat or eggs can be integrated with fish culture to reduce costs on fertilisers and feeds for fish and maximise benefits. Chicken (cock and hen) can be raised over or adjacent to the ponds and their excreta recycled to fertilise the fishponds. Chicken housing, when constructed above the water level using bamboo poles would fertilise fishponds directly. In fish-chicken integration, birds housed under intensive system are considered best. About 6-8 cm thick layer prepared from chopped straw, dry leaves, saw dust or groundnut shell is sufficient. Birds' droppings in the form of litter contain 3% nitrogen, 2% phosphate and 2% potash. Thus, it produces phytoplankton and zooplankton in fishpond. So application of extra fertiliser for fish is not required. This cuts the cost of fish production by 60%. In one year 25-30 birds can produce one tonne litter and based on that it is found that 500-600 birds are enough to fertilise one ha water spread area for good fish production. When phytoplanktonic bloom is seen over the surface water of pond then application of birds' litter to the pond should immediately be suspended. Chicken-fish integration also maximises the use of space and saves labour in transporting manure to the ponds.

Duck-fish system: A fishpond being a semi-closed biological system with several aquatic animals and plants provides excellent disease-free environment for ducks. In return ducks consume juvenile frogs, tadpoles and dragonfly, thus making a safe environment for fish. Duck dropping provides essential nutrients to stimulate growth of natural food for fish. This integration is highly profitable as it greatly enhances the animal protein production in terms of fish and duck per unit area. The duck dropping contains 25% organic and 20% inorganic substances with a number of elements such as carbon, phosphorus, potassium, nitrogen, calcium, etc. Hence, it forms a very good source of fish feed. Besides manuring, ducks eradicate the unwanted insects, snails and their larvae which may be the vectors of fish pathogenic organisms and water-borne disease-causing organisms infecting human beings. The number of ducks may be between 100 and 3000/ha depending on the duration of fish culture and the manure requirements. For culturing fish with ducks, it is advisable to release fish fingerlings of more than 10 cm size; otherwise the ducks may feed on the fingerlings. As the nitrogen-rich duck manure enhances both phytoplankton and zooplankton production, phytoplankton-feeding silver carp and zooplankton-feeding catla and common carp are ideal for duck-fish culture.

Livestock-crop production system

An integrated crop-livestock system is a form of mixed production that utilises crops and livestock in a way that they can complement one another through space and time. In livestock-crop system, the animal component is often raised on agricultural waste products while the animal is used to cultivate the land and provide manure to be used as fertiliser and fuel (Jayanthi et al., 2000). The animals like sheep, goats or cattle play important role in enriching the soil gradually and increasing soil organic matter to support the crop. Animal can also be used for farm operations and transport, while crop residues provide fodder for livestock and grain provides supplementary feed for productive animals. Animals also provide manure and other types of animal waste. Cow dung helps in the overall sustainability of the farming system. Dung contains macronutrients and micronutrients required for improving the soil fertility and crop growth. Cow dung is used for the production of biogas. Biogas is a source of renewable, alternative and sustainable energy (Godi et al., 2013).

Small ruminant-crop system: Small ruminants play a vital role in the economy of small, marginal and landless farmers. Integration of these enterprises can increase the total productivity, maintain ecological balance and economic sustainability. Small ruminants based integrated farming system also helps in improving the soil fertility. Small ruminants graze weeds in the crop field. They also directly graze fodder tree/shrubs. Thus, this system saves labours and spread of insect pests is checked. Boundary plantation with fodder shrubs/trees provides high quality leaf fodder for small ruminants.

Rice-duck farming: Rice-duck farming is an integrated type of farming technology. It is especially suitable for resource poor farmers to produce organic rice in low cost. The evidence from various countries including Japan, Bangladesh, Philippines, and Vietnam has proved the integration of ducks in rice field as a successful and productive farming technology. The integration of ducks in rice field creates symbiotic relationship between rice and ducks yielding mutual benefits to both entities. Ducks eat harmful insects and weeds averting the use of chemical pesticides and manual weeding in the rice field. Thus they get nutritious diet from eating insects and weeds in rice field. The droplets of ducks act as natural fertiliser to the rice crop preventing the use of chemical fertilisers. The continuous movement of ducks in the rice field provides natural stimulation and aeration which increases the availability of nutrients like nitrogen, phosphorous and potash to the rice crop. Rice-duck technology causes the reduction of emission of methane gas from rice field contributing to reduce the global warming. Against the traditional rice farming system, integrated rice-duck technology supersedes in terms of minimising the cost of production, increasing rice productivity, providing environmental benefits and increasing the income of farmers through sale of organic rice and duck meat.

Livestock-crop-fish based farming system: Livestock-crop-fish farming systems can be followed by integrating fish with livestock-crop farming system, without any additional feed to fish, rearing fish in the pond with the help of available dung from livestock. Integrated livestock-crop-fish farming can be carried out for increasing returns from a limited land area and reducing risk by diversifying crops. There should be ample supply of fresh and clean water available throughout the year to maintain water level for fish management purposes. The nutrients content in cow dung will help in growth of phytoplankton and zooplankton in the pond. The by-products of livestock can be used for aquaculture. Direct application of wastes of livestock is common practice. The by-products of livestock are manure and urine.

Fruit crops/vegetables on pond dykes: Pond acts as a mini water harvesting structure which holds a considerable amount of rain and runoff water after cessation of rainfall. Farmers can grow crops in the dyke with proper planning and adoption of suitable cropping sequence by utilising the stored water. Fruit crops like coconut, papaya, banana, pomegranate, lemon, and seasonal vegetables like ridge gourd, bottle gourd, bitter gourd, pumpkin, cucumber, etc. can be grown successfully on pond dykes.

BEEKEEPING IN IFS

Beekeeping has been evolved as an excellent enterprise for rural poor and a constituent part of integrated farming system. Beekeeping supplements income and employment generation and nutritional intake of rural population. Honeybees have been offering services to the society through ensured pollination in cross-pollinated crops as well as by providing honey and a variety of beehive products. Honeybees have also vital role in sustaining plants biodiversity resulting in environmental stability (Sarswat, 2017).

Benefits of beekeeping

The direct benefits from apiary are hive products like honey, bee wax, pollen, bee venom, royal jelly whereas indirect benefit is from pollination of different crops. The value of additional yield obtained due to pollination is 15-20 times more than the value of all the hive products put together.

1. Most suited to landless people or with small land holdings
2. Suitable for all irrespective of age bar and gender
3. Does not compete with any branch of agriculture
4. Does not require continuous labour
5. Improve crop yield through cross pollination
6. Unemployed youth can start this business with minimal funds
7. Proper utilisation of natural resources, nectar and pollen otherwise go waste
8. Different sectors and trades benefit from a strong beekeeping industry
9. Beekeeping encourages ecological awareness
10. Export of honey/beehive products attracts foreign exchange
11. It helps in rural development and promotes small village industry

Scientific beekeeping

For sustainable product of honey and other hive products, the following eight steps should be followed in scientific beekeeping.

Step 1: Training

- Honeybee is a highly organised social insect. It is very useful and lot of things are there to be learnt from honeybees.
- Scientific beekeeping helps to protect the bees and simultaneously harvest honey of good quality.
- Scientific beekeeping is very easy but it involves lot of skill, requires clean understanding of bee behaviour and their language. For this training is essential.

Step 2: Knowledge about honeybees

- The grower should get acquainted with various bee species like rock bees, little bees, Indian hive bees, Italian bees and stingless bees and the queen, drone and worker of their colonies.
- Knowledge must be acquired about the working style of the different honeybee species and their division of labour.

- The method of hiving the Indian hive bees, Italian honeybees and the stingless bees and their scientific management practices should be acquired.

Step 3: Selection of site

- Neat and clean area with plenty of fresh water, well ventilation, ample shade and sunshine, away from main road but well communicated and least disturbed area is very ideal for beekeeping.
- Site selected for apiary should have sufficient bee flora within a radius of 1 km and provide nectar and pollen to the bees round the year.
- Site should be away from water logged areas, cattle shade, kitchen/smoky area.

Step 4: Procurement of bee equipments

- ISI 'A' type beehives with 8 frames for Indian hive bees, langstroth hive with 10 frames for Italian honeybees and wooden box of 25 cm × 15 cm × 13 cm size for hiving the stingless bees are necessary.
- Smoker, honey extractor, queen gate, bee veil, gloves, queen excluder sheet, drone trap, knife, and hive tool should be procured prior to beekeeping.

Step 5: Procurement of colony

- To start the beekeeping a colony may be procured by dividing an old colony or capturing a swarm in the season. Beginners may also purchase the bee colonies from a progressive beekeeper.
- The colony should be of four frame colony with young (< 1 year) queen, free of drones and sufficient workers. The frame should have sufficient eggs, broods, good nectar and pollen store.
- Beekeeping should be initiated in the beginning of honey flow season. The ideal time either for procurement or purchase of colony is December-May.

Step 6: Seasonal management

- Regular inspection of the hive at 10 to 15 days interval round the year is required.
- Provisioning of heat in winter, protection from rainwater in monsoon and shade in summer are prime steps in seasonal management.
- Dividing strong colonies, swarm control, mounting super chamber over brood chamber, regular honey extraction, forcing bees for construction of new comb, uniting the weaker colonies is necessary during spring.

- Providing artificial diet, prevention of robbing and management of laying workers during dearth and queenless colony should be done immediately.

Step 7: Protection from enemies and bee disease

- Knowing the enemies of bees like wax moth, wasps, different species of mites, ants, spiders, viral diseases, microorganisms and their symptoms , appropriate preventative/curative method should be adopted,
- Maintenance of strong colony and keeping bottom board clean reduces occurrence of enemies and diseases in bees.

Step 8: Honey extraction, processing and storage

- Harvesting honey from the hive is the chief attraction of a beekeeper. This is normally done in honey flow season,
- Honey is extracted when 70-80% of comb in super chamber is filled with honey and sealed.
- Honey should not be kept exposed for a longer period during harvesting,

MUSHROOM CULTIVATION IN IFS

There are many reasons why mushroom agriculture is more sustainable than other types of agriculture. Farmers use by-products from other agricultural production systems such as cow manure, chicken litter, cotton hulls, paddy straw, etc. for substrate on which mushrooms grow. Then, throughout the growing process, their environmental impact is minimal. Mushrooms require less growing materials, water, and energy than other types of crops. Mushroom farming emits much less carbon dioxide. Mushrooms require less land than other agricultural practices because they are stacked vertically in growing facilities.

Button mushroom, oyster mushroom and paddy straw mushrooms are the three major types cultivated in India. Paddy straw mushrooms can grow in temperatures ranging from 35-40° C. Oyster mushrooms on the other hand, are grown in the northern plains while button mushrooms grow during the winter season. All these mushrooms of commercial importance are grown by different methods and techniques. Mushrooms are grown in special beds known as compost beds.

Management of agricultural waste through mushroom cultivation

Mushroom is a fruiting body of fungi. Mushrooms need substrate to grow, a mushroom substrate is a substance where mushroom mycelium will grow. Mushroom cultivation is landless farming using various substrates like paddy straw, wheat straw, sawdust, and other agricultural residues that are rich in carbon and nitrogen sources. Mushrooms need substrate either composted or non-composted if it is cultivated under the control condition for commercial production. The agricultural residue generated from the cropland after crop harvesting is generally burned in the crop field causing environmental problems. Therefore, this residue alternatively managed by using residues as a substrate for mushroom cultivations. Growing mushroom on agricultural residue generates additional income for the farmers.

Mushroom in supplementation to farmer income

Mushroom cultivation activities can play an important role in supporting the local economy by contributing to subsistence food security, nutrition, and medicine; generating additional employment and income through local, regional, and national trade; and offering opportunities for processing enterprises (Marshall and Nair, 2009). Mushroom cultivation contributes to strengthening the livelihoods through reducing vulnerability to poverty and generating reliable sources of income. Mushrooms are a good source of nutrition like protein, carbohydrate, minerals, vitamins, fat, ash and fibre. Mushrooms also contain a high amount of calcium, phosphorus, sodium, and potassium and low but available form of iron and magnesium which shows good health effects. It has also been examined that mushroom contains vitamins like thiamine, riboflavin, niacin, biotin, and vitamin C. Mushrooms have been utilised for food and medicine as they contain different primary and secondary metabolites, showing beneficial health effects. The demands of mushroom in pharmaceutical industry increase due to its nutraceutical application. Mushrooms are being used for the production of various value-added products such as mushroom papad, mushroom pickle, mushroom biscuits, mushroom beverage, etc.

Mushroom compost in horticulture

Spent mushroom compost (SMC) is the residual compost waste generated or left out after mushroom productions. This compost is readily available and its formulation generally consists of a combination of cereal straw, animal manure, ground chalk, etc. composted all together. SMC contains lots of salt and organic material along with enzymes and other nutrients that make it suitable habitats for different microbes including bacteria and fungi, synergistically they are found to be beneficial in disease suppression and plant growth promotion. Disease suppressive properties of composts rely on several factors including microbial activity, microbial population dynamics, nutrient concentrations, and other associated chemical and physical factors (Patil et al., 2018). Use of SMC as a soil amendment shows positive and significant effects on crop growth parameters. Spent mushroom substrates (SMS) are those residues which are left after the mushroom harvest from the non-composted substrate. Both SMS and SMC have a common role in the soil as a soil conditioner. These improve soil structure, provide plant nutrients, increase plant nutrient availability, increase soil microbial populations, increase soil cation exchange capacity, increase plant root structure, increase soil aeration, improve soil water status, and reduce soil compaction. Therefore, SMC and SMS are suitable to use in high-value horticultural crops (Singh et al., 2020).

Mushroom in animal husbandry

Some species of edible mushrooms as a feed additive to the broiler chickens and laying hens have dietary and health-promoting activities (Bederska-Lojewska et al., 2017). In poultry farm, antibiotic is common to increase the production which leads to human health hazards. Therefore, a natural source of the antibiotic substitute in the form of medicinal mushrooms shows positive effects on immunity on the broiler and laying hens (Mahfuz and Piao, 2019). The waste of oyster mushroom left from the mushroom industry may be used as supplements to the broiler chicken (Fard et al., 2014).

REFERENCES

Annadurai, K., Palaniappan, S. P., Chinnusamy, C. and Jayanthi, C. 1994. Integrated farming system: need of the hour. *Kisan World* **21**(9): 39.

Baishya, A., Kalita, M. C., Mazumdar, D. K., Hazarika, J. P. and Ahmed, S. 2007. Characterisation of farming systems in the Borpeta and Kamrup districts of lower Brahmaputra valley zone of Assam. *Journal of Farming Systems Research and Development* **13**(2): 168-175.

Bederska-Lojewska, D., Swiatkiewicz, S. and Muszynska, B. 2017. The use of Basidiomycota mushrooms in poultry nutrition - a review. *Animal Feed Science and Technology* **230**: 59-69.

Bedoussac, L., Journet, E. P., Hauggaard-Nielsen, et al. 2015. Ecological principles underlying the increase of productivity achieved by cereal-grain legume intercrops in organic farming - a review. *Agronomy for Sustainable Development* **35**(3): 911-935.

Behera, U. K. 2013. A Textbook of Farming Systems. Agrotech Publishing Academy, Udaipur. pp. 280.

Behera, U. K. and Mahapatra, I. C. 1999. Income and employment generation of small and marginal farmers through integrated farming systems. *Indian Journal of Agronomy* **44**(3): 431-439.

CARDI. 2010. A Manual on Integrated Farming Systems. Caribbean Agricultural Research and Development Institute, Ministry of Economic Development, Belize.

CGIAR. 1978. Farming systems research at International Agricultural research Centres. TAC Secretariat, Rome, Italy.

Chan, G. L. 2006. Integrated Farming System. What Does Integrated Farming System Do? http://www. scizerinm.org/chanarticle.htm.

Csavas, I. 1992. Regional review on livestock-fish production systems in Asia. In: Mukherjee, T. K., Moi, P. S., Panandam, J. M., and Yang, Y. S. (eds.). Proceedings of the FAO/IPT Workshop on integrated livestock-fish production systems, 16-20 December 1991, Institute of Advance Studies, University of Malaya, Kuala Lumpur, Malaysia.

Das, A. 2013. Integrated Farming: An approach to boost up family farming. *LEISA India* **15**(4): 1-5.

FAO. 1977. Recycling of organic wastes in agriculture. FAO Soil Bulletin 40. Food and Agriculture Organisation, Rome.

FAO. 2001. Mixed Crop-Livestock Farming: A Review of Traditional Technologies based on Literature and Field Experience. Animal Production and Health Papers, 152. Food and Agriculture Organisation of the United Nations. Rome.

Fard, S. H., Toghyani, M. and Tabeidian, S. A. 2014. Effect of oyster mushroom wastes on performance, immune responses and intestinal morphology of broiler chickens. *International Journal of Recycling of Organic Waste in Agriculture* **3**(4): 141-146.

Feliciano, D. 2019. A review on the contribution of crop diversification to sustainable development goal 1 'no poverty' in different world regions. *Sustainable Development* **27**(4): 795-808.

Fresco, L. O. and Westphal E. 1988. A hierarchical classification of farm systems. *Experimental Agriculture* **24**: 399-419.

Gangwar, B., Ravisankar, N. and Prasad, K. 2012. Agronomic research on cropping systems in India. *Indian Journal of Agronomy (Special Issue)* **57**: 171-181.

Garbach, K., Milder, J. C., DeClerck, F. A. J., et al. 2017. Examining multi-functionality for crop yield and ecosystem services in five systems of agroecological intensification. *International Journal of Agricultural Sustainability* **15**(1): 11-28

Godi, N. Y., Zhengwuvi, L. B., Abdulkadir, S. and Kamtu, P. 2013. Effect of cow dung variety on biogas production. *Journal of Mechanical Engineering Research* **5**(1): 1-4.

Hazra, C. R. 2001. Crop diversification in India. In: Papademetriou, M. K. and Dent, F. J. (eds.). Crop diversification in the asia-pacific region. Food and Agriculture Organisation of the United Nations Regional Office for Asia and the Pacific, Bangkok, Thailand.

http://www.agriculturesnetwork.org/magazines/india/family-farming/integrated-farming

https://vikaspedia.in/agriculture/livestock/income-generation-and-nutritional-security-through-livestock-based-integrated-farming-system

Hufnagel, J., Reckling, M. and Ewert, F. 2020. Diverse approaches to crop diversification in agricultural research - a review. *Agronomy for Sustainable Development* **40**(14): 1-17.

Jayanthi, C., Mythili, S. and Chinnasamy, C. 2002. Integrated farming systems – A viable approach for sustainable productivity, profitability and resource recycling under low land farms. *Journal of Ecobiology* **14**(2): 143-148.

Jayanthi, C., Rangasamy, A. and Chinnusamy, C. 2000. Water budgeting for components in lowland integrated farming systems. *Agricultural Journal* **87**: 411-414.

Kochewad, S. A., Meena, L. R., Kumar, D., Kumar, S., Meena, L. K., Singh, S. P. and Singh K. 2017. Livestock based integrated farming systems for livelihood improvement of small and marginal farmers. *South Asian Journal of Food Technology and Environment* **3**(1): 526-532.

Kumar, S. 2013. Integrated farming system models for food and nutritional security. In: Meena, M. S., Singh, K. M., Bhatt, B. P. and Kumar, U. (eds.). Gender perspective in integrated farming system, pp. 33-46. ICAR Research Complex for Eastern Region, Patna, Bihar.

Lal, R. and Miller, F. P. 1990. Sustainable farming for tropics. In: Singh, R. P. (ed.) Sustainable agriculture: issues and prospective Vol. 1, pp. 69-89, Indian Society of Agronomy, IARI, New Delhi.

Liu, K., Johnson, E. N., Blackshaw, R. E., Hossain, Z. and Gan, Y. 2019. Improving the productivity and stability of oilseed cropping systems through crop diversification. *Field Crops Research* **237**: 65-73.

Mahapatra, I. C. 1994. Farming system research - a key to sustainable agriculture. *Fertiliser News* **39**(1): 13.25.

Mahapatra, I. C. 2010. Policy and methodological issues: Road map and strategy for integrated farming systems research in India. pp. 12-19. In: Gangwar, B., Varughese, K., John, J., Rani, B. Vijayan, M. and Mathew, T. (eds.). Manual on integrated farming systems, KAU Cropping Systems, Thiruvananthapuram & Project Directorate of Farming Systems Research, Modipuram, Meerut.

Mahfuz, S. and Piao, X. 2019. Use of medicinal mushrooms in layer ration. *Animals* **9**(12): 1014.

Marshall, E. and Nair, G. 2009. Make Money by growing mushroom. Rural Infrastraucture and Agro-industries Division, Food and Agriculture Organisation of the United Nation. pp 64.

Meynard, J. M., Charrier, F., Fares, M., etc. 2018. Socio-technical lock-in hinders crop diversification in France. *Agronomy for Sustainable Development* **38**(5): 54-13.

Nishan, M. A. 2014. Crop Diversification for sustainability. *Popular Kheti* **2**(2): 20-24.

Okigbo, B. N. 1995. Major farming systems of the lowland savanna of SSA and the potential for improvement. In: Proceedings of the IITA/FAO workshop, Ibadan, Nigeria.

Panda, S. C. 2004. Cropping and Farming Systems. Agrobios (India), Jodhpur. 431 pp.

Panke, S. K., Kadam, R. P. and Nakhate, C. S. 2010. Integrated Farming System for suatainable rural livelihood security. In: 22nd national seminar on Role of Extension in Integrated Farming Systems for sustainable rural livelihood, 9-10th December, 2010, Maharashtra. pp. 33-35.

Patil, M. G., Rathod, P. K. and Patil, V. D. 2018. Compost: a tool for managing soil borne plant pathogens. *International Journal of Current Microbiology and Applied Sciences* **6**: 272-280.

Plucknett, D. L., Dillon, J. L. and Vallaeys, G. J. 1987. Review of concepts of farming system research: What, Why, and How. In: Proceedings of Workshop on Farming Systems Research, February 17-21, 1986, ICRISAT, Hyderabad. pp. 2-9.

Rana, S. S. and M. C. Rana. 2011. Cropping System. Department of Agronomy, College of Agriculture, CSK Himachal Pradesh Krishi Vishvavidyalaya, Palampur. 80 pp.

Reckling, M., Hecker, J. M., Bergkvist, G, etc. 2016. A cropping system assessment framework - evaluating effects of introducing legumes into crop rotations. *European Journal of Agronomy* **76**: 186-197.

Renard, D. and Tilman, D. 2019. National food production stabilised by crop diversity. *Nature* **571**: 257-260.

Rhoades, R. E. and Booth P. H. 1982. Farmer Back to Farmer: A Model for generating acceptable Agricultural Technology. *Agricultural Administration Journal* **11**(2): 127-137.

Sarswat, B. L. 2017. Beekeeping: vital input for sustainable agriculture. https://pib.gov.in/newsite/printrelease.aspx?relid=157604

Sasikala, V., Tiwari, R. and Saravanan, M. 2015. A review on integrated farming systems. *Journal of International Academic Research for Multidisciplinary* **3**(7): 319-328.

Shaner, W. W., Phillip, P .E. and Schmehl W. R. 1982. Farming system research and development: guidelines for developing countries. West view Press, Boulder, and Colorado, USA.

Singh, C., Pathak, P., Chaudhary, N., Rathi, A., Dehariya, P. and Vyas, D. 2020. Mushrooms and mushroom composts in integrated farm management. *Research Journal of Agricultural Sciences* **11**(6): 1436-1443.

Singh, J. P. 2010. Scenario of farming systems in different agroclimatic zones. pp. 20-41. In: Gangwar, B., Varughese, K., John, J., Rani, B. Vijayan, M. and Mathew, T. (eds.) Manual on integrated farming systems, KAU Cropping Systems, Thiruvananthapuram & Project Directorate of Farming Systems Research, Modipuram, Meerut.

Singh, K., Bohra, J. S., Singh, Y. and Singh, J. P. 2006. Development of farming system models for the north-eastern plain zone of Uttar Pradesh. *Indian Farming* **56**(2): 5-11.

Singh, R. P. and Ratan. 2009. Farming system approach for growth in Indian Agriculture. In: National seminar on enhancing efficiency of extension for sustainable agriculture and livestock production, December 29- 30, 2009, Indian Veterinary Research Institute, Izatnagar.

Soni, R. P., Katoch, M. and Ladohia, R. 2014. Integrated Farming Systems - A Review. *Journal of Agriculture and Veterinary Science* **7**(10): 36-42.

Suba Reddy, G. and Ramakrishna, Y. S. 2005. Farming system modules in rainfed agriculture. In: Singh, A. K., Gangwar, B. and Sharma, S. K. (eds.). Proceedings of Symposium on Alternative Farming Systems: Enhanced income and employment generation options for small and marginal farmers, September 16-18, 2004, Project Directorate for Cropping Systems Research, Modipuram. pp. 65-71.

Swaminathan, M. S. 2010. Enhancing the disaster resilience of agriculture. *The Hindu Survey of Indian Agriculture 2010*: 7-10.

Vijaya Bhaskar, A. V., Baresel, J. P., Weedon, O. and Finckh, M. R. 2019. Effects of ten years organic and conventional farming on early seedling traits of evolving winter wheat composite cross populations. *Scientific Reports* **9**(1): 9053.

13

Agroforestry
A Sustainable Land Use System

The general concept of agroforestry is to integrate trees and agriculture so as to create a more diversified landscape, while providing the producers with new environmental and economic benefits. In other words, agroforestry is a method of farming that allows trees and shrubs to grow along with crops and/or livestock, therefore blending agriculture and forestry in the same production system. The pressure on the agricultural lands has increased manifolds due to the increasing population, expansion of urban area and the industrialisation process. The environment has also been disturbed. Soil is losing its productivity and the biodiversity is threatened. Chemical fertilisers and pesticides are applied in higher proportion, causing environmental pollution hazards. Under all these circumstances agroforestry has shown that besides sustainable agriculture it can also help promote a better environment. Trees have always played an important role in mankind's survival.

CONCEPT OF AGROFORESTRY

A strictly scientific definition of agroforestry should stress two characteristics common to all forms of agroforestry and separate them from the other forms of land use.

1. The deliberate growing of woody perennials on the same unit of land along with agricultural crops and/or animals, either in some form of spatial mixture or in some temporal sequence.
2. There must be a significant interaction (positive and/or negative) between the woody and non-woody components of the system, either ecological and/or economical.

International Centre for Research in Agroforestry (ICRAF) in 1996 defined agroforestry as 'a dynamic, ecologically based natural resource management system that, through the integration of trees on farms and in the agricultural landscape, diversifies and sustains production for increased social, economic and environmental benefits for land-users at all levels'. Thus, in agroforestry systems there are both ecological and economical interactions between the different components (Lundgren and Raintree, 1982). This definition implies that

1. Agroforestry systems normally involve two or more species of plants (or plants and animals), at least one of which is a woody perennial.
2. An agroforestry system always has two or more outputs.
3. A cycle of an agroforestry system always requires more than one year for completion.
4. Even the simplest agroforestry system is more complex, ecologically (structurally and functionally) and economically, than monocropping system.

Additionally, there are three attributes which, theoretically, all agroforestry systems possess (Nair, 2008).

1. **Productivity:** Agroforestry systems aim to maintain or increase production of preferred commodities as well as productivity of the land.
2. **Sustainability:** Agroforestry can achieve and indefinitely maintain conservation and fertility goals by conserving the production potential of the resource base, mainly through the beneficial effects of woody perennials on soils.
3. **Adoptability:** Improved or new agroforestry technologies that are introduced into new areas should fit the social as well as environmental characteristics of the land-use system for which it is designed and should also confirm to local farming system.

BENEFITS FROM AGROFORESTRY

Agroforestry is the system of developing agricultural land in combination with forestry technologies. Through this system, land with shrubs and trees are used to grow crop and livestock to encourage profitability, productivity, diversity and sustainability. There are numerous benefits of agroforestry as it encourages the adaptation of natural ecological processes within the commercial system. It helps farmers in terms of controlling land degradation, sheltering crop and livestock, improving their landscape and enhancing wildlife habitat. Benefits from agroforestry can be grouped under three broad categories; environmental, economic or social benefits.

Environmental benefits

1. Reduction of pressure on natural forests
2. More efficient recycling of nutrients by deep rooted trees on the site
3. Better protection of ecological systems
4. Reduction of surface runoff, nutrient leaching and soil erosion through impeding effect of tree roots and stems on these processes
5. Improvement of microclimate, such as lowering of surface soil temperature and reduction of evaporation of soil moisture through a combination of mulching and shading
6. Increment in soil nutrients through addition and decomposition of litter
7. Improvement of soil structure through the constant addition of humus from decomposed litter

Economic benefits

1. Increment in outputs of food, fuelwood, fodder, fertiliser and timber
2. Reduction in incidence of total crop failure, which is common to single cropping or monoculture systems
3. Increase in levels of farm income due to improved and sustained productivity

Social benefits

1. Improvement in rural living standards from sustained employment and higher income
2. Improvement in nutrition and health due to increased quality and diversity of food outputs
3. Stabilisation and improvement of communities through elimination of the need to shift farm activities from one site to other

LIMITATIONS OF AGROFORESTRY

Agroforestry has different types of advantages but at the same time it also has some limitations.

1. Trees in agroforestry systems often compete with agricultural crops for light, water and nutrients from the soil which may reduce crop yields.
2. The use of farm machines is more difficult in the confined space in an agroforestry field.
3. This system is very difficult to manage and needs more accuracy with highly skillful management practices.

4. Increased susceptibility to pests and diseases often leads to dependence on potentially harmful pesticides.
5. Some of the ecological functions played by the trees in the natural forest may be lost when trees are used in an agroforestry system.
6. Damage to food crops during tree harvest operations.
7. Trees serve as hosts to insect pest and diseases that are harmful to agricultural crops.
8. Rapid regeneration by prolific trees which may displace food crops and take over the entire field.
9. Agroforestry systems are very labour-intensive which may cause labour scarcity at times of other farm activities.
10. Longer period is required for tree components to mature and acquire an economic value.
11. Farmers are usually unwilling to displace food crops with trees, especially where land is scarce.
12. Agroforestry is more complex, less understood and more difficult to apply as compared to monocropping.
13. There is no government policy for tree felling in private lands. Farmers are often not allowed to harvest certain tree species grown by them in their own land which discourage strongly to go for tree cultivation.
14. The marketing infrastructures for agroforestry produce are unavailable. As a result, it is largely a buyer's market and middlemen get major share in profit.
15. Usually, minimum support price for the tree products and other forest products is not fixed by any government agency. So, the growers do not get good price for the produce from the tree components of agroforestry systems.
16. Although tree farming requires high initial investment and return is usually delayed, there is no policy for financial support to the tree growers or agroforesters through nationalised banks.
17. There is shortage of quality planting materials of tree species for the diverse agroclimatic zones.

MAJOR AGROFORESTRY SYSTEMS

Different types of agroforestry systems exist in different parts of the world. These systems are highly diverse and complex in character and function. Several criteria can be used to classify and group these agroforestry systems and practices. The most commonly used ones are the system's structure (composition and arrangement of components), its function, its socioeconomic scale and level of management and its ecological spread. The structure of a system can be defined in terms of its components and their expected roles. This classification considers the composition of the components including spatial admixture of perennial woody component, vertical stratification of the component mix and temporal arrangement of the different components. Based on nature of components, agroforestry systems can be classified into agrisilvicultural, silvipastoral, agrisilvipastoral and other systems.

AGRISILVICULTURAL SYSTEMS

In these systems, annual agricultural crops and perennial woody trees/shrubs are integrated. Based on the nature of the components this system can be grouped into various forms.

Shifting cultivation

Shifting cultivation is an agricultural land use system in which a piece of forest land is cleared and cultivated until the soil loses fertility. Once the land becomes inadequate for crop production, it is left and allowed to rest longer than the period of cultivation to be reclaimed by natural vegetation. The ecological consequences are often deleterious, but can be partially mitigated if new forests are not invaded. This system is practised extensively in the hill region of north-eastern states and to some extent in hilly areas of Andhra Pradesh, Madhya Pradesh, Bihar, Odisha, Jharkhand, Chhattisgarh and Karnataka. The practice is called '*jhum*' in north-eastern states and '*podu*' in Andhra Pradesh and Odisha.

Taungya system

'*Taungya*' is a Burmese word, consisting '*taung*' means hill and '*ya*' means cultivation, i.e., cultivation in the hill. The *taungya* is a modified form of shifting cultivation in which cultivators are required to plant forest tree species and allowed to raise agricultural crops in between rows of tree species until the canopy closes. While shifting cultivation is a sequential system of growing woody species and agricultural crops, *taungya* consists of the simultaneous combination of the two components during the early stages of forest plantation establishment. Although wood production is the ultimate objective in the *taungya* system, the immediate motivation for practising it, as in shifting cultivation, is food production.

Taungya was reported to have started first in Burma (Myanmar) in 1850 by the British for replanting vast areas under teak plantation and in Java in 1856. It was introduced into India by Dietrich Brandis in 1856 and the first *taungya* plantations were raised in 1863 in north Bengal. It is practised in the states of

Kerala, West Bengal, Uttar Pradesh, Tamil Nadu, Andhra Pradesh, Odisha, Karnataka and the north-eastern states with an assured annual rainfall of over 1200 mm. A large variety of crops and trees, depending on the soil and climatic conditions are grown in India. The state forest department is the owner of the land and thus the trees. In fact, this system was introduced to raise forest plantation but finally became a recognised agroforestry system. *Taungya* systems are of three types.

Departmental taungya: In this system, labourers are employed on daily wage basis by the state forest department to grow seasonal agricultural crops along with perennial tree crops. The agricultural crops are raised mainly for suppressing weeds between the tree rows during the early years of plantation.

Leased taungya: The plantation land is leased by the forest department to the interested cultivators for a specified period who raise suitable crops in between the tree rows and take care of tree plantation.

Village taungya: In this system, people are allowed to settle down in a village inside the forest by the forest department and in return, they raise trees and agricultural crops for 3 to 4 years in the land they are allotted. It is the most successful practice of the three *taungya* systems

Alley cropping

Alley cropping or hedgerow intercropping is a management-intensive agroforestry practice in which perennial, preferably leguminous trees or shrubs are grown simultaneously with an arable crop. The trees, managed as hedgerows, are grown in wide rows and the crop is planted in the interspaces or 'alley' between the tree rows. The trees are pruned regularly during the cropping phase and allowed to grow freely to shade the inter-rows when there are no crops. Alley cropping retains the basic restorative attributes of the bush fallow through nutrient recycling, fertility regeneration and weeds suppression and combines these with arable cropping so that all processes occur concurrently on the same land, allowing the farmer to crop the land for an extended period.

Benefits from alley cropping

1. An important benefit of alley cropping is the addition of large amounts of organic materials from the prunings as mulch or green manure which can have favourable effects on soil physical and chemical properties, on microbiological activity and hence on soil productivity. This ultimately improves crop performance in alleys.

2. There is a reduction in the use of chemical fertilisers which decreases environmental pollution and maintains soil health.
3. There is an overall improvement in the physical nature of the soil environment. The addition of mulch moderates soil temperature, reduces evaporation and improves activity of soil fauna and soil structure resulting in better infiltration, reduced runoff and improved water use efficiency.
4. The tree rows on sloping land act as a physical barrier to soil and water movement, resulting in significant reductions in erosion losses. The presence of prunings applied as mulch in the alleys also controls soil erosion.
5. Alley cropping provides additional products such as forage, firewood or stakes when a multipurpose tree is used as the hedgerow.
6. During the fallow period, shading of the interspaces reduces weed growth, while in the cropping phase, the mulch inhibits germination and growth of weeds.

Management of alley cropping systems

The effectiveness of an alley cropping system depends to some extent on the soil type and agroecological zone in which the system is practised. But the success of the system is also very dependent on management strategies adopted. Factors such as choice of tree species, orientation, layout and manipulation of the hedgerows and crop husbandry practices are all important in determining the outcome of the alley cropping system. Trees are planted as hedgerows in farm fields to maximise the positive and minimise the negative effects of trees on crops. Trees compete with agricultural crops for nutrients, soil moisture and solar energy. However, right kind of tree species planted at right spacing, with proper management practices can reduce competition and actually produce a net increase in yields per unit area.

The position and spacing of hedgerow and crop plants in an alley cropping system depend on plant species, climate, slope, soil condition and the space required for the movement of people and tillage equipments. Ideally, hedgerows should be positioned in an east-west direction so that plants on both sides receive full sunlight. The spacing adopted is usually 2 m to 8 m between rows and 25 cm to 2 m between the trees within rows. The closer spacing is generally used in humid areas and the wider spacing in subhumid and semiarid regions. On sloping land, hedgerows are densely planted along the contours to form a barrier against soil erosion. Grass strips planted alongside hedgerows will create a more effective barrier. If the desirable east-west orientation of hedgerows is not maintained on the slopes then regular pruning of tree branches is required to prevent excessive shading to the agricultural crops in alleys.

The choice of tree species for alley cropping is extremely important and to a large extent determines the success or failure of the system. The tree species for an ideal alley cropping should have rapid growth rate, ability to withstand frequent cutting, good coppicing ability, easy to establish from seeds or cuttings, nitrogen fixing capacity, deep-rooted with a different root distribution to the crop, and ability to withstand environmental stresses. The suitable species for alley cropping are *Cassia siamea*, *Leucaena leucocephala*, *Gliricidia sepium*, *Calliandra calothyrsus*, *Sesbania sesban*, *Acacia mangium*, *Gmelina arborea*, *Albizia* spp., *Cajanus* spp., *Dalbergia sissoo*, *Tectona grandis*, *Chamaecytisus* spp., *Desmodium* spp., *Erythrina* spp., *Flemingia* spp., *Inga* spp. and *Tephrosia* spp.

Crop combinations with plantation crops

On large estates plantation crops such as tea, coffee, cashew, oil palm, rubber, cocoa and coconuts are usually grown in monoculture. But intercropping in these perennial tree crops on small holdings is possible or even desirable due to several reasons.

1. There are large spaces between tree crops during the early stages of growth.
2. The intercrops reduce soil erosion between widely spaced tree crops, especially during the early tree growth stages.
3. Roots of some tree intercrop combinations complement each other.
4. Shade tolerance of understorey crops such as arrowroot, pineapple, turmeric, ginger, *Aloe vera*, etc. favours growth under some plantation crops.
5. Complementary use of light, e.g., up to 8 years and after about 25 years, coconut palms allow a considerable amount of light to reach the intercrops. On the other hand, while oil palm has a high light requirement that is available in the upperstorey, cocoa in the understorey requires considerable shade, except at flowering stage.
6. Intercropping during early growth period of tree crops is economically viable while waiting for harvest of long maturity tree crops.
7. Some intercrops are compatible because of differing peak labour requirement time with that of plantation crops.
8. Some tree crops require an intercrop to serve as a nurse crop during the early growth period. For example, newly transplanted coffee seedlings benefit from the shade provided by taller intercrops, such as maize.

Shade trees for commercial plantation crops

Trees are often planted to provide shade for plantation crops such as coffee and tea. Young cocoa plants also need shade in the nursery and for the first 2-3 years in the field. The shade moderates the microenvironment so that excessive moisture stress is avoided during dry season. The use of shade trees in coffee plantations is common in the tropics and legume trees are mostly used for this purpose. In addition to providing shade, these trees contribute nutrients to low fertility soils where inorganic fertilisers are not used. The important shade providing tree species are *Erythrina poeppigiana*, *Cordia alliodora*, *Gliricidia sepium*, *Leucaena leucocephala*, *Albizia chinensis*, *Grevillea robusta*, etc.

Shelterbelts

A shelterbelt is a wide strip of vegetation consisting of several rows of trees that slows wind speed, thereby reducing wind erosion, evaporation and damage of field crops by the wind. It is sometimes referred to as windbreak, although the latter often implies a single strip of trees and other vegetation. A shelterbelt presents a mechanical barrier to the impact of the wind and separates two zones, the windward and the leeward zones. The windward zone refers to the side from which the wind blows, while the leeward zone relates to the side where the wind passes.

Characteristics of shelterbelts

1. Shelterbelts have a typical triangular shape acquired by raising tall trees in the centre.
2. A certain degree of penetration by wind is planned because by raising a solid wall of trees, the protection decreases very fast on the leeward side.
3. The ratio of height and width of a shelterbelt should be roughly 1:10. Shelterbelts up to 50 m width are considered ideal under Indian conditions.
4. Orientation of shelterbelts depends upon the direction and velocity of the prevailing winds. Rows of trees are established at right angles to the prevailing wind.
5. The minimum length of a shelterbelt should be about 25 m.
6. Tree species selected for shelterbelt establishment should have ability for wider adaptability to climatic conditions (drought, frost, extreme temperatures, etc.), high growth rate, quick crown formation, developing a strong and deep root system (so that they do not compete for moisture and nutrients with agricultural crops), evergreen foliage, increased land productivity.

Benefits from shelterbelts

1. **Reduces soil erosion by wind:** A shelterbelt modifies the microclimate, mostly in its leeward side. This modified microclimate includes reduced wind speed and, therefore, reduced soil erosion. A significant reduction of

wind speed occurs on leeward side for a distance extending to approximately 20 times the height of the shelterbelt and also 3 to 5 times its height on the windward side. Therefore, a shelterbelt 5 m in height will provide a degree of protection for soils and crops for a total distance of up to 25 times its height, i.e., 125 m. Since the zone of protection provided by a single shelterbelt is limited, a series of shelterbelts is required to protect the whole field. The shelterbelts must be planted perpendicular to the direction of the prevailing wind to provide complete protection.

2. **Increases moisture for crop growth:** Shelterbelts reduces evaporation, thus providing more moisture for crop growth. It also protects the leeward areas from the desiccating effects of hot wind. Shelterbelts use moisture and nutrients from a greater depth than most annual crops.
3. **Reduced wind damage to crops:** Crops benefit from the reduced wind speed in the protected zone. The crop plants are less likely to be twisted by the wind.
4. **Potential for increased farm productions:** Most of the research conducted around the world reports crop yield increases due to shelterbelts. However, crops vary in their yield response to shelterbelt protection. Drought tolerant crops show the lowest response, forage crops are moderately responsive and weather-sensitive crops such as vegetables show the highest response. Besides, shelterbelts provide potential source of income for farmers (e.g., biomass, timber and non-timber products).

Windbreaks

Windbreaks are narrow strips of trees, shrubs and/or grasses planted to protect crop fields, homes, canals and areas from wind and blowing sand. Where wind is a major cause of soil erosion and moisture loss, windbreaks can make a significant contribution to sustainable production. From the farmers' point of view, gentle winds are advantageous as they can contribute to pollination of crops and seed dispersal. Strong winds, on the other hand, are damaging and could be detrimental to agricultural crops, human life and properties. Below are some of the effects of strong winds on crops.

1. Wind may increase transpiration rate, and this may lead to soil moisture deficits.
2. It can spread pests and diseases because disease causing spores and insects are dropped whenever wind speed is reduced.
3. It can cause crop lodging and deform plants.
4. It can increase loss of topsoil through wind erosion especially in semiarid areas.

5. It can cause drifts during herbicide and insecticide spray, hence, leading to wasteful application and ecotoxicity.

Benefits from windbreaks

1. It protects crops and pastures from hot, drying winds and cold wind, frost.
2. It reduces/prevents soil erosion.
3. It improves the microclimate by moderating temperature.
4. It reduces evaporation from farmlands, also checks transpiration.
5. It prevents grass fire.
6. It provides fencing and boundary demarcation.
7. It increases overall farm productivity.

Design and management

The basic design of a windbreak should include permeability, the appropriate orientation, placement, length and height, number of rows, spacing, density and continuity to provide effective protection.

1. **Permeability:** The effectiveness of the windbreak is influenced by its permeability. If it is dense, like a solid wall, the airflow will pass over the top of it and cause turbulence on the leeward side due to the lower pressure on that side. This gives a comparatively limited zone of effective protection on the leeward side compared to the zone that a moderately permeable shelter creates. Windbreaks must be semipermeable, ideally filtering 50-60% of the wind to reduce its strength. The desired permeability can be obtained by carefully selecting tree species. Species such as *Eucalyptus* and *Casuarina* will form effective windbreaks. Permeability of dense windbreaks can be improved by pruning lower branches at 0.5-0.8 m from the soil level.
2. **Orientation:** When designing a windbreak, the direction of the wind must be considered. A barrier should be established perpendicular to the direction of the prevailing wind for maximum effect. To protect large areas, a number of separate barriers can be created as parts of an overall system. When the prevailing winds are mainly in one direction, a series of parallel shelterbelts perpendicular to that direction should be established. Before establishing windbreaks or shelterbelts, it is important to make a thorough study of the local winds and to plot on a map the direction and strength of the winds.
3. **Height and length:** While selecting species for windbreak more importance is given to the height than the thickness. It is generally accepted that a windbreak protects an area over a distance up to 3-5 times its own height on the windward side and up to 20 times its height on the leeward side.

Thus, the species selected should be erect and tall growing, hardy and drought resistant, and should occupy less space as far as possible. Windbreaks are more effective when they stretch without major gaps for distances exceeding 12 times the mature height of the trees.

4. **Tree rows and spacing:** In reducing wind speeds, narrow barriers can be as effective as wide ones. Although in theory, one-row barrier should suffice, in practice three to five rows are more effective. If a single row windbreak is to be planted, tree species that retain their foliage to the ground and give a fairly dense growth should be selected. *Eucalypltus* trees are generally unsuitable as single row windbreaks because of their habit of losing the lower branches. The main disadvantage of a single row is that if one tree is lost, gap is created which reduces the efficiency of the entire windbreak. Distance between trees varies with the relative importance of the protective versus productive purposes of the windbreak. Initial spacing of 3 m between the rows, with trees 1-2 m apart in the rows is desirable.
5. **Species:** In general, tree or shrub species selected for windbreaks should have the characteristics of rapid growth, straight stems, wind firmness, good crown formation, deep tap root system and resistance to drought. Some of the tree genera used for windbreaks are *Cassia*, *Acacia*, *Casuarina*, *Leucaena*, *Grevillea*, *Cupressus*, *Pinus*, *Dalbergia*, *Syzygium*, *Erythrina*, *Mangifera*, *Bambusa*, etc.

Windbreak for fruit orchards

Fruit orchards when exposed to strong wind usually incur heavy losses. Heavy wind increases the losses of moisture both by increasing transpiration and surface evaporation. The high winds also cause damage to fruit trees by breaking of branches, destruction of blooms and dropping of immature fruits. The growth and yield in protected orchard is definitely better than the exposed orchard. Hence, establishment of a tall growing windbreak is necessary to protect an orchard. Planting of windbreaks should be done at least two years after planting of fruit trees. The first row of it should be planted 10 m away from fruit plants at a close spacing to form a thick screen. The windbreak trees sometimes may compete with the fruit trees for water and nutrient. To prevent this competition, a trench may be dug about 1 m deep and 5 m away from the row of windbreak trees and all the roots of windbreak trees exposed in the trench are cut off periodically.

Multipurpose windbreaks

Although planting windbreaks is an investment that takes some land out of production, well designed windbreaks protect the health and productivity of crops enough to make the overall return positive. A multipurpose windbreak is designed to provide multiple functions and/or products, in addition to wind protection. Multiple products from a windbreak can include yields such as fruit, timber, animal fodder, mulch, wildlife habitat and other economic farm products. Adding multiple functions or products to a windbreak plan can make the installation and management more satisfying and economically viable for the farmer. Multipurpose windbreaks require special care in planning and management to maintain the primary function of wind protection while maximising secondary yields. Once the form and position are carefully determined, then multiple functions or products can be added.

Since planting trees for a windbreak involves a long-term investment, trees having timber value may also be included. The main drawback of having timber as a secondary yield from a windbreak is that wind damage may produce timber of poor quality. Also, since windbreak trees require no or very little pruning, the lack of pruning may reduce timber quality and quantity in certain species that require a lot of pruning for optimal timber production. Planning for timber harvest requires the most careful effort as the entire trees are removed. The planting, harvesting and replanting must be coordinated to avoid the creation of gaps in the tree rows. Integrating timber trees with permanent rows of non-timber windbreak trees will help maintain the effectiveness of the windbreak. *Grevillea robusta*, *Pterocarpus indicus* and *Azadirachta indica* are some of the best suitable species for this purpose.

Nitrogen fixing trees (NFTs) can also be integrated in a multi-row windbreak and pruned regularly to provide nutrient-rich mulch for crops, or a nutritious fodder to supplement the diet of farm animals. Although pruning should be avoided for most windbreak trees, the practice of cutting back NFTs and allowing them to resprout can be integrated with windbreak management. Pruned NFTs are much more susceptible to wind damage if they are allowed to regrow to a large size, but if they are cut regularly and the regrowth kept small, they will be effective as a short row. To maintain the windbreak's primary function with this practice, it is essential to prune the NFTs regularly. These species are planted on the most sheltered side of the windbreak. The important species for this purpose are *Leucaena leucocephala*, *Sesbania sesban*, *Calliandra calothyrsus*, and *Gliricidia sepium*.

Soil conservation hedges

Trees and shrubs are planted on physical soil conservation works such as grass strips, bunds and terraces wherein they play two roles.

- to stabilise the soil conservation structure through the root system
- to make productive use of the land they occupy

The risers or terraces in steeply sloping landscapes are densely planted with multipurpose and/or fruit trees. The primary role of multipurpose/fruit trees and agricultural species is soil conservation and provision of various tree products. The important tree species used for soil conservation include *Grevillea robusta*, *Acacia catechu*, *Pinus roxburghii*, *Acacia modesta*, *Prosopis juliflora*, *Alnus nepalensis*, *Leucaena leucocephala*, etc.

SILVIPASTORAL SYSTEMS

Silvipasture as an agroforestry practice is specifically designed and managed for the production of trees, tree products, forage and livestock. Silvipasture results when forage crops are deliberately introduced or enhanced in a timber production system, or timber crops are deliberately introduced or enhanced in a forage production system. Integrating livestock production in the farming system promotes farm product diversification, improves food security and offers a diversified marketing opportunity that can stimulate rural economic development. The interactions among timber, forage and livestock are managed intensively to simultaneously produce timber commodities, a high-quality forage resource and efficient livestock production. Overall silvipastoral systems can provide economic returns while creating a sustainable system with many environmental benefits.

Before new silvipastoral systems are established, implications of merging forestry and agricultural systems should be explored thoroughly for economic and environmental considerations. Factors like availability of potential markets, soil type, climatic conditions and species compatibility should be considered while choosing tree and forage crop species. The timber component should be marketable, of high quality, fast growing, deep rooted, drought tolerant, and capable of providing the desired products and environmental services. The forage component should be a perennial crop that is suitable for livestock grazing, compatible with the site (soil, temperature, precipitation), productive under partial shade and moisture stress, responsive to intensive management and tolerant of heavy utilisation or grazing. A successful silvipastoral system requires understanding forage growth characteristics and managing the timing and duration of grazing to avoid browsing of young tree seedlings. Improper management of silvipastures can reduce desirable woody and herbaceous plants by overgrazing and soil compaction.

Benefits from silvipastoral systems

1. Economic risk is reduced because the system produces multiple products, most of which have an established market. Production costs are reduced and comprehensive land utilisation in silvipastoral systems provides a relatively constant income from livestock sale and selective sale of trees

and timber products. Well managed forage production provides improved nutrition for livestock growth and production.

2. Grazing can enhance tree growth by controlling competition of grass for moisture, nutrients and sunlight. Well managed grazing provides economical control of weeds. Fertiliser applied for forage is also used by trees. In addition, livestock manure recycles nutrients to trees and forage.
3. Some forage species tend to be lower in fibre and more digestible when grown in a tree-protected environment. Trees that provide shade or wind protection can have a climate-stabilising effect to reduce heat stress on livestock. Protection by trees can cut the direct cold effect by 50% or more and reduce wind velocity by as much as 70%.
4. The forage protects the soil from water and wind erosion, while adding organic matter to improve soil properties. Some forage species fix nitrogen.
5. Silvipastures provide an attractive landscape with an aesthetic value. It can also increase wildlife diversity.

Within the broad category of silvipastoral system there are several types of practices which can be identified depending on the role of the tree/shrub component. These practices include protein banks (fodder tree banks), trees and shrubs on rangeland or pastures, live fences of fodder trees and shrubs and plantation crops with pastures and animals.

Protein bank

Protein banks are blocks of forage plants deliberately planted to alleviate fodder shortages in arid, semiarid and mountainous regions, especially during the dry seasons. The forage plants may be leguminous trees and shrubs or herbaceous legumes, and they may be grown in combination with suitable grasses. The fodder trees are pruned regularly to feed livestock. When based on legumes, the fodder banks become important sources of protein and are referred to as protein banks. The important tree and shrub species for this system are *Acacia nilotica*, *Albizia lebbeck*, *Leucaena leucocephala*, *Gliricidia sepium*, *Sesbania grandiflora*, *Artocarpus* spp., *Bombax malabaricum*, *Cordia dichtotoma*, etc.

Trees and shrubs on rangelands or pastures

The primary use of trees in range and pasture land is to provide shade. Besides this vital function, trees on rangeland also provide other benefits like fodder and wood. Normally such trees are scattered at random but sometimes they may be arranged according to a systematic pattern. The number of species with potential for this practice is great. They include *Acacia nilotica*, *Acacia mangium*, *Acacia auriculiformis*, *Gliricidia sepium*, *Bauhinia* spp., *Leucaena leucocephala*,

Derris indica, *Azadirachta indica*, *Emblica officinalis*, *Psidium guajava*, *Prosopis* spp. and *Tamarindus indica*.

Live fences of fodder trees and shrubs

Live fences are lines of trees or shrubs planted on farm boundaries or on the borders of farmyards, pasture plots or animal enclosures. Sometimes they are also used around agricultural fields. They serve mainly as field boundaries to protect the property from stray animals. They can be made of densely planted single or multiple rows. The shrubs or trees are regularly pollarded and trimmed. The live fences provide shade, protection and privacy for the livestock integrated in this silvipastoral system. The trees can also serve as windbreaks that produce wood and foliage products. The foliage can be used as fodder. Legumes are especially valued as they have high protein content. In arid and semiarid zones, live fences are often made of thorny species of the *Acacia* or *Prosopis* genera. In the humid and subhumid tropics, leguminous species such as *Gliricidia sepium*, *Sesbania grandiflora* and *Erythrina berteroana* are used.

AGRISILVIPASTORAL SYSTEMS

Agrisilvipastoral system is the combination of agricultural crops, woody perennials and livestock. Agroforestry practices under this system include agrisilvicultural system converted to silvipastoral systems, multistorey system with free grazing, homegardens and alley cropping system using pasture grasses/fodder crops and agricultural crops.

In agrisilvicultural system converted to silvipastoral systems, the initial cropping combinations include tree seedlings and annual agricultural crops. As the trees grow and close their canopies, the yield recovery of annual agricultural crops gradually decreases and it will no longer be possible to grow these crops. Instead, shade tolerant grasses and vines will be more compatible to land use system where animals are allowed to graze freely, thus converting agrisilvicultural system to silvipastoral system. Multistorey system with free grazing is similar to the multistorey under agrisilvicultural system, except that in this case, grazing animals are an added component. Alley cropping with pasture grasses and agricultural crops is similar to hedgerow cropping with seasonal agricultural crops. However, instead of all alleys planted with agricultural crops, some alleys in between the hedgerows are grown with improved pasture grasses and/or fodder trees or shrubs which are regularly cut and fed to livestock. In woody hedgerows, various woody hedges especially fast growing and coppicing fodder shrubs and trees are planted for the purpose of browse, mulch, green manure, soil conservation, etc. The species for this system are *Erythrina* spp., *Leucaena luecocephala*, *Sesbania grandiflora*, etc.

Homegardens

This is one of the oldest agroforestry practices found extensively in the high rainfall areas in tropical south and south-east Asia. The word 'homegarden' has been used to describe diverse practices from growing vegetables behind houses to complex multistorey systems. It is the intimate association of multipurpose trees and shrubs with seasonal and permanent arable crops and livestock within the compounds of individual houses. Mostly the whole system, i.e., crop-tree-animal unit is managed by family labour. These systems are common in all ecological regions in the tropics and subtropics, especially in humid lowlands with high population density. The average size of a homegarden is usually much less than one hectare. In India, every homestead has around 0.2 to 0.5 ha land for personal production. In many parts of the world the fruit, vegetable and other foodstuff produced from homegardens provide a substantial part of the household food requirement. Food production is the primary function and role of most of the homegardens. The combination of crops with different production cycles results in a relatively uninterrupted supply of food products throughout the year. In most cases, animals and birds are kept in the homegardens. Fodder and legumes are widely grown to meet the daily fodder requirement of the animals. The waste materials from crops and homes are also used as fodder for animals and feed for birds. Cattle are mostly kept for dairy products and land cultivation. Goats, chickens and fish are kept for household consumption. Products from animals or the animals themselves can also be sold in the market.

Structure of homegardens

The layered configurations and combination of compatible species are the most important characteristics of all homegardens. The gardens are carefully structured systems where each component has a specific place and function. Every homegarden usually consists of an herbaceous layer near the ground, a tree layer at the upper level, and intermediate layers in between. The lower layer is usually divided into two, with the lowermost (less than 1 m height) dominated by different vegetable and medicinal plants, and the second layer (1-3 m height) being composed of food plants such as banana, papaya, yam and so on. The upper tree layer is also partitioned into two, consisting of emergent, fully grown timber and fruit trees occupying the uppermost layer of over 25 m height, and medium-sized trees of 10-20 m height occupying the next lower layer. The intermediate layer of 3-10 m height is dominated by various fruit trees.

Benefits of homegardens

1. Production is very diverse and continuous with less risk and there is a regular flow of products such as food, fuel, fodder, fruit, spices and timber.

2. The protective function of homegardens is very high with soil conservation, water retention and favourable microclimate for humans and animals.
3. The use of labour for the farm is very efficient due to proximity of the garden to the house, within the walking distances.
4. Valuable crops or animals in need of protection can be given extra care by the farm family because of the short distance from the house.
5. Homegardens ensure a pleasant living environment, shade and provision of privacy.
6. They can also provide aesthetic values and open patches in the homegarden often serve as family or village gathering places.

Constraints of homegardens

1. The high diversity of ecological niches in a homegarden may provide a habitat for species that can become pests like snakes, insects and fungi.
2. During tree harvesting other plants are often damaged.
3. Proximity to homesteads makes houses prone to damage from falling branches and extensive lateral roots from trees.

OTHER AGROFORESTRY SYSTEMS

Other minor agroforestry technologies include aquaforestry, agrisilviaquaculture, apisilviculture, sericulture, etc.

Aquaforestry

The aquaforestry system comprises of composite fish culture in farm ponds, and various trees and shrubs (*Leucaena leucocephala*, *Morus alba*, *Gliricidia sepium*, *Moringa olifera*, etc.), leaves of which are preferred by fish are planted on the boundary and around fishponds. Leaves of these trees are used as feed for fish. Inland fish such as *catla*, *rohu*, *mrigal*, common carp, silver carp and grass carp can be grown in the ponds. In the coastal regions, farmers are cultivating fish and prawn in saline water and growing coconut and other trees on bunds of ponds. Now fish culture in the mangroves is also advocated, which form a rich source of nutrition to the aquatic life and breeding ground for fish and prawn. Other terms that have been used for this practice are silvipisciculture, agrisilviaquaculture and aquasilviculture. Some other modified aquaforestry systems are

1. A well-balanced system of animal husbandry including goatery, poultry, duck farming, turtles and fishes in the small ponds in homegardens make a balanced system of high moisture, energy and nutrient use efficiency per unit area.

2. In paddy field, fish can easily be reared by planting trees on field bunds or boundary to provide leaves used as fish feed. This system can be practised in high rainfall areas.
3. Coconut plants can also be successfully planted on raised paddy field bunds with an alley space of 5 m width depression which is utilised for pisciculture purpose.

Apisilviculture

In this system, various honey or nectar producing trees frequently visited by honeybees are planted on the boundary of the agricultural field. Bees benefit the trees and trees in turn provide series of benefits to bees. The primary purpose of this system is to produce honey. Apisilviculture with *Eucalyptus*, *Gliricidia*, *Grevillea*, *Gmelina*, *Leuceana* and *Albizia* species were more remunerative and a good source of generating additional farm income in rural areas. Fruit trees can also be integrated with apisilviculture for more profits.

AGROFORESTRY FOR SOIL FERTILITY AND PRODUCTIVITY

Agroforestry systems have the potential to make use of marginal and degraded lands through the soil improving effects of trees. Trees have a different impact on soil properties than annual crops, because of their longer residence time, larger biomass accumulation and more extensive root systems. 'Trees improve soils' – this hypothesis is substantiated by the following facts (Young, 2005).

1. The soil under natural forest is fertile. It is well structured, has a good water holding capacity and has a store of nutrients bound up in the organic matter. From time immemorial, farmers have known that they will get a good crop by planting in forest clearings.
2. In natural forests, nutrients are efficiently cycled with very small inputs and outputs from the system. A forest ecosystem is a relatively closed system in terms of nutrient transfer, storage and cycling in contrast to agricultural systems.
3. Reclamation of degraded land through afforestation or a similar type of tree-based land-use system is found to be best to restore soil fertility in many areas.
4. The practice of shifting cultivation demonstrated the power of trees to restore fertility lost during cropping.
5. The conversion of natural forest ecosystems to arable farming systems leads to a decline in soil fertility and a degradation of other soil properties unless appropriate corrective measures are taken.

Nitrogen fixation and a high biomass production are the two most desirable characteristics of a tree. However, many properties are specific to particular objectives of systems in which the trees are used. The following properties are likely to make a tree suitable for soil fertility maintenance or improvement.

1. A high rate of production of leafy biomass
2. A high rate of nitrogen fixation
3. The existence of deep roots and a dense network of fine roots with a capacity for abundant mycorrhizal association
4. A high and balanced nutrient content in the foliage; litter with high nitrogen, low lignin and polyphenols
5. An appreciable nutrient content in the root system
6. Rapid litter decay, where nutrient release is desired, and a moderate rate of litter decay, where maintenance of a soil cover is required
7. Absence of toxic substances in the litter or root residues
8. A capacity to grow on poor soils for reclamation
9. Absence of severe competitive effects with crops, particularly for water
10. Low invasiveness
11. Productive and service functions other than soil improvement

Beneficial effects of trees on soil

Trees improve soil fertility by processes of increasing additions to the soil, reduction of losses from soil, and improving soil physical, chemical and biological conditions. Most of the beneficial effect of trees on soils is primarily due to their root, crown, litter fall and N-fixing ability and the effects are directly proportional to the density of the trees. Understanding the magnitude and rate of beneficial as well as adverse effects of trees on soil is the key to successful design and management of agroforestry systems.

Maintenance or increase in soil organic matter: Trees are believed to increase, or at least maintain, the organic-matter levels of the soil. This is mainly through litter fall and continuous degeneration or sloughing-off of roots of standing trees. Pruning materials from the woody perennials used in agroforestry systems add a huge quantity of organic matter to the soil, and this is generally high in nitrogen. The root biomass of trees is usually 20-30% of total plant biomass (Armson, 1977), but it may be as low as 15% in some rain forests or as high as 50% or more in semiarid and arid climate (Nair, 1989). This biomass consists of structural roots, fine roots and associated mycorrhizae. The ability of the root system to improve soil organic matter is a crucial factor in low-input agricultural systems with low productivity levels. Roots also store considerable quantities of nutrients.

Nitrogen fixation: Biological nitrogen fixation takes place through symbiotic and non-symbiotic means. Symbiotic fixation occurs through the association of plant roots of many leguminous plant species with N_2-fixing microorganisms, *Rhizobium* or *Bradyrhizobium.* A few non-leguminous tree species also nodulate with a genus of actinomycetes, *Frankia.* About 200 non-leguminous plant species belonging to 19 genera nodulate with *Frankia. Alnus*, *Casuarina* and *Allocasuarina* are the main genera of nodulating non-leguminous plants (Bond, 1983). The N-fixing ability of trees and shrubs in agroforestry system varies from 50-500 kg/ha/annum depending upon the tree species and growing conditions (Jhon, 1988).

Nutrient uptake: Trees are more efficient than herbaceous plants in taking up nutrients released by the weathering of soil at deeper horizons. Potassium, phosphorus, calcium, magnesium and micronutrients are released by rock-weathering, particularly in the B/C and C soil horizons which tree roots often penetrate. Thus, nutrients in deeper soil horizons, that are unavailable to shallow rooted crops, are taken up by deep rooted trees.

Atmospheric input: Atmospheric deposition makes a significant contribution to nutrient cycling. It consists of nutrients dissolved in rainfall (wet deposition) and those carried in dust (dry deposition). Trees reduce wind speed considerably and thus favour dry deposition.

Increased water infiltration: Trees facilitate water infiltration along the root channels due to higher organic matter content in the soil. The leaf litter fall from trees also absorbs a large quantity of water. This substantially reduces runoff and increases water input to the soil.

Water retrieval: The tap roots of trees draw water from soil well below the depth ordinarily reached by roots of crop plants. This process increases water input in plant-soil system.

Protection from erosion: Soil erosion causes loss of soil organic matter and nutrients, and thus results in reduction in crop yield. Agroforestry systems reduce erosion through ground litter cover by tree leaves and understorey vegetation. The tree canopy may also reduce the erosion to a little extent by slashing raindrops. Trees and shrubs through their proper planting arrangement and management can act as effective barriers to control soil erosion.

Nutrient retrieval and recycling: The tree-root systems in an agroforestry system intercept, absorb, and recycle nutrients in the soil that would otherwise be lost through leaching, thereby making a more closed nutrient cycle. The mycorrhizal systems associated with tree roots are an agent in the nutrient cycling process. They penetrate a large proportion of the soil, facilitating the uptake of nutrients, which can move only short distance by diffusion. Thus, the nutrient-use efficiency in an agroforestry system is greatly enhanced. The leaf litter after decomposition forms humus, releases nutrients and improves soil properties. The nutrient contents of leaf litter may vary from 150 to 300 kg N, 10 to 20 kg P, 75 to 150 kg K and 100 and 300 kg Ca/ha/year which means that nutrient cycling can substantially reduce the fertiliser requirement of the associated crops grown in agroforestry system (Negi and Sharma, 1984).

Reduction of water loss through evapotranspiration: Tree canopies lower temperature and increase relative humidity in an agroforestry system which leads to low transpiration rate. Tree litter and prunings act as mulch on ground surface that reduces the evaporative loss of soil moisture. The water storage capacity of soil is also increased due to improved soil organic matter.

Reduction in the decomposition rate of humus: The rate of decomposition of organic matter is comparatively low under trees than in agricultural use through lower topsoil temperature and less soil disturbance.

Maintenance or improvement of physical properties: The soil physical properties such as structure, porosity, moisture retention, and erosion resistance are much improved in an agroforestry system than that of an open area or area with seasonal field crops. There is much evidence of the influence of physical properties of tropical soils on crop growth, independent of nutrient or other effects (Lal and Greenland, 1979).

Modification of extremes of soil temperature: A cover of tree leaf litter and prunings greatly reduce the surface temperatures of ground in the tropics, which sometimes exceed 50° C and adversely affect crop growth. In temperate regions, there is some protection from ground frost by the 'blanketing' effect of litter.

Penetration of compact layers by roots: Tree roots are capable of penetrating the layers of laterite, calcrete or hard pans in the soil, so drawing nutrients and water beneath.

Reduction of acidity: Trees tend to moderate the effects of leaching through the addition of bases to the soil surface. The bases released by litter decay can help to check acidification caused by the application of fertilisers to the crops in an agroforestry system.

Reduction of salinity and sodicity: Perennial woody species in an agroforestry system can reclaim saline and alkaline soils by lowering salinity and reducing sodium saturation of the exchange complex. Trees as such cannot contribute to reduction of salinity, but they assist in water management that leads to soil amelioration and leaching out salts. Afforestation with *Acacia nilotica* and *Eucalyptus tereticornis* in Karnal region successfully reduced the soil *p*H from 10.5 to 9.5 over five years (Dagar, 2009).

Reduction of soil toxicities caused by pollution: The trees do not extract pollutants from a chemically polluted soil, but some species are tolerant to them. The trees build up soil organic matter and ameliorate fertility which results in reducing pollution levels in the soil.

Improvement in the activity of soil fauna: The presence of tree litter in soil is frequently associated with higher rates of activity of soil fauna.

Effects of shading: Shade caused by the tree canopy in an agroforestry system lowers ground-surface temperatures, which reduce the rate of loss of soil organic matter by oxidation. Shading is allowed to a certain extent while trees are grown with crops. In most of the cases, the branches are pruned up to 1/3rd height of the trees with advancement of age of the trees.

Root nodulation: Roots of some nitrogen fixing trees have more nodules where they are in close contact with roots of non-nitrogen fixing plants, the latter stimulating the former.

Exudation of growth-promoting substances into the rhizosphere: Root exudates by a few tree species are believed to work as growth-promoting substances for other plant species.

Adverse effects of trees on soil

Removal of organic matter and nutrients through tree harvest: Trees accumulate large quantities of carbon and nutrients in their biomass, part of which is removed in harvest. The problem is greater where there is whole-tree harvesting, with gathering of fine branches and litter by local people after a timber harvest. From a soil management point of view, all branches and litter should be allowed to decay *in situ*, although such a practice appears unreasonable to the local people.

Competition between trees and crops for nutrients: Usually an established root system of trees or shrubs dominates that of newly planted annual crops. Ideally, trees in agroforestry systems should have a deep penetrating taproot system with limited lateral spread. Whereas lateral spread of the canopy can be controlled by pruning, root pruning is generally too expensive to be practical.

Competition between trees and crops for moisture: In the semiarid and arid zones, this is possibly the most serious problem encountered in agroforestry.

Production of substances which inhibit germination or growth: Some *Eucalyptus* species produce toxins which can inhibit the germination or growth of some annual herbs. Some trees release organic compounds to the soil which are phytotoxic and may cause problems in agroforestry systems.

Nutrient cycling in agroforestry systems

Nutrient cycling occurs to varying degrees in all land use systems. Agroforestry systems show a high potential to improve soil fertility with more efficient nutrient cycling because of the beneficial effects of woody perennials on the soil. The woody perennials have more extensive and deeper root systems than herbaceous plants and thus have a potential to capture and recycle a larger amount of nutrients. Their litter contribution to the soil is also more than that of herbaceous plants.

Forest ecosystems have high rates of turnover and low rates of losses from the system and these are usually self-sustaining through closed and efficient nutrient cycling systems. But agricultural systems are often open and the turnover within the system is relatively low and losses are comparatively high per unit area and unit time. Nutrient cycling in agroforestry systems falls between these two extremes and as compared to agricultural systems more nutrients in these systems are reused by plants before being lost from the system.

Agroforestry systems provide an opportunity for modifying nutrient cycling through management which results in more efficient use of nutrients in the soil. There are several management options for exploiting the advantages of efficient nutrient cycling in agroforestry systems.

1. The deep root systems of trees reach deeper soil horizons, which are not often accessed by roots of common agricultural crops. Thus, there is an enhanced nutrient uptake as a result of rock weathering or percolation past herbaceous plant roots. This process, which is commonly called nutrient pumping, is a significant factor of soil fertility improvement in agroforestry systems.
2. Gains from symbiotic N_2-fixation by trees can be enhanced through selection of tree species and admixture.
3. Different trees and shrubs used in alley cropping systems vary in quantity, quality and decomposition rate of leaf biomass. The timing of hedge pruning can be regulated in such a way that the nutrient release through decomposition of biomass is synchronised with the peak period of crop's nutrient demand.
4. Better management practices which improve organic matter status of the soil in an agroforestry system will also improve nutrient cycling and, thus better soil and crop productivity.
5. A major management consideration in agroforestry is to reduce nutrient loss through soil conservation.

Soil organic matter in agroforestry systems

Woody perennials differ from herbaceous crops in the quantity, quality and time of addition of organic matter to the soil. Trees usually add organic matter in a steady rate whereas herbaceous crops add biomass mostly at harvest. Biomass added to the soil from trees contains more lignin than herbaceous crops, which affect the rate of decomposition and humus formation. Majority of the organic materials are of plant origin, others include microbial tissue and dead biomass of soil fauna. Soil organic matter consists of two parts, fully decomposed organic matter or humus and plant and microbial remains that are in various stages of decomposition, commonly called litter. Organic matter decomposition refers mainly to the conversion of litter to humus through the activity of soil fauna and microflora.

Plant litter with high nitrogen is decomposed rapidly and is considered to be of high quality, whereas woody residues and other materials like cereal straw which have high lignin content are more resistant to decomposition and are considered to be of low quality. Materials with low carbon to nitrogen ratios (C: N) decompose rapidly and release relatively larger quantities of nitrogen. But materials with high C: N ratios provide a source of energy (C) for the microbes to multiply rapidly and utilise the N reserves from the soil. This causes a temporary unavailability of nitrogen for the plants as the added material is very low in N. Subsequently when the C source is depleted, the microbial population declines and the N that have been temporarily incorporated in microbial tissues are once again released to the soil and available for plant uptake.

Prunings of many of the woody perennials used in agroforestry systems are generally high in nitrogen. Fresh leaves of some leguminous trees and shrubs such as *Leuceana leucocephala, Gliricidia sepium,* and *Erythrina* spp. decompose relatively fast and release a major part of their nutrients from the applied mulch within about four weeks under humid tropical conditions. But nutrient-rich and easily decomposable material may not always be desirable because nutrient release may exceed plant nutrient demands resulting in asynchrony between supply and demands. Mulches of species such as *Cassia siamea*, *Flemingia macrophylla* and *Dactyladenia barteri* that are also commonly used in agroforestry systems as hedgerow planting are generally slow to decompose. Suppression of weed growth or reduction of soil moisture loss through evaporation may be sometimes more desirable than providing a quick supply of nitrogen to the crops. In such situations these slow decomposing mulches which provide a good ground cover for a longer period of time are preferred. Thus, better nutrient use efficiency in an agroforestry system can be achieved by selecting tree species with differing rates of litter decomposition, adjusting the timing of pruning to regulate the time of addition of the mulch and modifying the method of application of the mulch, i.e., surface addition or soil incorporation. The roots of trees add a substantial amount of organic matter to the soil. The roots of crops function on a seasonal basis but tree roots function all year round. Mycorrhizae absorb carbohydrates from the host plant and in turn, function as an expanded root system which increases nutrient absorption.

Effect of agroforestry on soil erosion

Agroforestry improves or maintains soil productivity through its effect on soil conservation. Soil conservation addresses not only erosion, but also other forms of physical, chemical and biological deterioration of soil. The adverse effect of erosion is expressed in terms of loss of soil organic matter and plant nutrients which leads to reduction of crop yields.

While construction and maintenance of physical barriers like bunds and terraces to control runoff involves huge investment and causes irreparable damage to valuable topsoil, agroforestry provides continuous ground cover formed of living and dead plant materials, crop residues and tree prunings which effectively check raindrop impact and runoff and thus, conserves soil. The presence of multiple layers of canopy in agroforestry systems can considerably reduce the velocity of falling raindrops and thus reduce the severity of their impact. Barrier hedges such as in alley cropping are effective in reducing runoff. Woody perennials also act as runoff barriers.

Agroforestry practices influence the soil erodibility factor through their effect on soil physical properties and soil organic matter. Soil structure under agroforestry has higher stability, lower detachability and higher water infiltration capacity than that of pure cropping. Agroforestry systems can reduce the rainfall erosivity. However, it may not be true in all agroforestry situations. Low and dense canopy usually reduce erosivity. Thus, well managed agroforestry systems can reduce overall erosion losses.

Where wind is a major cause of soil erosion and moisture loss, windbreaks and shelterbelts can make a significant contribution to sustainable production. Trees and shrubs on the farmlands are traditionally used for soil erosion control in a number of ways. These include both direct use of these species for reducing erosion and their supplementary use for stabilising physical structures that are created specifically for erosion control. Direct use involves use of trees and shrubs in agroforestry systems for increasing soil cover, providing live or dead barriers and enhancing soil's resistance to erosion through maintenance or build-up of organic matter and desirable physical properties. The supplementary use consists of use of trees and shrubs to stabilise physical structures such as bunds, risers or embankments.

AGROFORESTRY IN MITIGATING CLIMATE CHANGE

Sustainable diversification of agricultural systems

Diversification of agricultural enterprises is one of the best practices to reduce the risks associated with variable climate. It is an adjustment of the farm enterprise pattern to increase farm income or reduce income variability by reducing risk and exploiting new market opportunities. At the farm level it is the adoption of multiple production activities that are complementary in economic and/or ecological dimensions involving crops, trees and livestock. Agroforestry systems are a suitable pathway for sustainable diversification of agricultural systems. Agroforestry ensures food security by generating direct benefits to farmers such as food, fodder, feed, fuelwood, live fences and other products. The diversity of crops provides multiple harvests at different times of the year, thereby reducing the risk of crop loss and food shortage. Quality of life of farmers is improved by increasing income due to multiple harvests and sale of products from different components, thereby providing regular income throughout the year.

Improvement in rainwater use efficiency

Climate change has both direct and indirect impacts on water availability. The direct impacts include changes in precipitation patterns while the indirect ones are increases in losses through runoff and evapotranspiration. The presence of trees has positive effect on water budget of the soil and crops growing between or beneath them. There are several mechanisms whereby agroforestry may use available water more effectively than the annual crops. Agroforestry systems with a perennial tree component can make use of the water remaining in the soil after harvest of annual crops and the rainfall received outside the crop season. A mulch or litter layer increases the infiltration of rainwater while simultaneously reducing evaporation from the soil. The changes in microclimate also reduce the evaporative demand and make more water available for transpiration.

Soil and water conservation

There are several ways by which climate change manifests soil degradation. Higher temperatures and drier conditions lead to lower organic matter accumulation in the soil resulting in poor soil structure, reduction in infiltration of rainwater and increase in runoff and erosion. The occurrence of extreme rainfall events has adverse impacts on the severity, frequency and extent of erosion. Trees improve soil fertility by processes of increasing additions to the soil, reduction of losses from soil, and improving soil physical, chemical and biological conditions. Trees have a different impact on soil properties than annual crops in terms of their longer residence time, larger biomass accumulation and more extensive root systems. Trees increase or at least maintain the organic matter levels of the soil. This is mainly through litter fall and continuous degeneration or sloughing-off of roots of standing trees. Pruning materials from the woody perennials used in agroforestry systems add a huge quantity of organic matter to the soil and this is generally high in nitrogen.

Potential to limit carbon emissions and sequester carbon

The tree component of the agroforestry systems can be a significant sink for atmospheric carbon. An efficient agroforestry system not only maximises the benefit it provides but also ensures the link to climate change mitigation. Trees are important carbon warehouses that filter massive quantities of carbon from the atmosphere, trapping it in their biomass. Through the natural process of photosynthesis, plants absorb carbon dioxide from the atmosphere and release oxygen. They store this carbon in their leaves, branches, trunks and roots. An average tree can remove about 23 kg of carbon dioxide from the atmosphere annually. In agroforestry systems, the amount of carbon sequestration is further increased. The interactions of the different components of agroforestry systems can help absorb and sequester carbon dioxide and other greenhouse gasses from the atmosphere. Thus, trees in an agroforestry system make it a potential strategy in mitigating climate change.

Microclimate amelioration

The full genetic potential of many crops can only be exploited under optimum environmental conditions. Any change in these conditions, especially during the reproductive stage, will have a direct impact on the production and economic viability of certain crops. In an agroforestry system, the presence of trees reduces heat and light. Trees bring about favourable changes in the microclimatic conditions by influencing radiation flux, air temperature, wind speed all of which will have a significant impact on modifying the rate and duration of photosynthesis and subsequent plant growth, transpiration and soil water use.

Interventions for drought mitigation

Tree canopies lower the temperature and increase relative humidity in an agroforestry system which leads to a lower transpiration rate. Tree litter and pruning materials act as mulch on ground surface that reduces the evaporative loss of soil moisture. The water storage capacity of soil is also increased due to improved soil organic matter. Higher survival and biomass generation under agroforestry is possible due to greater efficiency for photosynthesis, tapping of nutrients and water from deeper layers, efficient utilisation of off-season rainfall by the perennial tree component and better environmental conditions.

Protection and stabilisation of ecosystem

Trees give protection from adverse effects of different kinds of pollutants such as dust, dirt and other physical air pollutants. The dust level in the air can be as much as 75% lower on the sheltered side of the tree compared to the windward side. It is estimated that one hectare of a close forest filters about 50 t of dust and dirt every year. Trees also absorb some of the chemical air pollutants.

AGROFORESTRY AND THE FUTURE

Agroforestry is bound to play a major role in future, not only for its importance in food and livelihood security, but also for its role in combating the environmental challenges. It is estimated that by 2050, the demand for fodder, food grain, timber and fuel wood in the country would increase by 1.5, 2, 3 and 2 times, respectively (CAFRI, 2015). Agroforestry has the potential to meet this demand for food, fodder, firewood and timber. With the modern-day crisis of shortage of land for forestry and agriculture, agroforestry is well positioned to provide a perfect balance and a viable solution.

Table 13.1. Total domestic demand for various commodities and agroforestry contribution in 2050

Items	Projected for 2050	Contribution from agroforestry in 2050	Contribution (%) from agroforestry in 2050
Food grains (millions t)	457.1	41.14*	9.00
Fruits (millions t)	305.3	47.74*	15.64
Fodder (millions t)	1545	154.50	10.00
Fuel wood (millions t)	629	308.00	48.97
Timber (millions t)	347	295.00	85.01
Biodiesel (millions t) required for 20% blending of diesel	37.92	30.34	80.01
Area (millions ha) required for tree-borne oilseeds	21.67	17.34	80.02
Agroforestry (millions ha)	53.32		

*Food-grains/fruits production from systematic agroforestry systems, viz., agrisilviculture/ agrihorticulture only considered

Source: CAFRI, 2015

Agroforestry today has become a sustainable method to manage forest and agriculture together, while being economically and environmentally viable. This has the potential to reduce regional disparity, bring desirable peace, prosperity and happiness and ensure an optimistic future for the generations to come. Thus, the need of the hour is to invest in further research and development in this relatively new science. Agroforestry research should focus on development and demonstration for the adoption of various agroforestry models linked with market for enhancing productivity and profitability for small holding farmers. The state governments should evolve state agroforestry policy in consonance with the National Agroforestry policy 2014, to plan, prioritise and develop agroforestry action plan suitable for each agroclimatic zone and focus on strengthening research and activities through public-private partnership mode for developing profitable, ecologically and socioeconomically viable agroforestry models for all farmers (Dhyani, 2014). The promotion of sustainable agroforestry practices on a large scale in future only possible through amalgamation of proactive farmer policies of government, involvement of the industries, support services from non-government organisations and willingness of farmers (Verma et al., 2017).

REFERENCES

Armson, K. A. 1977. Forest Soils: Properties and Processes. University of Toronto Press, Toronto, Canada.

Bond, G. 1983. Taxonomy and distribution of non-legume nitrogen-fixing systems. *In*: Gordon, J. C. and Wheeler, T. C. (Eds.) Biological Nitrogen Fixation in Forest Ecosystems: Foundations & Applications. The Hague, the Netherlands .pp. 55-87.

CAFRI. 2015. Vision 2050. ICAR-Central Agroforestry Research Institute, Jhansi, India.

Dagar, J. C. 2009. Opportunities for alternate land uses in salty and water scarcity area. *International Journal of Ecology and Environmental Science* **35**: 53-66.

Dhyani, S. K. 2014. National Agroforestry Policy 2014 and the need for area estimation under agroforestry. *Current Science* **107**(1): 9-10.

Jhon, B. 1988. Litter production and nutrient cycling in coffee plantation with shade. *Agroforestry systems* 7(2): 103-114.

Lal, R. and Greenland, D. J. (eds.). 1979. Soil Physical Properties and Crop Production in the Tropics. Wiley, Chichester, UK.

Lundgren, B. O. and Raintree, J. B. 1982. Sustained agroforestry. In: Nestel, B. (ed.). Agricultural Research for Development: Potentials and Challenges in Asia. ISNAR, the Hague, the Netherlands.

Nair, P. K. R. (ed.). 1989. Agroforestry systems in the tropics. Kluwer Academic Publishers, Dordrecht, the Netherlands.

Nair, P. K. R. 2008. An Introduction to Agroforestry. Springer (India) Pvt. Ltd., New Delhi.

Negi, J. D. S and Sharma, D. C. 1984. Distribution of nutrients in age series of *Eucalyptus globulus* plantation in Tamil Nadu. *Indian Forester* **100**(9): 944-953.

Verma, P., Bijalwan, A., Dobriyal, M. J. R., Swamy, S. L. and Thakur, T. K. 2017. A paradigm shift in agroforestry practices in Uttar Pradesh. *Current Science* **112**(3): 509-516.

Young, A. 2005. Agroforestry for soil management. CAB International, Walling Ford, U.K.

14

Organic Farming for Sustainable Agriculture

Organic farming is one of the several approaches found to meet the objectives of sustainable agriculture. It is often associated and sometimes confused with 'sustainable farming'. However, policy and ethics-wise 'organic farming' is different from 'sustainable farming'. Organic farming is based on various laws and certification programmes whereas sustainable agriculture is the practice of farming using principles of ecology. Organic farming through sustainable agriculture meets not only the food requirements of present generation in an environment friendly way but also the requirements of future generations and maintains our environment. Organic farming is a production system which avoids or largely excludes the use of synthetic fertilisers, growth regulators, livestock feed additives and pesticides. It relies on agronomic management practices like crop rotation, crop residues, animal manures, green manuring, organic wastes, biological pest control, etc.

The United State Department of Agriculture defines organic farming as a system which avoids or largely excludes the use of synthetic inputs such as fertilisers, pesticides, hormones, feed additives, etc. and to the maximum extent feasible rely upon crop rotations, crop residues, animal manures, off-farm organic waste, mineral grade rock additives and biological system of nutrient mobilisation and plant protection. In another definition Food and Agriculture Organisation suggests that organic agriculture is a unique production management system which promotes and enhances agroecosystem health, including biodiversity, biological cycles and soil biological activity and this is accomplished by using on-farm agronomic, biological and mechanical methods in exclusion of all synthetic off-farm inputs. Organic agricultural methods are internationally regulated and legally enforced by many nations, based on the standards set by the International Federation of Organic Agriculture Movements (IFOAM). IFOAM defines organic farming as a production system that sustains the health of soils, ecosystems and people. It relies on ecological processes, biodiversity and cycles adapted to local conditions, rather than the use of inputs with adverse effects. Organic agriculture combines tradition, innovation and science to benefit the shared environment and promote fair relationships and a good quality of life for all involved.

HISTORY OF ORGANIC FARMING

The origin of organic farming goes back to 1940s. During this period, the path breaking literature on the subject published by Jerome Irving Rodale in the United States, Lady Balfour in England and Sir Albert Howard in India contributed to the cause of organic farming. In Germany, Rudolf Steiner's development, biodynamic agriculture, was probably the first comprehensive organic farming system. This began with a lecture series Steiner presented at a farm in Koberwitz (now in Poland) in 1924. Steiner emphasised the farmer's role in guiding and balancing the interaction of the animals, plants and soil. Healthy animals depend upon healthy plants (for their food), healthy plants upon healthy soil, and healthy soil upon healthy animals (for the manure). He characterised a farm as 'an organism' (Paull, 2011a).

The British botanist Sir Albert Howard is often referred to as the father of modern organic agriculture. From 1905 to 1924, he worked as an agricultural adviser in Pusa, Bengal, India where he documented traditional Indian farming practices and came to regard them as superior to the conventional agriculture science. His research and further development of these methods are recorded in his book, 'An Agricultural Testament' published in 1940, which influenced many scientists and farmers of the day. In 1939, influenced by Sir Albert Howard's work, Lady Eve Balfour launched the Haughley Experiment on farmland in England. It was the first scientific, side-by-side comparison of organic and conventional farming. Four years later, she published 'The Living Soil', based on the initial findings of the Haughley Experiment.

The term 'organic farming' was coined in 1940 by Lord Northbourne and first appeared in his book 'Look to the Land' (Paull, 2006). From Steiner's characterisation of the farm as 'an organism' and the practices of biodynamic farming, he described a holistic, ecologically balanced approach to farming (Paull, 2011b). During the 1950s, sustainable agriculture was a topic of scientific interest. In the USA, J. I. Rodale began to popularise the term and methods of organic growing, particularly to consumers through promotion of organic gardening.

In Japan, Masanobu Fukuoka, a microbiologist working in soil science and plant pathology, began to doubt the modern agricultural movement. In 1937, he quit his job as a research scientist, returned to his family's farm in 1938, and devoted the next 60 years to developing a radical no-till organic method for growing grain and many other crops, now known as 'Nature Farming' or 'do-nothing' farming. In 1975, Fukuoka published his first book, 'The One-Straw Revolution', with a strong impact in certain areas of the agricultural world. His approach to small-scale grain production emphasised a meticulous balance of the local farming ecosystem, and a minimum of human interference and labour. The International Federation of Organic Agriculture Movements was founded in 1972 in Versailles, France (Paull, 2010).

The negative effects of modern chemical based farming system were first experienced by those countries, which introduced it initially. So, naturally, it was in those countries organic farming was adopted in relatively large scales. In the 1980s, around the world, farming and consumer groups began seriously pressuring for government regulation of organic production. This led to legislation and certification standards being enacted through the 1990s. Since the early 1990s, the retail market for organic farming in developed economies has been growing by about 20% annually due to increasing consumer demand. Concern for the quality and safety of food, and the potential for environmental damage from conventional agriculture, are apparently responsible for this trend.

In March 1999, organic agriculture was officially included into the agenda of Food and Agriculture Organisation as a way to support sustainable progress. The interest of FAO in organic agriculture stems from its potential to contribute to rural development and global food security. FAO member countries in the International Conference on Organic Agriculture and Food Security in Rome (2007) emphasised the organic agriculture's capability to create a more stable food supply, increased access to food in rural areas, and maintenance of natural resources (Morgera et al., 2012).

The organic agriculture movement in India received inspiration and assistance from IFOAM. All India Federation of Organic Farming (AIFOF) is a member of IFOAM and consists of a number of NGOs, farmers' organisations, promotional bodies and institutions.

AIMS AND PRINCIPLES OF ORGANIC FARMING

The aims and principles of organic farming concisely expressed in the statement of basic standards for production and processing of the International Federation of Organic Agriculture are as follows (IFOAM, 1998). As the statement makes clear, the scope of the principles extends beyond simple biophysical aspects to matters of justice and responsibility (Rigby and Caceres, 2001).

1. To produce food of high quality in sufficient quantity
2. To interact in a constructive and life-enhancing way with natural systems and cycles
3. To consider the wider social and ecological impact of the organic production and processing system
4. To encourage and enhance biological cycles within the farming system, involving microorganisms, soil flora and fauna, plants and animals
5. To develop a valuable and sustainable aquatic ecosystem
6. To maintain and increase long-term fertility of soils

7. To maintain the genetic diversity of the production system and its surroundings, including the protection of plant and wildlife habitats
8. To promote the healthy use and proper care of water, water resources and all life therein
9. To use, as far as possible, renewable resources in locally organised production systems
10. To create a harmonious balance between crop production and animal husbandry
11. To give all livestock conditions of life with due consideration for the basic aspects of their innate behaviour
12. To minimise all forms of pollution
13. To process organic products using renewable resources
14. To produce fully biodegradable organic products
15. To produce textiles which are long-lasting and of good quality
16. To allow everyone involved in organic production and processing a quality of life which meets their basic needs and allows an adequate return and satisfaction from their work, including a safe working environment
17. To progress towards an entire production, processing and distribution chain which is both socially just and ecologically responsible

A conventional farmer needs a farming model that can leave the environment better than before, the forest healthier, the land more diverse, wildlife more prolific and soil more fertile resulting in clean water, clean air, and healthy animals and plants (Deshmukh, 2010). Practice of organic farming would fulfil these. The IFOAM definition of organic agriculture is based on the four principles. The principles of health, ecology, fairness, and care are the roots from which organic agriculture grows and develops. They express the contribution that organic agriculture can make to the world and a vision to improve all agriculture in a global context. The IFOAM has elaborates these four principles as follows (Luttikholt, 2007).

The principle of health

Organic Agriculture should sustain and enhance the health of soil, plant, animal, human and planet as one and indivisible. Health is the wholeness and integrity of living systems. It is not simply the absence of illness, but the maintenance of physical, mental, social and ecological well-being. Immunity, resilience, and regeneration are key characteristics of health. The role of organic agriculture, whether in farming, processing, distribution, or consumption, is to sustain and enhance the health of ecosystems and organisms from the smallest in the soil to human beings. In particular, organic agriculture is intended to produce high quality, nutritious food that contributes to preventive health care and well-being. In view of this, it should avoid the use of fertilisers, pesticides, animal drugs and food additives that may have adverse health effects.

The principle of ecology

Organic Agriculture should be based on living ecological systems and cycles, work with them, emulate them and help sustain them. This principle states that production is to be based on ecological processes, and recycling. Nourishment and well-being are achieved through the ecology of the specific production environment. For example, in the case of crops this is the living soil; for animals it is the farm ecosystem; for fish and marine organisms, the aquatic environment. Organic farming, pastoral and wild harvest systems should fit the cycles and ecological balances in nature. These cycles are universal but their operation is site-specific. Organic management must be adapted to local conditions, ecology, culture and scale. Inputs should be reduced by reuse, recycling and efficient management of materials and energy in order to maintain and improve environmental quality and conserve resources. Organic agriculture should attain ecological balance through the design of farming systems, establishment of habitats and maintenance of genetic and agricultural diversity. Those who produce, process, trade, or consume organic products should protect and benefit the common environment including landscapes, climate, habitats, biodiversity, air and water.

The principle of fairness

Organic agriculture should build on relationships that ensure fairness with regard to the common environment and life opportunities. Fairness is characterised by equity, respect, justice and stewardship of the shared world, both among people and in their relations to other living beings. This principle emphasises that those involved in organic agriculture should conduct human relationships in a manner that ensures fairness at all levels and to all parties, farmers, workers, processors, distributors, traders and consumers. Organic agriculture should provide everyone involved with a good quality of life, and contribute to food sovereignty and reduction of poverty. It aims to produce a sufficient supply of good quality food and other products. This principle insists that animals should be provided with the conditions and opportunities of life that accord with their physiology, natural behaviour and well-being. Natural and environmental resources that are used for production and consumption should be managed in a way that is socially and ecologically just and should be held in trust for future generations. Fairness requires systems of production, distribution and trade that are open and equitable and account for real environmental and social costs.

The principle of care

Organic agriculture should be dealt with in a responsible and precautionary manner in order to guard the health of the present and future generations and for the well-being of environment. Organic agriculture is a living and dynamic system that responds to internal and external demands and conditions. Practitioners of organic agriculture can enhance efficiency and increase productivity, but this should not be at the risk of jeopardising health and well-being. Consequently, new technologies need to be assessed and existing methods reviewed. Given the incomplete understanding of ecosystems and agriculture, care must be taken. This principle states that precaution and responsibility are the key concerns in management, development and technology choices in organic agriculture. Science is necessary to ensure that organic agriculture is healthy, safe and ecologically sound. However, scientific knowledge alone is not sufficient. Practical experience, accumulated wisdom and traditional and indigenous knowledge offer valid solutions, tested by time. Organic agriculture should prevent significant risks by adopting appropriate technologies and rejecting unpredictable ones, such as genetic engineering. Decisions should reflect the values and needs of all who might be affected, through transparent and participatory processes.

BENEFITS OF ORGANIC FARMING

Organic agriculture is a holistic way of farming. Its primary objective is to conserve natural resources like soil, water, climate, biodiversity and non-renewable energy for achieving sustainability in agricultural production. The basic approaches to organic farming are as follows (Walia et al., 2011).

1. To improve and avoid overexploitation and pollution of natural resources
2. To minimise the consumption of non-renewable energy sources
3. To exploit synergies those exist in natural ecosystems
4. To maintain and improve soil health by stimulating activity of soil organisms with organic manures and avoid harming them with pesticides
5. To obtain optimum economic returns, within a safe, secure and healthy working environment
6. To acknowledge the virtues of indigenous knowledge and traditional farming systems

Organic farming provides long-term benefits to people as well as the environment. The environmental advantages of organic farming include increased biodiversity, improved soil quality and reduced pesticide and fertiliser pollution and packaging waste, and water usage (Ziesemer, 2007). Besides its potential to alleviate poverty and economic gains, organic farming proves to be valuable in a number of ways.

Environmental benefits

Organic farming is based on ecological cycles and procedures, using low external inputs. Thus, it protects the environment from harmful effects of synthetic inputs such as pesticides, herbicides, fertilisers and hormones. Fertilisers and pesticides release dangerous toxic chemicals into soil and water. Some pesticides can cause harm to the environment or on direct exposure, they can prove to be toxic and dangerous to human health. Organic farming is a system based on generating food with negligible destruction to ecosystems, humans, or animals.

Organic agriculture more or less constantly supports more biodiversity and usually has a positive environmental impact per unit of land (Siddique et al., 2014). Organic farming system provides greater biodiversity as compared to the traditional farming system due to decreased soil changes and chemical application (Nascimbene et al., 2012). All non-crop species exhibit higher inclination for organic farming system in terms of both diversity and abundance. An average of 30% more species reside in organic farms. Butterflies, birds, beetles, spiders, earthworms, mammals, soil microbes, and vegetation are particularly affected (Gabriel et al., 2006). Agricultural biodiversity consists of all shapes of life directly significant to agriculture, and can subsist equally in a farm and crossway farms. This amplifies crop resilience and ecological firmness.

Organic farms are more capable of withstanding harsh weather conditions as compared to the conventional farms. In drought conditions, their yield is about 70% more than conventional farms (Lotter, 2003). Organic farms are more cost effective in the drier regions because of their better drought performance. These can also check loss of topsoil due to water or wind erosion. Hence, organic farming contributes positively to the reduction in soil, air, and groundwater pollution. Moreover, it is also an excellent solution to nitrate pollution. In addition, it improves soil fertility, structure, and soil fauna (Fan et al., 2005). The organic matter is globally recognised to enhance soil fertility. In addition, improving the soil's chemical, physical, and organic matter has the prospective to add to climate change alleviation by impounding carbon from the atmosphere.

Economic benefits and profitability

Organic agriculture consumes zero pesticides and less energy (Mader et al., 2002). High prices that consumers disburse for organic products along with decreased cost of pesticide inputs and synthetic fertilisers add to increased profits. Organic farms in long-run have always been found to be more profitable as compared to the conventional farms (Lotter, 2003). In the majority of cases, 30-50% less energy is consumed by organic farming as compared to the traditional farming. Organic agriculture typically uses energy more competently than conventional agriculture. Organic agriculture often needs about one-third additional manual labour hours as a substitution for energy-intensive inputs used in traditional agriculture (Ziesemer, 2007).

Organic agriculture is about 30% more efficient to produce the same amount of food as compared to the traditional farming. Conventional farming is capital intensive, which requires more manufactured inputs and energy as compared to the organic farming which is knowledge and labour intensive (Alroe et al., 2006). The system engages large skilled and semiskilled/unskilled labour for various tasks to be performed (sowing, planting, cultivating, rearing, maintenance, aftercare, harvesting, cleaning, washing, grading, bar coding, labelling, packing, transporting, and marketing) in order to follow a strict code of organic farming (Pimentel, 2006).

Health benefits

Pesticide residues in foodstuff have adverse health effects. The organically produced vegetables and fruits contain less agrochemical residues than their conventionally grown alternatives (Magkos et al., 2006). It is believed that organic products are healthier than traditionally grown products. Organic products contain less nitrate content. Larger amount of nitrate in food and drinking-water can cause cancer of the alimentary tract and methemoglobinemia in infants (Forman and Silverstein, 2012). The organically grown vegetables and fruits have been linked with lesser risk of chronic human health harms like hypertension, cancer and cardiovascular diseases because of their elevated phytochemicals. The health advantages of vegetables and fruits have been endorsed to the antioxidant characteristics of phytochemicals (Siddique et al., 2014).

Social benefits

Organic farming has an important social effect on local communities. It provides employment opportunities to the local people. More manual labour is often required in organic agriculture to compensate for pesticide and synthetic fertiliser loss, thus providing more jobs in local communities. Commonly, the labour required to run an organic farm is 10-20% higher as compared to the traditional farms. Organic farmers also expand their crops and widen their planting schedules throughout the year in order to enhance soil health and maintain biodiversity. This establishes year-round employment opportunities and may lessen the problems related to migrant labour. More job opportunities will increase the population of local communities and also halt migration to urban areas. Thus, organic agriculture can increase the local communities and support rural development. In order to stay competitive, farmers must adjust to the local conditions by managing land, labour, and resources so that the production can be increased. Farmers also depend on their neighbours to sustain certain principles in order to guarantee the reliability of their own water, soil, and air. Ties within the community are strengthened by association on these issues, leading to greater association among organic farmers and also partnerships. Cooperatives or organised groups can thus gain power in trade negotiations, gather their resources, and enjoy greater access to markets. Increased collaboration may result in new businesses among local communities. Consumer protection is another keystone of organic farming. The well-built regulatory frameworks, whereby the government verifies organic certifications, are essential for consumers to trust the food that they buy (Morgera et al., 2012).

IMPACTS OF ORGANIC FARMING ON GLOBAL WARMING

Organic agriculture increases water and crop quality by removing external inputs, pesticides, herbicides, synthetic fertilisers and genetically modified crops. During the production of synthetic fertilisers, carbon dioxide is released as a side product. In case of organic agriculture, only on-farm compost, animal manure, crop residue, etc. are used as plant nutrients. Production and transportation of organic manures also contribute to greenhouse gas emissions, but this is much less than that of the off-farm mineral fertilisers. Fossil fuels are usually the source of energy for the manufacturing and shipment of artificial pesticides. Thus, the global warming potential of organic pest control is less than that of traditional methods (Siddique et al., 2014).

Change in rainfall patterns due to global warming may lead to drought conditions in many areas. Extenuating the emissions of GHGs is thus an important challenge to improve food security. One way to achieve this is by reducing CO_2 emission due to combustion of fossil fuel, but it can also be done by farming (Moreau, 2007). Organic agriculture supplies more GHGs in the soil than conventional agriculture. Carbon is stored in the soil instead of heating up the atmosphere, which acts like a carbon sink. Thus, more carbon can be stored in organically farmed soil than any other farming system depending on the use of plant protection chemicals & artificial nitrogen fertiliser. Farmers across the globe may help to solve the climate disaster by switching to organic methods, along with avoiding the use of synthetic pesticides and fertilisers (Isaacs, 2012). It is estimated that at least 30% of global warming is due to emission of GHCs, mainly CH_4, N_2O & CO_2 by agricultural practices. Fertiliser industry mainly emits CO_2. A significant contributor for CO_2 emission by agriculture is deforestation. CH_4emissions from livestock are mostly from enteric fermentation. Rice fields also emit CH_4.

About half of the energy used in agriculture in developed countries is in the manufacturing of fertilisers, mostly nitrogen fertilisers. However, there is an increased amount of energy utilised by the fertilisers due to minor mechanisation and less efficient use of fertiliser plants. Since organic agriculture does not use artificial nitrogen fertiliser, less energy is consumed than conventional agriculture, and therefore less CO_2 is emitted. It has been estimated that in Europe for key crops, organic farming uses about half the energy used in traditional agriculture systems (Siddique et al., 2014). Despite lower yield in organic agriculture, the amount of crops produced remains a significant factor. In European livestock production, the energy consumed in the production of organic milk is 25% of conventional milk production. This is because organic cows are mainly grazing, whereas the feed of conventional ones is based on grain and soybean cake (Moreau, 2007). It has been observed that the organic carbon content in many areas of the world has decreased progressively due to the increase in agricultural practices with deep ploughing and use of chemical fertilisers. Deforestation in tropics has led to a faster decline in the organic matter content. Organic farming increases or at least maintains carbon content of soil. This ability to seize carbon, contributes to alleviate the contribution of agriculture to the greenhouse effect (Moreau, 2007).

Soil mainly emits nitrous oxide (N_2O). The Intergovernmental Panel on Climate Change has evaluated that about 1.25% of the amount of nitrogen applied as fertiliser is represented by these emissions. The amount of nitrogen applied normally is lower in organic than conventional agriculture. The emissions of N_2O increase radically when nitrogen fertilisation exceeds the needs of the crop, which happens habitually in conventional farming. Thus, the emission of N_2O is less in organic farming as compared to conventional farming. (Moreau, 2007).

There are three main origins of methane from agriculture; anaerobic fermentation of flooded rice, fermentation of animal dejections and enteric fermentation of ruminants. The production of methane per animal is about the same in conventional and organic breeding. But, the emission per kilo of meat or milk is more in intensive production. This increase can be compensated by better endurance of organic cows. Methane emission by fermentation of compost is less in organic than conventional breeding, because composting is aerobic, whereas storing compost in heaps or slurry is mainly anaerobic (Moreau, 2007).

ORGANIC FARMING IMPROVES SOIL HEALTH

Soil is a living system and soil fertility is the key to agricultural productivity. Any input that destroys this living system and undermines soil health basically undermines the agricultural productivity. The maintenance of the fertility of soil is the primary step in any permanent system of agriculture. The soil carries a lot of fertility of its own. Some soils can be highly productive without any addition of external inputs. However, chemicals destroy the life in soils by making them inert. Healing the dead soil requires reviving life in the soil (Shiva et al., 2004).

Soil is given more care by organic farming. Organic farming enhances the status of micronutrients. Green manuring and other organic practices, and vesicular-arbuscular mycorrhiza (VAM) fungi, enhance the uptake of macronutrients such as phosphorus and nitrogen. An increase in the organic matter of soil occurs, pH of soil becomes stable, and organic carbon enhances, which are important for good biological environment of soil. Organic manure speeds up the nutrition transformation processes. The enhancement of soil's physical properties occurs in terms of aggregation stability, porosity, nutrient retention, soil aeration, and water-holding ability (Deshmukh, 2010). In organic production systems, the soil health is maintained and improved through stimulation of soil microorganisms' activity with organic manures and by avoiding harming them with the use of synthetic pesticides and fertilisers.

Application of organic manures reduces bulk density and improves water holding capacity of soil (Parkash et al., 2002; Prasad and Gill, 2009). Infiltration rate is also increases with higher organic matter content in the soil (Patil et al., 2003). Organic farming ameliorates soil structure with an increased A horizon depth, organic matter content, porosity, earthworm abundance and activity with coarser and better developed aggregates than conventional farming.

The organic manures results in slow mineralisation rates during the crop period with synchronisation with crop uptake and results in reduced mineral N losses from the rooting zone during the growing season. In contrast, application of inorganic N leads to substantial losses of soil mineral N during the growing season. In organic agriculture where the crop residues are back to the system the soil may increase availability of P by decreasing its adsorption (Ohno and Erich, 1997). Cereal straws contain large amounts of K, 75% of which is water soluble and its incorporation increases K content of soil (Bhat et al., 1991). The organic manures enhance the availability of the native soil zinc and other micronutrients which are effectively utilised by the crop plants.

Some of the important aspects of organic farming in the context of soil health as summarised by Walia et al. (2011) are given below.

1. Organically managed soils have a high potential to counter soil degradation, erosion, and desertification, as they are more resilient both to water stress and nutrient loss.
2. Organic farming protects the long-term fertility of soils by maintaining organic matter levels, fostering soil biological activity, giving due importance to the

basic principles of crop rotation and intercropping, and providing crop nutrients indirectly by using sources that are not readily soluble.

3. In organic farming, the supply of nutrients is more balanced, which helps keep the plants healthy and improve the biological activity of soil, which in turn improves nutrient mobilisation from organic matter and native soil reserves and minimises the losses of nutrients, thus protecting the environment.
4. Microorganisms have a good feeding base and create a stable soil structure.
5. Inclusion of legumes or cover crops, mulching, intercropping and agroforestry plays an important role in protection against soil erosion and degradation.
6. Organic agriculture technologies increase the organic matter content of the soil, which has a positive effect on soil aggregation and water-holding capacity.
7. Animal manure, green manure and composts favour the composition processes and can replenish nutrients required by the crops and supply the soil with essential organic matter.
8. Legume crops are a highly valuable source of nitrogen. Closed nutrient cycles and efficient use of local resources are important for subsistence farmers depending on few and limited assets.
9. Water-retention capacity of the soil is increased due to their higher level of organic matter content and permanent soil cover. Due to the resulting capacity for higher moisture retention, the amount of water needed for irrigation is reduced.
10. Organic agriculture helps mitigate the climate change, because it reduces emission of greenhouse gases, especially nitrous oxide.

NUTRIENT MANAGEMENT IN ORGANIC AGRICULTURE

Nutrient management on organic farms requires long-term planning and a diverse combination of cultural practices and inputs. Usually, nutrient needs of crops are met through organic sources such as green manuring, crop residues, composts and farm yard manures, biofertilisers, etc. The inherent soil fertility should also be conserved through agronomic practices like crop rotation, cover cropping, organic mulching, etc. There are an increasing number of commercially available organic fertilisers and biofertilisers, but the most profitable organic farms typically source nutrients on or very near the farm by using organic wastes, scavenging residual soil nutrients, and biological fixation of nitrogen.

Animal manure

Manure application is often valuable to organic production and enriches the soil with natural components that originated from animals, either raw or composted. The method has restrictions though, as the material must not contain any synthetic additives, and manures are allowed at least three months prior to harvesting. Composted forms are preferable since they are more compact in volume and contain fewer potential pathogens and contaminants. Sometimes, applying manure to meet all of the crop nitrogen demand can lead to excessive soil phosphorus because crops remove more nitrogen than phosphorus. The excessive soil phosphorus is not likely to be harmful to crops but contributes to phosphorus loss in runoff and erosion and contamination of waterbodies. Therefore, manure nitrogen needs to be complemented by biological nitrogen fixation or other nitrogen sources in organic systems (Wortman et al., 2017). A nutrient budget can be a useful tool for long-term accounting of phosphorus addition via manure and removal via harvested crops. This approach to nutrient management can help to ensure organic farmers do not unsustainably mine inherent soil nutrient reserves, and on the other end of the extreme, do not build up soil nutrient levels of high risk for environmental contamination.

Compost

Composting processes organic waste into material of higher nutrient concentration, and reduces the bulk of organic materials through carbohydrate and water loss during decomposition. Compost is often easier to handle than the bulk organic material, and the composting process kills some pathogens and weed seeds. Compost has less odour and fewer microbial pathogens, with less risk of microbial contamination of fresh produce than with raw manure.

Nitrogen mineralisation for finished composts with a carbon: nitrogen ratio (C: N) less than 20:1 occurs readily with much of the organic nitrogen released to the first crop following application. Composts with C: N greater than 30:1 will result in net immobilisation of soil nitrogen with reduced nitrogen availability to the first crop following application. Nearly all of the nitrate and ammonium in the compost will be available to the plants the first season after application, but only 15% of the organic nitrogen in the compost may be available during that season if the C: N is high. In addition to supplying nitrogen, compost is an important source of other macro- and micronutrients. As with other organic materials, compost application can improve soil organic matter content, cation exchange capacity, soil porosity, aggregate stability, and water-holding capacity; although the magnitude of improvements will depend on current soil organic matter levels. Similarly, soil biological properties, including microbial biomass, microbial enzymatic activity, and nitrification potential are often increased following compost application (Wortman et al., 2017).

Biofertilisers

Biofertilisers are products designed to provide enhanced nutrient availability and uptake, stimulation of crop growth, biological nitrogen fixation, and protection against insect pests and disease. Depending on the purpose, biofertiliser products can be applied to soil, seeds, or foliar tissue. Biofertilisers may be most beneficial in soils of low to moderate soil organic matter and nutrient availability. In contrast, biofertiliser applications to soils of more than 3% organic matter rarely result in measurable crop or soil benefits. Introduced microorganisms often fail to compete and survive with already well-established and resilient microbial communities.

Several common categories of biofertilisers include nitrogen-fixers, phosphorus-solubilisers and phosphorus-absorbers. Nitrogen-fixers such as *Rhizobium*, *Azospirillum*, and *Azotobacter* convert atmospheric nitrogen into ammonia. *Bacillus* and *Pseudomonas* are examples of microbes found in phosphorus-solubilising biofertilisers that lower the soil *p*H to dissolve soil-bound phosphate for plant availability and may be most effective for calcareous soils. Arbuscular mycorrhizal fungi take up soil phosphorus, zinc and copper and transfer these to plant roots (Wortman et al., 2017).

Cover crops

Cover crops can improve soil physical properties, nutrient cycling and soil microbial activity. In addition, cover crops can scavenge residual nitrogen mineralised from soil and organic amendments before it is lost to volatilisation, runoff or leaching. However, like other sources of organic nitrogen, nitrogen contained in cover crop biomass is not entirely available to the next crop. It is important to consider the C: N of the cover crop residue. Species with high C: N (> 20:1) result in net immobilisation of soil nitrogen in the short-term, whereas nitrogen will be more readily available following decomposition of species with low C: N (< 20:1). Nitrogen fixation by legume cover crops can be useful for increasing soil nitrogen availability and balancing the abundant phosphorus and potassium supplied through manure or compost application (Wortman et al., 2017). Cover crops tackle soil erosion; improve water filtration and aeration with their roots. They also suppress weeds.

Green manuring

Green manuring is the practice of incorporating undecomposed green plant tissue into the soil. The function of a green manure crop is to add organic matter to the soil. As a result of the addition, the nitrogen supply of the soil may be increased and certain nutrients made more readily available to crop plants. It also increases moisture levels and adds nutrients for microorganisms, thus improving the soil quality and productivity. Green manuring plants usually belong to leguminous family. The most important green manuring crops are sunnhemp (*Crotalaria juncea*), dhaincha (*Sesbania aculeata*), *cowpea (Vigna unguiculata),* cluster beans (*Cyamopsis tetragonoloba*) and sesbania (*Sesbania rostrata, Sesbania speciosa).*

Crop rotation

Crop rotation can contribute to improved soil physical properties, pest management, nutrient availability, nutrient use efficiency, and crop yield. Legumes in the rotation can result in a nitrogen credit for subsequent crops due to biological nitrogen fixation and less nitrogen immobilisation compared with a non-legume as the preceding crop. However, the nitrogen credit for forage legumes is greater than for grain or vegetable legumes because most of the nitrogen is removed during harvest of the grain. In contrast, forage legumes contribute nitrogen from vegetative biomass, which can be comparable to cover crop nitrogen contribution if the forage has time to regrow between the last cutting and termination. Root architecture is important to sustainable crop rotation. Deep-rooted crops like alfalfa can scavenge immobile nutrients like phosphorus and leached nutrients like nitrate-nitrogen from deep soil layers, which may be released near the soil surface with decomposition of the crop residue for availability to subsequent crops in rotation (Wortman et al., 2017).

WEED MANAGEMENT IN ORGANIC AGRICULTURE

Organic weed management is an approach to weed removal and prevention that does not involve the use of synthetic chemicals. Some organic weed control strategies are cultural and mechanical methods, focusing on prevention, crop rotation and cultivation. Weeds can also be managed in several ways such as through careful planning and planting and by using mulches. For effective weed management in organic systems, soil management, machinery, weather and time should also be considered. The main principle behind organic weed control is that healthy soil promotes plant growth that controls the weeds. The weeds can also be managed by biological methods.

Cultural methods

Cultural weed control includes diversifying rotations, delaying planting, changing planting rate, timing of nutrient application and using cover crops. Cultural methods are the first line of defence in weed management. Good agronomic practices that result in vigorous, competitive crop plants are the real key to successful weed control. Many agronomic procedures that encourage healthy soil conditions with a diverse microbial population can also reduce weed pressure. Cultural practices may not prevent all weed growth, and some mechanical follow-up will be necessary, but cultural practices can improve soil conditions, permitting more effective mechanical control, they can adjust weed species to ones that are easier to control. Cultural weed management practices produce high-quality, vigorous, high-yielding organic crops.

Sanitation: Using clean seed will prevent the introduction of new weed problems. The use of mowing weeds around the edges of fields or after harvest prevents weeds from going to seed. Hand-rouging weeds in problem areas and thoroughly composting manure can reduce the spread of weed seeds and difficult weed species. Planting clean and high-quality seeds is essential to avoid weed seeds. Some other sanitation factors include thorough cleaning of any machinery which might have been used in weedy fields, and the establishment of hedgerows to limit windblown seeds.

Crop competition: The most effective way to control weed growth is to have highly competitive crops. A vigorously growing crop is less likely to be adversely affected by weed pressure. It is imperative to create conditions where the intended crop can establish dominance quickly. Using high-quality, vigorous seed, well-adjusted planting equipment, adapted varieties, optimal soil fertility, good soil drainage and tilth, and proper soil preparation will usually result in rapid, vigorous crop growth (Howell and Martens, 2002).

Soil fertility and condition: An active and diverse soil microbial population is the key to growing healthy, high-yielding organic crops. Successful organic fertility management should primarily feed the soil ecosystem in a long-term manner, rather than simply feeding the plants. Soil organic matter is a tremendous source of plant nutrients and water-holding capacity. Careful attention to the balance of key nutrients can often reduce weed problems and enhance crop plant growth. The improper application of manure or improperly finished compost can throw off the balance of certain soil nutrients and microbial life and can often increase weed growth. Some soil fertility amendments like gypsum can increase the looseness and tilth of the soil. This improves mechanical cultivation operations, and also reduces the pressure from certain weed species that are favoured by hard, tight soils (Howell and Martens, 2002).

Crop rotation: Diverse crop rotations are essential to build a healthy, sustainable organic system and break weed cycles. The variation in crop species and planting times is essential, as crop rotation creates a changing environment and prevents weed dominance of certain species. When making a crop production plan, rotations should be designed for each field keeping the weed species and potential problems in mind. Crops that are fast growing and can out-compete weeds should be added to the rotation.

Green manures: Green manuring crops can be useful for organic weed management. They can be grown where there is a weed problem allowing the weeds to grow with them. Both are then grazed off by livestock or incorporated into the soil before seeding occurs. Thus, the weed seed bank is reduced and fewer weeds germinate in the following crops. Green manures are very competitive and can weaken perennial weeds. Several green manuring crops coupled with the right cultivations can rapidly clean up a land and improve nutrient levels.

Cover crops: Cover crops are being used to help control weeds after harvest and before planting the next crops. A healthy cover crop can help suppress weed growth by changing the soil dynamics, becoming established and growing quicker than weeds, and also by smothering weed seedlings. Also, some cover crops produce chemicals that are toxic to certain germinating seeds.

Allelopathy: Some plant species compete with each other by releasing chemical substances from their roots that inhibit the growth of other plants. This allelopathy is one of nature's most effective techniques of establishing plant dominance. Selecting allelopathic crops can be useful in particularly weedy fields with reducing overall weed pressure (Howell and Martens, 2002).

Variety selection: Careful selection of crop varieties is essential to limit weeds in crop fields. It is important to consider planting disease-resistant varieties if certain pathogens are prevalent in the area. The crop variety that grows more rapidly than the weeds and is able to quickly shade inter-row space restricts weed growth effectively. Deep shading crops, which intercept most of the sunlight that strikes the field and keeps the ground dark, will prevent the growth of many weed species (Howell and Martens, 2002).

Timing and spacing: Planting dates can be manipulated to escape weed germination windows. Seeding density of crops can also be increased to suppress late-germinating weeds. When perennial weeds are a problem, seasonal cash crops can be rotated with perennial crops.

Mulching: Organic mulches are effective at stopping weeds from germinating. The ground to be mulched must be completely free of established weeds before applying the mulch, including dormant perennial weeds. If the un-mulched areas contain creeping and spreading weeds, these will often invade the mulched area and then spread very rapidly, due to the lack of competition. Such spreading weeds should be removed from the un-mulched area. Many organic materials like straw, leaves can be effective mulches. Straw and other materials that are easily decomposed are applied to vegetables during the growing season.

Solarising: Solarisation is the method of covering an area with clear plastic to heat the soil and kill weeds and weed seeds in the top 15 cm of soil. This works effectively in warm weather conditions, where sufficient heat is collected under the sheet to kill weeds. The soil may be tilled to bring weed seeds to the surface to sprout just before solarising. Solarisation can be an effective process of controlling many annual weeds.

Mechanical methods

Mechanical weed control involves tillage, harrowing and the cutting and pulling of weeds. The mechanical weed control method is critical for managing weeds in organic systems. It includes these of pre-plant tillage such as ploughing, disking, and field cultivating. Then, these types of primary and secondary tillage can help reduce the rate and spread of certain perennial weeds and kill emerged weed seedlings and bury weed seeds below the germination zone.

Mechanical weed control becomes more effective by using a diversity of equipment that provides options to eliminate weeds at different stages of crop growth. Lack of favourable weather or soil conditions to perform a mechanical weed control operation in a timely manner is one of the biggest reasons for failure; thus, the availability of different implements that allow operation under different conditions can reduce risk. Though some methods become difficult after the cotyledon stage and their selectivity decreases with increasing crop and weed age. So, if the weeds have become too large, an intensive and aggressive adjustment of the implements is necessary to control the weeds, and doing this increases the risk of damaging the crop severely.

Blind cultivation: Blind cultivation is the easiest and best opportunity to destroy the weeds that would be growing within the rows and presenting direct competition to the crop. In blind cultivation, the entire field is tilled shallowly with the implement, paying little attention to where the rows are. The point of blind cultivation is to stir the top 2-3 cm of soil, adding air and causing the millions of tiny germinating weed seeds to dry out and die. The larger crop seeds germinate below the level of the cultivation and are not usually damaged by this operation. Weed seedlings are very vulnerable to drying out and to burying at this stage. Blind cultivation also breaks a soil crust, allowing crop seedlings to emerge. Usually, the first blind cultivation is done right before crop emergence, with a second one done about a week later, depending on conditions. The most effective blind cultivation is done when the soil is fairly dry and the sun is shining, a wind also improves the effect (Howell and Martens, 2002).

Biological methods

Use of living organisms such as insects, plant pathogens, herbivorous fish, snails or even competitive plants for the control of weeds is called biological control. The objectives of biological control are not eradication, rather the reduction and, regulation of weed population below the levels of economic injury. This method is not useful to control all types of weeds. Introduced weeds are best targets for biological control.

A successful bioagent is host specific, rapid destroyer of the target weed, effective on several species of the weed in question, adjustable to new environment and easy to multiply. A bioagent kills the weed plants by exhausting plant food reserves, defoliation, boring and weakening structure of the plant. The bioagents should possess the following qualities.

1. The bioagent must feed or affect only one host and not other useful plants.
2. It must be free of predators or parasites.
3. It must readily adapt to environment conditions.
4. The bioagent must be capable of seeking out itself to the host.
5. It must be able to kill the weed or at least prevent its reproduction in some direct or indirect way.
6. It must possess reproductive capacity sufficient to overtake the increase of its host species, without too much delay.

Larvae of *Coctoblastis cactorum*, a moth borer, control *Opuntia* sp. by tunnelling through the plants and destroy it. Cochineal insects (*Dactylopius indicus* and *D. tomentosus*) are also used to control this weed by sucking the plant sap. *Lantana camara* is controlled by larvae of *Crocidosema lantana*, a moth bores into the flower, stems, eat flowers and fruits. *Cuscuta* spp. is controlled by *Melanagromyza cuscutae*. *Bactra verutana*, a moth borer can control *Cyperus rotundus*. *Ludwigia parviflora* is completely denuded by *Altica cynanea* (steel blue beetle). Herbivorous fish *Tilapia* controls algae. Common carp, a non-herbivorous fish controls submerged aquatic weeds. It is apparently due to uprooting of plants while in search of food.

The biological method of weed control has the least harm to the environment with no residual effect. This is relatively cheaper and has comparatively long lasting effect. Usually bioagents do not affect non-targeted plants and safer in usage. However, the biological method of weed control has some demerits like costlier multiplication, slow control process, limited success of control and non-availability of host-specific bioagents.

Bioherbicides: The use of plant pathogens which are expected to kill the targeted weeds are called bioherbicides. These are native pathogen, cultured artificially and sprayed just like post-emergence herbicides each season on target weed, particularly in crop areas. Fungal pathogens of weed have been used to a larger extent than bacterial, viral or nematode pathogens, because, bacteria and virus are unable to actively penetrate the host and require natural opening or vectors to initiate disease in plants. Pathogenic organisms damage the host plants through enzymatic degradation of cell constituents, production of toxins, obstruction in the translocation of food materials and minerals and malfunctioning of physiological processes. A bioherbicide can often live in the environment and wait for the next growing season when there will be more weeds to infect. This reduces the farmer's cost of applying herbicides year after year.

PEST MANAGEMENT IN ORGANIC AGRICULTURE

Organic farming does not intend to destroy the pests completely as it may cause changes in ecosystem. This concept of agriculture strongly relies on the alternative methods of pest control, i.e., prevention, constructing natural barriers, physical removal, and tackling pest invasions with their biological enemies (predators). To design and implement best practices of pest management in organic farming, it is necessary to have accurate knowledge of crop and pest biology, ecology, phenology and pest-crop interactions, while following strictly regulated rules of organic farming.

Cultural practices

Cultural practices to protect the crop from pests include crop rotation, crop isolation, mixed cropping, intercropping, etc. Crop rotation is the potential method of preventing pests from getting used to any single type of plants that are being cultivated. This method uses cultivation of alternate species of crops every year. Intercropping involves simultaneous cultivation of two or more crops in the same farm. The definite distance remains between crops of the same species. So, this technique is efficient in attracting pests away from their target host plant. The timing of sowing and planting can also be used to allow young plants to establish to a tolerant stage before an attack occurs and to reduce the susceptible period of attack. Use of trap crops may divert insect attack away from the crop.

Sticky traps

These are particular colour traps and a sticky substance is spread over them. These traps are used to catch the insect which is attracted towards a particular colour. For cucumber beetles, a white colour mobile trap can be used. A large object can be painted white and moved in crop rows, cucumber beetles will be attracted towards that. For white flies, fruit flies and all the common insects, yellow trap can be used. These yellow traps are placed at a distance of 1-1.5 m in the crop fields.

Mechanical or physical pest control

This includes cutting, mulching, mowing, tillage and organic soil coverage and barriers. This method also includes simply hand picking of insects. Some devices can also be used to protect crops from pests. It includes row covers, protective nets with varying mesh size according to the pests. Insect vacuums can be used to eliminate insects from the plant surface and collect them into a collection box. Water can also be sprayed to remove some insects from the plant surface. Installation of bird perches in cotton and chickpea fields has proved effective in checking the bollworm infestation.

Biological pest control

Biological control in organic plant protection is a method of controlling insect pests and diseases using other beneficial organisms which rely on predation, parasitism, or some other natural mechanisms. This method is of three types. First one is the importation of pests' natural enemies, which do not occur naturally there, but are released in the farm. Second is, boosting the naturally occurring population of beneficial microorganisms. And the third one is conservation of existing natural enemies of pests in the environment. *Trichogramma* spp. are the most popular parasitoids being applied on a number of host crops. A number of microorganisms such as *Trichoderma* spp., *Verticillium* spp., *Aspergillus* spp., *Bacillus* spp. and *Pseudomonas* spp. that attack and suppress the plant pathogens have been exploited as biological control agents.

Biopesticides

Biopesticides are certain types of pesticides derived from animals, bacteria and trees. These have a narrow target range and a very specific mode of action. These suppress, rather than eliminate a pest population. Biopesticides have no risk on environment. These are slow acting pesticides, and have relatively critical application times. But, biopesticides have no residue problems, so they do not damage the crops.

Biochemical pesticides

These are naturally occurring substances that control pests by non-toxic mechanisms. Biochemical pesticides include substances that interfere with growth or mating, such as growth regulators, or substances that repel or attract pests, such as pheromones. Pheromones are often used to detect or monitor insect populations, or in some cases, to control them.

Chemical control

Organic standards are designed to allow the use of naturally occurring substances such as pyrethrin and rotenone. There are also few synthetic substances allowed in organic farming, such as fixed coppers (copper hydroxide, copper oxide, copper oxychloride and copper sulphate), hydrated lime, hydrogen peroxide, lime, sulphur, and potassium bicarbonate.

CONSTRAINTS TO THE ADOPTION OF ORGANIC FARMING

If organic agriculture is to play a role in providing sustainable food security and sustainable livelihoods, it needs to be accessible to poor farmers. The adoption or non-adoption of an agricultural technology or practice is influenced by the demographic and socioeconomic characteristics of the farmer, by the policy, institutional and economic context, as well as by the nature of the technology itself (Seufert, 2012).

Lack of awareness and information

Good management practices are more important in organic than in conventional systems of farming for achieving high yields (Seufert et al., 2012). Information and education about organic agriculture is thus a necessary precondition for the diffusion of organic agriculture and the lack thereof has often been shown to hinder the adoption and success of organic agriculture (Mendez et al., 2010).

Non-availability of labour

In organic farming inputs are usually replaced with labour. Instead of using herbicides, it relies on weeding and ploughing for weed management, and instead of using chemical fertilisers, it relies on incorporation of crop residues, cover cropping and compost for nutrient management. This high labour requirement has often been identified as an important barrier to the adoption of organic farming systems (Moser and Barrett, 2003; Lee, 2005). The increased labour requirements, especially during the periods of peak labour demands, can often not be met by household resources and thus require additional cash expenditure (Bray et al., 2002; Bolwig et al., 2009). This is particularly problematic in areas where availability of agricultural labour is limited due to emigration of workmen or due to movement of labour to non-agricultural sectors. On the other hand, the increased labour requirement of organic production can also be an asset by creating employment in rural areas with limited job opportunities (Goldberger, 2008).

Market related constraints

Organic agriculture in developing countries is an export-oriented farming system. Organic farmers are thus dependent on access to international markets, they need to undergo and pay for an extensive certification process, and they are subject to international regulations dictated often by foreign certification bodies. This dependence on international markets can pose substantial obstacles to small farmers and can potentially increase farmers' vulnerability. In many developing countries, there can be substantial delays before farmers receive the full payments for their organic produce and organic farmers often sell parts of their organic produce on lower-paying conventional markets to receive immediate payments (Bacon, 2005). They also bring produce to the conventional market that they cannot sell to the organic exporting companies, who often have a rather low demand and high quality-requirements (Valkila, 2009; Chongtham et al., 2010).

Non-availability of organic inputs

The content of plant nutrients in organic manure is very less. Thus to meet the nutrient requirement of crops a voluminous organic manure is required. This increases the transport and application costs. The NPK contents in various organic manures vary and very often depend on the method of preparation, storage and incorporation. This confuses the farmers to replace the NPK requirements from different sources. Adequate organic pest control inputs like biopesticides, biocontrol agents (parasitoids, predators and pathogens), etc. are not timely available to the farmers which strongly discourage them to shift to organic practices.

REFERENCES

Alroe, H. F., Byrne, J. and Glover, L. 2006. Organic agriculture and ecological justice: ethics and practice. In: Halberg, N., Alroe, H. F. and Knudsen, M. (eds.). Global development of organic agriculture: challenges and prospects. CAB International. pp. 75-112.

Bacon, C. 2005. Confronting the coffee crisis: can fair trade, organic, and specialty coffees reduce small-scale farmer vulnerability in northern Nicaragua? *World Development* **33**(3): 497-511.

Bhata, A. K., Beri, V. and Sidhu, B. S. 1991. Effect of long-term recycling of crop residues on soil productivity. *Journal of Indian Society of Soil Science* **39**: 380-383.

Bolwig, S., et al. 2009. The economics of smallholder organic contract farming in tropical Africa. *World Development* **37**(6): 1094-1104.

Bray, D. B., et al. 2002. Social dimensions of organic coffee production in Mexico: lessons for eco-labelling initiatives. *Society & Natural Resources* **15**(5): 429-446.

Chongtham, I. R., et al. 2010. Assessment of the strategies of organic fruit production and fruit drying in Uganda. *Journal of Agriculture and Rural Development in the Tropics and Subtropics* **111**(1): 23-34.

Deshmukh, S. M. 2010. Organic farming: Principles, Prospects and Problems. Agrobios (India), Jodhpur.

Fan, T., Stewart, B., Payne, W., Yong, W. and Luo, J. 2005. Long-term fertiliser and water availability effects on cereal yield and soil chemical properties in northwest China. *Soil Science Society of America Journal* **69**: 842-855.

Forman, J. and Silverstein, J. 2012. Organic foods: health and environmental advantages and disadvantages. *Pediatrics* **130**(5): 25-79.

Gabriel, D., Roschewitz, I., Tscharntke, T. and Thies, C. 2006. Beta diversity at different spatial scales: plant communities in organic and conventional agriculture. *Ecological Applications* **16**(5): 2011-2021.

Goldberger, J. R. 2008. Diffusion and adoption of non-certified organic agriculture: a case study from semiarid Makueni District, Kenya. *Journal of Sustainable Agriculture* **32**(4): 531-564.

Howell, M. and Martens, K. 2002. Organic weed control: cultural and mechanical methods. *Acres U.S.A.* **32**(8): 1-6.

https://blog.agrivi.com/post/best-practices-for-pest-management-in-organic-farm-production

https://eos.com/blog/organic-farming/

https://krishijagran.com/agripedia/6-organic-techniques-for-effective-pest-management/

https://rodaleinstitute.org/blog/weed-management-the-organic-way/

https://www.agrifarming.in/weed-management-in-org

https://www.ecofarmingdaily.com/eco-farming-index/organic-weed-control/

https://www.ifoam.bio/why-organic/shaping-agriculture/four-principles-organic

IFOAM. 1998. Basic Standards for Organic Production and Processing. International Federation of Organic Agriculture Movements. Tholey-Theley, Germany.

Isaacs, J. R. 2012. Organic farming keeps carbon out of the atmosphere. http://news.mongabay. com/2012/1128-isaacs-organic-carbon.

Lee, D. R. 2005. Agricultural sustainability and technology adoption: Issues and policies for developing countries. *American Journal of Agricultural Economics* **87**(5): 1325-1334.

Lotter, D. W. 2003. Organic agriculture. *Journal of Sustainable Agriculture* **21**(4): 1-63.

Luttikholt, L. 2007. Principles of organic agriculture as formulated by the International Federation of Organic Agriculture Movements. *NJAS - Wageningen Journal of Life Sciences* **54**(4): 347-360.

Mader, P., Fliessbach, A., Dubois, D., Gunst, L., Fried, P. and Niggli, U. 2002. Soil fertility and biodiversity in organic farming. *Science* **296**: 1694-1697.

Magkos, F., Arvaniti, F. and Zampelas, A. 2006. Organic food: buying more safety or just peace of mind? A critical review of the literature. *Critical Reviews in Food Science and Nutrition* **46**(1): 23-56.

Mendez, V. E., et al. 2010. Effects of fair trade and organic certifications on small-scale coffee farmer households in Central America and Mexico. *Renewable Agriculture and Food Systems* **25**(3): 236-251.

Moreau, T. L. 2007. International conference on organic agriculture and food security. Rome, Italy, 3–5 May 2007

Morgera, E., Caro, C. B. and Duran, G. M. 2012. Organic agriculture and the law. Food and Agriculture Organisation of the United Nations, Rome.

Moser, C. M. and Barrett, C. B. 2003. The disappointing adoption dynamics of a yield-increasing, low external-input technology: the case of SRI in Madagascar. *Agricultural Systems* **76**(3): 1085-1100.

Nascimbene, J., Marini, L. and Paoletti, M. 2012. Organic farming benefits local plant diversity in vine yard farms located in intensive agricultural landscapes. *Environmental Management* **49**: 1054-1060.

Ohno, T. and Erich, M. S. 1997. Inhibitory effects of crop residue derived organic ligands on phosphate adsorption kinetics. *Journal of Environ Quality* **26**: 889-895.

Parkash, Y. S., Bhadoria, P. B. S. and Rakshit, A. 2002. Comparative efficacy of organic manures on the changes in soil properties and nutrients availability in an Alfisol. *Journal of Indian Society of Soil Science* **50**: 217-221.

Patil, P. V., chalwade, P. B., Solanke, A. S. and Kulkarni, V. S. 2003. Effect of fly ash and FYM on physico-properties of vertisols. *Journal of Soils and Crops* **13**: 59-64.

Paull, J. 2006. The farm as organism: The foundational idea of organic agriculture. *Journal of Bio-Dynamics Tasmania* **80**: 14-18.

Paull, J. 2010. From France to the World: The International Federation of Organic Agriculture Movements (IFOAM). *Journal of Social Research & Policy* **1**(2): 93-102.

Paull, J. 2011a. Attending the first organic agriculture course: Rudolf Steiner's agriculture course at Koberwitz, 1924. *European Journal of Social Sciences* **21**(1): 64-70.

Paull, J. 2011b. The Betteshanger summer school: Missing link between biodynamic agriculture and organic farming. *Journal of Organic Systems* **6**(2): 13-26.

Pimentel, D. 2006. Impacts of organic farming on the efficiency of energy use in agriculture. The Organic Center, Cornell University, Ithaca.

Prasad, K. and Gill, M. S. 2009. Development and strategies perspective for organic farming in India. *Indian Journal of Agronomy* **54**(2): 186-192.

Rigby, D. and Caceres, D. 2001. Organic farming and the sustainability of agricultural systems. *Agricultural Systems* **68**: 21-40.

Seufert, V. 2012. Organic Agriculture as an Opportunity for Sustainable Agricultural Development. Policy Brief No. 13. Institute for the Study of International Development. Montreal, Canada.

Seufert, V., et al. 2012. Comparing the yields of organic and conventional agriculture. *Nature* **485**: 229-232.

Shiva, V., Pandey, P. and Singh J. 2004. Principles of Organic Farming: Renewing the Earth's Harvest. Navdanya, New Delhi.

Siddique, S., Hamid, M., Tariq, A. and Kazi, A. G. 2014. Organic farming: the return to nature. In: Ahmad. P., et al. (eds.). Improvement of Crops in the Era of Climatic Changes, Volume 2. Springer Science & Business Media, New York. pp. 249-282.

Valkila, J. 2009. Fair trade organic coffee production in Nicaragua: Sustainable development or a poverty trap? *Ecological Economics* **68**(12): 3018-3025.

Walia, U. S., Walia, S. S., Kler, D. S. and Singh, D. 2011. Science of Agronomy. Scientific Publishers, Jodhpur, India. pp. 482-483.

Wortman, S. E., Wortmann, C. S., Pine, A. L., Shapiro, C. A., Thompson, A. A. and Little, R. S. 2017. Nutrient Management in Organic Farming. Nebraska Extension. Institute of Agriculture and Natural Resources, University of Nebraska-Lincoln, USA. pp. 1-7.

Ziesemer, J. 2007. Energy use in organic food systems. Natural Resources Management and Environment Department, Food and Agriculture Organisation of the United Nations, Rome.

15

Sustainable Livestock Farming

The livestock sector has emerged as a vital sector for ensuring a more inclusive and sustainable agriculture system. Livestock production can be an important component of a sustainable agricultural system because it can provide a quality source of plant nutrients, be an income generator, and provide an environmentally sound use of certain lands. Some farmland is just not suitable for crop production, but may be utilised sustainably for livestock production.

Sustainable livestock production depends on the following six important factors.

1. Sustainable animal housing and husbandry systems
2. Animal health and welfare
3. Integration of livestock farming into the landscape and the community
4. Contributions in the areas of energy, climate and the environment
5. Business opportunities and a viable market
6. Responsible consumption

BENEFITS OF LIVESTOCK FARMING

- Livestock farming is a vital activity in the development of humanity and continues to occupy a prominent place among the primary activities of the world economy.
- It generates high-quality food products such as meat, egg, milk cheese, etc.
- Other local economic sectors benefit directly or indirectly from its activity; food processing industries, handicrafts, tourism and hospitality
- It is one of the few human productive economic activities that are truly sustainable.
- It generates employment opportunities and serves as a source of income.
- It can also serve as a hobby for some people.
- It can help a country to generate foreign exchange earnings through the export of livestock products. This will further strengthen the local currency value.

- Bigger animals such as cattle, horse and donkeys can be used for some special farm operations such as ploughing, harrowing and even beasts of burden.

SUSTAINABLE DAIRY FARMING

India ranks first among the world's milk producing nations since 1998 and has the largest bovine population in the world. Dairying has become an important secondary source of income for millions of rural families and has assumed the most important role in providing employment and income generating opportunities particularly for marginal and women farmers. Most of the milk is produced by animals reared by small, marginal farmers and landless households. Thus, the future of the Indian dairy sector depends on the sustainability of family dairy farms. Dairy farming requires large amounts of cattle feed, meal supplements, antibiotic drugs and other inputs. A chronic shortage of cattle feed coupled with the poor quality of fodder has become the major constraint in dairy farming in India. In the current system of intensive livestock production, there is high emphasis on concentrate feeding, which has increased the cost of milk production and has substantially decreased the profits for farmers. The increasing cost of feed ingredients and its seasonal variability also adds to the gravity of the situation (Quazi, 2016).

Sustainable dairy farming in India is a new concept that prophesies the proper and efficient usage of the resources, without being over-exploitative. Small scale dairy farming can also be sustainable with good care of animals and the use of farm inputs. Buffalo have significant potential for meeting India's dairy demand in a sustainable way. Buffalo are more adaptable to India's increasingly dry, hot climate. They also convert the low-quality indigenous grasses into milk more efficiently than dairy cows, improving productivity and reducing methane emissions. Consumers are increasingly choosing buffalo over dairy cow milk because they prefer the higher fat content of buffalo milk, while farmers prefer the higher returns they receive from buffalo milk compared to cow milk. Improving genetics, feed and animal care practices for cows and buffalo can provide more milk using fewer animals, while also reducing methane emissions. Increasing access to mechanisation for small and medium-scale farmers would reduce reliance on cattle for draught power, allowing investments in milk production. Increasing the productivity and sustainability of milk production is essential if India is to reach its targets for reducing food insecurity, malnutrition, poverty and climate impacts.

The objective for good dairy farming practice is that safe, quality milk should be produced from healthy animals using management practices that are sustainable from an animal welfare, social, economic and environmental perspective. Food and Agriculture Organisation of the United Nations and the International Dairy Federation have issued the following guidelines to achieve the desired outcome from a sustainable dairy farming (FAO & IDF, 2011).

- Animal health
- Milking hygiene
- Animal welfare
- Environment
- Socioeconomic management

Animal health

Establishment of the herd with resistance to disease

1. **Breeds well suited to the local environment and farming system:** Different dairy species and breeds have different requirements. Selecting dairy animals that are suited to the local environment will greatly reduce the risks to productivity posed by animal health and welfare problems. The animals' ability to adapt to climatic extremes, feed quality, local parasites and their acquired resistance to endemic disease must be considered before selecting the breeds. The demands on the animals also vary with the farming system.
2. **Herd size and stocking rate:** Larger herds and higher stocking rates generally require a higher level of organisation, infrastructure and skill to manage. Disease burdens may be higher and individual animals requiring intervention may be more difficult to identify and treat. Good planning and management skills are required to manage risks that have serious consequences. Thus, herd size and stocking rate should be based on management skills, local conditions and the availability of land, infrastructure, feed, and other inputs.
3. **Vaccination of animals:** Vaccination is a useful tool to limit the impact of disease by increasing the immunity of the animal population to specific pathogens.

Prevention of entry of disease onto the farm

1. **Procurement of animals of known health status:** The most effective way to prevent the spread of infectious diseases is to keep a closed herd. This means no new animals enter the herd and previously resident animals do not re-enter after they have left the herd. Increased risk of disease may also occur when animals share grazing or other facilities. Prior to being introduced to the farm, all dairy animals should be screened for diseases

that are significant to their area of origin and new location. Where the animals' health status is unknown, they should be kept under quarantine or separate to the existing animals for an appropriate length of time.

2. **Safe transportation of animals:** Potential buyers of animals should always ask and be told if the animals are sick or diseased. No sick animals should be transported. The disposal of diseased and dead animals should be done in a way that minimises the risk of disease spread.
3. **Monitoring risks from neighbours:** The farmer should be aware of local endemic diseases and/or exotic diseases which have the potential to affect the health of the herd or flock, especially from neighbouring farms. The animals should be appropriately contained to ensure there is no risk of disease spread between farms and within farms.
4. **Limit access of people to the farm:** People visiting a number of farms may spread disease between the farms. Visitors to the farm should wear clean protective clothing and clean, disinfected footwear if entering areas that pose a high risk of transferring disease onto or from the farm.
5. **Vermin control programme:** Vermin breeding sites should be eliminated, especially if those sites also harbour disease pathogens, such as manure heaps, livestock disposal sites, etc. Vermin control measures may also be required in the milking shed, feed and water storages and animal housing areas. Vermin species vary geographically but can include indigenous animals, rodents, birds and insects.
6. **Use of clean equipment from a known source:** All agricultural and veterinary equipment introduced onto the farm should be clean and steps should be taken to prevent the introduction of disease by procuring these from a known source.

Effective herd health management programme in place

1. **Use of identification system for all animals:** An identification system that allows all animals to be identified individually from birth to death should be introduced so that they are easily identifiable by all people who come in contact with them. The systems used should be permanent, allowing individual animals to be uniquely identified by systems like ear tagging, tattooing, etc.
2. **Herd health programmes:** Herd health programmes should include the farm's practices for the diagnosis, treatment, prevention and control of relevant animal diseases, including internal and external parasites. It is important to ensure a consistent approach to herd health, so all staff should be aware of and understand the farm's herd health programme. Effective

herd health programmes should be developed in consultation with appropriately skilled people such as veterinarians.

3. **Regular checking of animals for signs of disease:** All animals should regularly be checked for detection and accurate diagnosis of infectious disease. Detailed breeding and reproductive records should be kept and animals observed at appropriate stages as many diseases are associated with reproduction. Clinical diseases should be investigated to determine the underlying causes so that animals can be treated and further cases prevented.
4. **Quick treatment of sick animals:** The sick animals should be treated quickly by proven methods after accurate diagnosis.
5. **Isolation of sick animals:** The sick animals should be kept isolated on the farm to minimise the spread of contagious disease. Prompt treatment can limit the spread of infectious agents.
6. **Separation of milk from sick animals:** Appropriate procedures should be followed to separate milk from sick animals and animals under treatment. This milk is not suitable for human consumption and if stored on farm should be clearly labelled as such. Milking equipment and utensils should be thoroughly cleaned to avoid cross contamination.
7. **Documentation of all treatments:** It is important that people involved with handling dairy animals on the farm know what treatments have been given to which animals. An appropriate system should be introduced to readily identify treated animals.
8. **Management of animal diseases that can affect public health:** Through appropriate animal management and hygienic practices direct transmission of animal diseases to people may be avoided. The animal waste and carcasses should be safely disposed. Milk from sick animals should not be consumed.

Use of chemicals and veterinary medicines

1. **Use of chemicals approved under relevant legislation:** The chemicals that have been assessed and registered for use in dairy production by the relevant authority should only be used.
2. **Use of chemicals according to directions and dosages:** Using agricultural and veterinary chemicals for the purpose of which they were approved, and in accordance with label directions, gives a predictable outcome while managing the potential risks. Dairy farmers should manage the use of all chemicals to prevent the chemicals adversely affecting animal health and productivity, the health and safety of the user, the environment or the safety and quality of milk and milk products.

3. **Use of medicines as prescribed by veterinarians:** Only approved veterinary medicines, at the recommended dose or as prescribed by a veterinarian should be used. Relevant withholding periods must be observed.
4. **Secured storage of chemicals and veterinary medicines:** The chemicals and veterinary medicines should be stored securely to ensure they are not used inappropriately or do not unintentionally contaminate milk and feed. Chemicals and their containers should also be disposed of in a way that will not cause contamination to animals or the farm environment.

Milking hygiene

Milking is the defining activity of dairy farming. Consumers demand high standards of milk quality, so milking management aims to minimise microbial, chemical and physical contamination. Milking management covers all aspects of the process of obtaining milk from animals quickly and effectively, while assuring the health of the animals and the quality of the milk.

Prevention of injuries to the animals during milking

1. **Identification of individual animals that require special milking management:** Individual animals should be easily identifiable by all people who come in contact with them. The system used should be permanent. Additional temporary identification systems should be in place on farms to manage animals that require special handling at milking, such as treated or diseased animals, or animals producing milk that is not suitable for human consumption.
2. **Appropriate udder preparation for milking:** Dirty teats should be washed and dried before milking. Udder and teats are checked for any abnormalities which may indicate clinical mastitis.
3. **Consistent in milking techniques:** Good milking technique should be consistently applied. Incorrect or variable milking techniques may result in a higher mastitis risk and injury to the animal. Over-milking should be avoided.
4. **Appropriate disposal of milk from sick or treated animals:** Animals whose milk is unfit for human consumption should be milked last with a separate bucket. This abnormal milk may be discarded in a manner that would not pose a risk to people, animals and the environment.
5. **Installation and maintenance of milking equipment:** Materials used for milking equipment that come into contact with milk and with cleaning and disinfecting fluids should be made from adequately resistant materials and should not impart a taint to milk. Only approved cleaning and disinfecting agents should be used. These chemicals should be used in a way that ensures they do not have an adverse effect on the milk or milking equipment.

6. **Supply of sufficient clean water:** A sufficient supply of clean water should be available for milking operations, for cleaning the equipment that comes into contact with milk and for cleaning the milking area. The quality of the water should be suitable for its intended use.

Hygienic conditions for milking

1. **Clean housing environment:** A high standard of cleanliness should be maintained at all times in housing areas to decrease soiling of the udder and so protect udder health. The housing area should be designed to provide good drainage and ventilation and to avoid animal injury.
2. **Cleanliness of milking area:** The milking area should be designed to allow it to be kept clean and tidy. It should be easy to clean with a clean water supply.
3. **Hygiene rules for milkers:** The milker should wear suitable and clean working clothes; keep hands and arms clean especially when milking. He should not have any infectious disease transmissible via milk.
4. **Clean milking equipment:** The milking equipments are to be cleaned before each use. Approved chemicals should be used for cleaning and disinfecting of milking equipments.

Proper handling of milk after milking

1. **Ensure milk is cooled or delivered for processing within the specified time:** Milk should be cooled as soon as possible after milking to the required storage temperature and within the specified time. Limits on the time taken between milking and delivery to the milk collection centre may exist where the cooling or processing of milk is undertaken off the farm.
2. **Clean and tidy milk storage area:** The milk storage area should be clean and clear of accumulated rubbish, any products or chemical substances not in constant use and any feedstuffs.
3. **Storage of milk at the specified temperature:** The storage equipment should be capable of holding milk at the required temperature until collection, and be constructed of materials that do not taint the milk.
4. **Clean milk storage equipment:** The milk storage equipment should be cleaned before each use. Milk contact surfaces should be sanitised, if required.

Nutrition

The quantity and quality of the feed and water provided largely determines the dairy animal's health and productivity, and the quality and safety of its milk.

Feed and water supplies from sustainable sources

1. **Herd's feed and water requirements:** Budgeting the herd's feed and water requirements in advance reduces risk and may help the dairy farmer identify less expensive sources of feed. Planning feed and water requirements ahead can enhance the sustainability of the farm enterprise.
2. **Sustainable nutrient, irrigation and pest management practices for growing feed:** Many farming systems rely on home grown feed for their livestock. Implementing good dairy farming practice includes managing the flow of nutrients on the farm, including the appropriate use of effluent and fertilisers for growing feed. Resources such as water for irrigation must also be managed sustainably. Implementing integrated pest management strategies can reduce chemical use.
3. **Farm inputs from suppliers implementing sustainable systems:** Dairy farmers should consider sourcing farm inputs such as feed, water, fertiliser and energy from suppliers adopting sustainable practices, and so reduce the environmental impact of their own enterprise.

Feed and water of suitable quantity and quality

1. **Nutritional needs of animals:** Dairy animals should be provided with sufficient feed and water daily, according to their physiological needs. The quality and quantity of the feed, including appropriate fibre, should reflect the animal's age, bodyweight, stage of lactation, production level, growth, pregnancy, activity and climate.
2. **Impact of feed on the quality or safety of milk:** Dairy farmers should ensure that the feed fed to cows does not contain chemical residues, toxins or other contaminants that pose a risk to animal health or the safety or quality of milk.
3. **Maintenance of quality water supply:** Water supplies should be of suitable quality and free of excrement. Many contaminants can enter water supplies and threaten the health or safety of people, livestock and the milking equipment rinsed with the contaminated water. The most common contaminants include pathogenic microorganisms and their toxins, as well as toxic chemicals such as pesticides, petroleum, solvents and nitrates.
4. **Separate equipment for handling chemicals and feed stuffs:** Separate equipment should be used for handling agricultural and/or veterinary

chemicals and feed stuffs and/or water for animals. Chemical residues may remain on equipment and contaminate.

5. **Appropriate use of chemicals on pastures and forage crops:** Records of all chemical applications to crops and pastures must be maintained and grazing withholding periods are closely observed. Adequate precautions should be taken when allowing stock to drink the water after spraying herbicides and pesticides as spray drift may be a potential source of residues.

6. **Chemicals for treatment of animal feeds and withholding periods:** Only chemicals approved for use in dairy operations should be used. Chemicals should be managed in a manner that avoids their accidental introduction into the feed and water and, as a result, into milk. Withholding periods may also apply to pastures, forage crops and stored grains if they have been treated with an agricultural chemical.

Storage conditions of feed

1. **Storage conditions to avoid feed spoilage or contamination:** Feed should be protected from contaminants. Pesticides, treated seeds, medicated feed and fertilisers should be stored properly. Hay and dry feeds should be protected from a moist environment. Silage and other fermented crops should be kept under sealed conditions.

2. **Rejection of substandard feed:** Feeding any mouldy or substandard feed to dairy animals should be avoided. A wide range of feeds can contain poisonous fungal toxins that can be transferred to milk, particularly if they have not been stored correctly.

Traceability of feedstuffs brought on to the farm

1. **Source of animal feed from suppliers having an approved feed quality assurance programme in place:** If feed is bought, it is to be ascertained that the feed supplier has an assurance programme in place, and he can monitor appropriate residues and diseases and can trace the ingredients used back to their source.

2. **Records of feed received on the farm:** There should be an appropriate system to record and trace all feed or feed ingredients received in the farm so that the farmer can identify and trace all treatments applied to feeds including crop and grain treatments.

Animal welfare

Animal welfare is the application of sensible and sensitive animal husbandry practices to the livestock on the farm. Animal welfare is primarily concerned with the well-being of the animal.

Free from thirst, hunger and malnutrition

1. **Sufficient feed and water for all animals every day:** Dairy livestock should be given sufficient feed, based on their physiological needs. Their requirements will vary according to their age, bodyweight, stage of lactation, production level, growth, pregnancy, activity and environment. The quality (palatability and nutrient content) of the feed also needs to be considered, based on the animal's dietary requirements. Dietary supplements need to be considered if the ration is unable to meet the animal's nutrient requirements. Animals should be fed a balanced diet and have unrestricted access to clean water.
2. **Protection from toxic plants and other harmful substances:** Animals should be protected from access to toxic plants and contaminated areas.
3. **Provision of good quality water:** Animals should have free access to a clean fresh water supply. The water supply should be adequate to meet peak requirements. Drinkers should fill sufficiently quickly to avoid any animals in a group remaining thirsty.

Free from discomfort

1. **Hazard free shelter:** Consideration should be given to the free flow of animals when designing and building animal housing and/or milking sheds. Dairy buildings are to be safely wired and properly earthed.
2. **Provision of adequate space allowances and clean bedding:** Overcrowding of animals, even for short periods should be avoided. Adequate space should be provided to each animal for resting on comfortable bedding. These areas should be kept clean.
3. **Protection from adverse weather conditions:** Animals should be protected from adverse weather conditions. This includes stress factors such as weather extremes, forage shortages, unseasonal change and others causing cold or heat stress.
4. **Provision of adequate ventilation:** All animal housing should be adequately ventilated allowing sufficient supply of fresh air to remove humidity, allow heat dissipation and prevent build-up of gases such as carbon dioxide, ammonia or slurry gases.

5. **Provision of suitable flooring and safe footing in housing and animal traffic areas:** Floors should be constructed to minimise slipping and bruising due to slippery or uneven floors. Excessively rough concrete or surfaces with sharp protrusions and stones can cause excessive wear or penetrations to the sole of the hoof, resulting in lameness. Protective floor coverings can be used on walkways to reduce hoof abrasions that lead to secondary hoof infections.
6. **Protection from injury and distress:** Transport can pose risks to the welfare of dairy animals. The loading and unloading facilities should be appropriate. For longer journeys feed, watering and resting requirements are to be ensured.

Free from pain, injury and disease

1. **Effective health management programme:** Animals should be regularly checked to detect injury and/or disease. Treatment and preventative health management programmes should be in place.
2. **Avoidance of procedures and processes that cause unnecessary pain:** People carrying out veterinary related tasks should be able to demonstrate competency, especially for procedures that could cause suffering. Good hygiene is essential for surgical-type procedures.
3. **Appropriate birthing and weaning practices:** An appropriate birthing plan should be developed that provides safe birthing facilities and regular checking of animals to ensure prompt, experienced help is provided if required. New born animals should be fed colostrum soon after birth.
4. **Appropriate procedures for marketing young dairy animals:** Calves should not be offered for sale until sufficiently hardy to be transported. Adequate bodyweight and dry navel are good indicators.
5. **Protection against lameness:** Laneways, yards, milking stalls and housing should be constructed to minimise the incidence of lameness. Regular hoof care management practices should be implemented and the animals' diets adjusted to minimise lameness. Lameness should be investigated to determine underlying causes and treated appropriately.
6. **Milking of lactating animals:** The lactating animals should be milked regularly. A regular milking routine appropriate to the stage of lactation should be established that does not overly stress the animals.
7. **Avoidance of poor milking practices:** Poor milking practices can affect animal well-being and production. Milking equipment should be well maintained and regularly serviced.

Free from fear

1. **Provision of competent stock handling and husbandry skills and appropriate training:** Good stock handling and husbandry skills are key factors in animal welfare. Without competent, diligent care of animals their welfare will be compromised. A competent operator should be able to understand the significance of a change in the behaviour of the animals and anticipate potential problems and take the necessary preventive action.
2. **Facilities and equipment suitable for stock handling:** The facilities and equipments used to handle the animals should be appropriate for the purpose, well designed and maintained. This can avoid injury to both people and the animals. Careful use of equipment can reduce fear in animals and make them easier and safer to handle.

Environment

Increasingly, consumers are concerned that the production of food is sustainable and is undertaken in harmony with the environment. To meet these concerns it is important that dairy farmers produce milk in a way that meets the wider community's expectations, by using natural resources efficiently and minimising any adverse impact on the environment. Every farmer can play a role in protecting the future of his enterprise by adopting management practices that enhance the environmental sustainability of his farming system.

Implementation of an environmentally sustainable farming system

1. **Use of farm inputs such as water and nutrients efficiently and sustainably:** Some dairy farming systems rely heavily on external inputs such as irrigation water, fertiliser and feed. Dairy farmers need to consider the short and longer term impacts of using these resources on the land they farm. There may also be wider impacts on the local environment. Good dairy farming practice is aimed at improving productivity by using farm inputs efficiently and in a sustainable manner.
2. **Minimum production of environmental pollutants from dairy farming:** Practices that manage the risks of environmental pollutants protect the surrounding environment and can also be good for the farm enterprise. The careful application of fertilisers and an effective effluent management plan will minimise off-farm impacts and can boost productivity. Practices that improve productivity such as feeding a more digestible diet and applying fertilisers appropriately can reduce methane and nitrous oxide gas emissions respectively. Ensuring the nutrient requirements of both the plants and animals are met appropriately will help reduce/manage any potential nutrient losses.

3. **Minimum production of environmental pollutants from dairy farming:** Grazing and housing strategies that minimise environmental impacts may be implemented. Waterways and other sensitive areas may be fenced-off from livestock.

4. **Selection and use of energy resources:** Dairy farmers should examine their farm's energy consumption and consider options to reduce it. Useful energy sources can be harnessed from dairy waste streams, such as effluent. Non-renewable sources of energy such as electricity and fossil fuels can be supplemented by renewable sources such as solar, biofuels and wind energy, etc.

Appropriate waste management system

1. **Practices to reduce, reuse or recycle farm waste as appropriate:** Reusing and/or recycling are good dairy farming practices. Many farms can reduce their water and energy consumption by properly maintaining equipment and infrastructure, or implementing minor changes to their work routines. Opportunities to recycle plastics, drums and other consumables should also be investigated.

2. **Disposal of wastes to minimise environmental impacts:** Waste storage areas such as manure heaps, slurry stores and farm dumps, should be sited appropriately, considering the local amenity with regard to sight and smell, and the risk to the environment from pollution. Other wastes such as waste milk, dead livestock, farm chemicals and fertilisers, etc. are to be disposed appropriately to prevent the environmental pollution. Potential breeding sites for flies and other disease carrying vermin should be eliminated. A simple waste management plan may be developed to identify when, where and at what rate to spread manures, slurry and other organic wastes to minimise the risk of causing pollution. All organic manure wastes including slurries should be spread or incorporated into soil when there is minimal risk to the environment.

Impact of dairy farming practices on the local environment

1. **Containment of dairy runoff on-farm:** Dairy farmers should adopt systems that avoid the potential for the contamination of the local environment. Storage facilities for oil, soiled water and other polluting substances must be located in a safe place. Disposal of agricultural and veterinary chemicals should be done in such a way that these do not enter the local environment.

2. **Overall appearance of the dairying operation:** To limit the potentially adverse impact of dairying on the landscape and to provide a positive image

of dairy production, dairy farmers should ensure that access roads to their farms and the farm surroundings are clean, sheds are correctly maintained and that access roads used by dairy animals are free of effluent.

Socioeconomic management

Socially responsible and economically sustainable are integral elements to good dairy farming practice. Managing and looking after the farm's human resources is critical to the sustainability of the dairy enterprise. Dairy farmers must also consider the role their enterprise has in the wider community; as an employer, consumer of natural resources and its potential impact on neighbours. Like all businesses, dairy farm businesses must be financially viable to have a long-term future.

Effective and responsible management of human resources

1. **Sustainable work practices:** Managing a dairy farm is a very demanding job. Even on a small landholder scale the dairy farmer is responsible for producing quality food from livestock that require daily care and attention. Dairy farmers need to consider their own health and welfare, and that of their families and farm workers. The workload of all should be reasonable and sustainable. Society has a duty to dairy farmers in this regard too, providing a fair return for farm produce.
2. **Employment of farm staff:** Farm staff should be engaged on clearly defined and documented terms. Labour productivity is enhanced when staffs properly understand their duties, so a documented job description can be useful. Employment contracts must be lawful.
3. **Management of human resources:** Dairy farmers and staff should ensure that they are able to cope with the physical and mental demands required by dairy farming. Dairy farmers need to have realistic expectations of their staff. Working hours must be sustainable and not exceed legislated limits.
4. **Farm working environment:** Farm equipment and infrastructure should not be a risk to the health and safety of farm staff and visitors. Good design and maintenance will remove many risks. Personal protective equipment, toilets and washing facilities should be provided where required.

Carrying out farm tasks safely and competently

1. **Appropriate procedures and equipment in place for undertaking dairy farming tasks:** It is important to ensure tasks undertaken on the farm are done safely, correctly and consistently by all farm staff. It is the dairy farmer's responsibility to ensure that farm staffs are aware of and understand the procedures specific to their enterprise. They also need to clearly identify who is responsible for particular tasks.

2. **Training for staff:** Farm staffs need to be properly trained to work productively and safely. This includes being formally introduced to the working environment and their specific role. Training opportunities for existing staff can also improve productivity and increase work satisfaction. Competent and qualified people should be chosen to develop and deliver staff training.

Management of the enterprise to ensure its financial viability

1. **Financial management systems:** Most decisions that support good dairy farming practice have financial implications for the farm enterprise. Book keeping systems can provide the dairy farmer with up-to-date information about the farm's financial position. Access to this information underpins good decision-making.
2. **Adoption of agricultural practices that contribute to the productivity and profitability:** Dairy farmers operate in constantly changing physical and market environments. Implementing good dairy farming practice is about being adaptive to change. Dairy farmers should implement new technologies and practices that are consistent with their goals. Feed and labour are significant costs to most dairy farm businesses and so improvements in these areas have the largest impact on the financial sustainability of the enterprise.
3. **Management of financial risks:** Forward budgeting of both farm income and expenses should be used to manage the financial risks to the farm enterprise. Budgeting can help identify and manage future cash flow issues, as well as minimise any financing costs.

ROLE OF SMALL RUMINANT IN SUSTAINABLE FARMING

Small ruminants play an important role in Indian economy and source of livelihood and employment to millions of rural households especially the landless, marginal and small farmers. The socioeconomic value of small ruminant rearing as compared to other livestock species, for poor farmers is immense. Goat and sheep are also among the main meat-producing animals in India, whose meat is readily preferred irrespective of caste, creed and religion. They produce a variety of products - meat, milk, skin, wool and manure.

The goat and sheep are mainly dependent on common property resources (CPRs) for meeting their feed and fodder requirements. The CPRs comprise barren and uncultivable lands, cultivable wastes, permanent pastures and other grazing land and land under miscellaneous trees, crops and other fallow land. The rural households have often developed highly efficient agricultural and livelihood systems that make the most rational and conservative use of the scarce resources available to them. The rural poor who cannot afford to maintain a cow or a buffalo find goat/sheep as the best alternative source of supplementary income and milk. Unlike a cow or buffalo, a few goats can be maintained easily and can be easily sold at the time of need (Kumar and Roy, 2013).

Benefits of goat farming

1. The initial investment needed for goat farming is low.
2. Due to small body size and docile nature, housing requirements and managerial problems with goats are less.
3. Goats are friendly animals and enjoy being with the people.
4. Goats are prolific breeders and achieve sexual maturity at the age of 10-12 months. Gestation period in goats is short and at the age of 16-17 months it starts giving milk. Twinning is very common and triplets and quadruplets are not rare.
5. In drought prone areas risk of goat farming is very much less as compared to other livestock species.
6. Unlike large animals in commercial farm conditions both male and female goats have equal value.
7. Goats are ideal for mixed species grazing. The animal can thrive well on wide variety of thorny bushes, weeds, crop residues, agricultural by-products unsuitable for human consumption.
8. Under proper management, goats can improve and maintain grazing land and reduce bush encroachment without causing harm to the environment.
9. No religious taboo against goat slaughter and meat consumption prevalent in the country.
10. Slaughter and dressing operation and meat disposal can be carried without much environmental problems.
11. The goat meat is more lean (low cholesterol) and relatively good for people who prefer low energy diet and sometimes goat meat is preferred over sheep meat because of its chewability.
12. Goat milk is easy to digest than cow milk because of small fat globules and is naturally homogenised. Goat milk is said to play a role in improving appetite and digestive efficiency.
13. Goat milk is non allergic as compared to cow milk and it has antifungal and antibacterial properties and can be used for treating urogenital diseases of fungal origin.

14. Goats are 2.5 times more economical than sheep on free range grazing under semiarid conditions.
15. Goat creates employment to the rural poor besides effectively utilising unpaid family labour. There is ample scope for establishing cottage industries based on goat meat and milk products and value addition to skin and fibre.
16. Goat is termed as walking refrigerator for the storage of milk and can be milked number of times in a day.

Benefits of sheep farming

1. Sheep do not need expensive housing facility and on the other hand require less labour than other kinds of livestock.
2. The foundation stock is relatively cheap and the flock can be multiplied rapidly.
3. Sheep are economical converter of grass into meat and wool.
4. Sheep will eat varied kinds of plants compared to other kind of livestock. This makes them excellent weed destroyer.
5. Unlike goats, sheep hardly damage any tree.
6. The production of wool, meat and manure provides three different sources of income to the shepherd.
7. The structure of their lips helps them to clean grains lost at harvest time and thus convert waste feed into profitable products.
8. Mutton is one kind of meat towards which there is no prejudice by any community in India and further development of superior breeds for mutton production will have a great scope in the developing economy of India.

Constraints faced by goat and sheep farmers

In spite of several advantages the small ruminant sector faces a lot of constraints. The following constraints and limitations are identified by Chakraborty and Gupta (2017).

1. **Absence or weak livestock keeper organisations:** The small ruminant farmers are not organised and thus they are unable to express their demands and requirements.
2. **Transportation:** Major constraint faced by commercial small ruminant farmers is the lack of proper transportation guidelines for live animals. While transporting breeding stock from long distances farmers are charged lump sum by police in the name of well-being of animals which discourages the farmers for this enterprise.

3. **Disease:** Mortality rate of goats are high and growth in kids is also poor. High mortality is due to diseases like peste des petits ruminants (PPR), diarrhoea, pneumonia, tetanus, etc. There is dearth of knowledge of better methods of goat farming, low level of prophylaxis, lack of vaccines and lower accessibility to veterinary doctors. Furthermore, the funds allocated for veterinary services for small ruminant sector are less. Majorly small ruminants are located in remote areas and are migratory in nature; therefore the services available mostly do not reach them.
4. **Breeding stock and its breeding:** Farmers are unable to identify pure breed animals because of lack of knowledge. They also face difficulty in accessing good quality breeding animals. Finest breeds (mostly males) are usually sold to slaughterers, hence leading to shortage of better quality breeds.
5. **Availability of inputs and services:** There is unavailability of vaccines especially PPR and cost effective tagging materials.
6. **Marketing:** Goat and sheep farmers receive very low price for their animals since bodyweight is not considered in livestock markets. More often, there is no consistency in price of small ruminant products and it fluctuates as per the demand during certain occasions. Major share of the profits goes to middlemen or the slaughterhouse owner/meat seller.
7. **Reduced access to credit and insurance:** Low income farmers have meagre capital assets to finance their livestock enterprises for activities such as acquiring finest breeds, health care, feeding and management practices, etc. Furthermore, insurance support to livestock, particularly small ruminants has largely been neglected by insurance providers and banks.
8. **Poor synergy between departments:** There is a lack of convergence between policies/programmes of different departments in small ruminant development. In addition, no consultation is done while formulating policies/ programmes with community, elected village council and related government departments such as animal husbandry, forest and agriculture. Due to lack of interstate coordination, migratory men and women farmers face constraints during their transit to nearby states in search of grazing resources.
9. **Processing and abattoirs constraints:** Abattoirs are facilities proposed for animal meat processing which includes slaughtering, chilling, packaging for exporting meat, and cooking into different dishes.
10. **Unhygienic meat processing:** Animal meat processing units in many cases lack proper flooring, ventilation, and water supply. As a result, they have poor hygiene standards, posing threats to public health and environment. The waste from slaughterhouse is also not disposed as per guidelines.

Sustainable development of small ruminant production

The potential for goats and sheep to contribute to the attainment of food security, economic development and environmental sustainability is tremendous. The broad genetic variability of Indian goat and sheep breeds enables them to survive under stressful environmental conditions, including high disease incidence, poor nutrition and harsh temperatures. Environmental pressure also maintains a wide range of genotypes, each adapted to a specific set of circumstances. Under on-station management, indigenous sheep and goat breeds have shown good growth and reproductive performances, indicating their potential and ability to respond to improvements in management, particularly nutrition and disease control (Kumar and Roy, 2013).

The following recommendations were given by Kumar and Roy (2013) and Chakraborty and Gupta (2017) for sustainable development of small ruminant production as a means of livelihood security of resource-poor rural people in India.

1. Scarcity of feed and fodder being the major constraint in the existing small ruminant production system, better management of common feed resources is crucially important for improved fodder supplies in rainfed areas. Each watershed should essentially have a component to develop common and private fodder resources as part of it and its compliance by the project implementation agencies should be monitored. The available fodder resources should be efficiently utilised. Chopping of coarse cereal stover (sorghum/ ragi, etc.) needs to be promoted to reduce wastage and improve digestibility.
2. The panchayats may be given incentives to develop common grazing resources in the village commons through convergence of various schemes under the departments of soil conservation and watershed management, horticulture, animal resources, etc. An interdepartmental committee may be formed to explore possibilities of convergence in policies/programmes.
3. 'Small Ruminant Policy' may be formed at national and state level under the department of animal husbandry in consultation with stakeholders including livestock keeper organisations. A state level 'Small Ruminant Cell' should coordinate the implementation of the Small Ruminant Policy with all major stakeholders.
4. With the involvement of animal husbandry experts and other major stakeholders (farmers, local veterinary doctor, dairy cooperatives, etc.), the potential indigenous breeds of different species need to be identified at micro level considering the yield potential, resource situation and socioeconomic factors and should be used for grading up programmes.

5. The poor quality breeding males of small ruminants should be weeded out. Once the breeds of animals to be promoted at block and village level are decided, the farmers may be asked not to keep breeding males of non-descript poor breeds. The poor breeding males may be replaced as part of breed improvement programme.
6. For improving fodder availability for the small ruminants, there is need to institutionalise linkages between goat farmers in fodder scarce regions and areas with better fodder availability.
7. Production and mortality losses in small ruminants due to diseases and parasites are huge. Awareness campaigns should be carried out through mass media focusing on importance of prevention programme for diseases and parasites in small ruminants.
8. Deworming against parasites and vaccination against major diseases should be taken up.
9. Farmers should be encouraged and supported for collective action in livestock production and marketing. Linkages need to be created between smallholder producers, processors, and exporters. Farmers should not sell their livestock when need arises, instead they should be sensitised on market demand to help realise better price in domestic market and produce healthy pure breed goats for export.
10. The insurance of livestock is the best safeguard for minimising the risk of loss of production and animals due to diseases and feed scarcity. Though the insurance companies have provision for insurance of animals throughout the country, only a very small share of small ruminants has been insured. There is need to increase the farmers' awareness and make mandatory provision for the companies to undertake livestock insurance of interested farmers.
11. Because of the constraints inherent to the situation, the level of technology adoption in small ruminants is very low. Package of practices and technological guidelines for rearing small ruminants should be provided to the farmers.
12. Farmers must be trained and retrained to develop their management skills for proper feeding including fodder development and conservation, proper breeding skills, disease control and prevention, basic on-farm processing methods to add value to their goat farming, simple record-keeping and the exploitation of synergies between livestock and crops.
13. Animal husbandry department should undertake the function of enhancing infrastructure facilities required for small ruminants. Many functions such as input distribution and marketing could be organised through public private partnership.

14. Credit facility should be provided by both central and state government to private investors for livestock development and meat production. Incentives should be provided for establishing new urban or rural abattoirs and for upgrading existing ones. Financial scheme for supporting goat and sheep enterprises should be formulated by central and state government. Bank loans should be made easily available for livestock farming.
15. Women livestock producer associations may be formed for training and empowering women in livestock development. They should be able to take advantage of credit, insurance, inputs, and marketing services from such association. They should be empowered to take decisions on small ruminant issues at household and organisation level.

SUSTAINABLE POULTRY FARMING

Poultry, defined as domesticated birds kept either because of their meat, feathers or eggs, constitute a fascinating and diverse group of animals, which can be integrated into many farming systems throughout the world. Poultry, including laying hens, meat chickens, turkeys, and ducks, are key in sustainable food and agriculture. Their ability to convert many types of feed, such as residuals from agricultural activities, households and food processing industries, into animal products and protein sources is more efficient than most other animal species. Their contribution to food security, protein supply, and peoples' livelihood makes them valuable animals on a global level (Vaarst et al., 2015). Poultry is a remunerative and popular module in any integrated farming system that fit into urban and peri-urban agricultural production systems, and in many parts of the world they generate income over which women have control. Rural poultry is a good occupation for the rural masses for economic sustainability and nutritional security. However, poultry are raised on large, mid, and small-scales around the world.

Aspects of sustainability of poultry production

Environmental aspects: The current global growth of the livestock sector is achieved at substantial environmental cost (Gerber and Steinfeld, 2008). Livestock activities have been estimated to contribute about 18% of total anthropogenic greenhouse gas emissions (Steinfeld et al., 2006). Considering the agricultural and forestry sector as a whole, the livestock share alone is over half. These activities are linked to multiple factors related to production of feed and deforestation, N fertiliser use, processing and transport. However, poultry is the livestock species with generally the least impact in terms of land size, water use, environmental stress and footprints both for poultry meat and eggs (De Vries and De Boer, 2010). Poultry farming, especially broiler production is the most resource-efficient meat production in agriculture, mostly due to genetic development and nutritional efficiency (Williams et al., 2009). For climate mitigation, depending on how birds and feed are raised, poultry production can be carbon neutral. Compared to grazing animals, poultry do not produce significant amounts of the greenhouse gas, methane, during digestion.

Resource management: The type of feed has an important role in sustainability. Poultry are typically fed grains and pulses. Manure from the birds can be used to fertilise feed crops, especially when raised on the same farm as the birds. Or poultry can be fed by-products, such as fruit pulp or cull crops, to increase nutrient cycling. Insects raised on crop and food residues can also provide sustainable feeds for poultry. The water footprint of poultry and other meat production activity is mainly determined by feed conversation ratio (FCR), feed composition and ingredient origin. Concentrated feed has a larger water footprint than roughage though it has a higher FCR (Gerbens-Leenes et al., 2013). A relevant aspect of resource management related to feed and food is the potential risk of poultry competing with humans for protein sources, e.g., if more of a protein-rich feedstuff is fed to birds. The 100% organic feeding can be based on increased amounts of locally-sourced feed stuffs, which do not compete with human food or needs transport, and therefore can be regarded as more sustainable. However, to meet the birds' requirement of methionine, there may be protein overfeeding which increases N-excretion to the environment with the risk of disease and impaired production. Thus, 'adequate feeding' with a systems approach, giving more emphasis on resilience related to feeding will lead to a sustainable poultry farming (Steenfeldt et al., 2013).

Pollution: In a well-integrated farming system, animals are part of a nutrient cycle. Ideally, the animals are sustained from products on the farm, with minimum use of labour and resources, in particular non-renewable resources. The manure of the animals contributes to soil fertility. This cycle does not exist in industrial animal production, which relies on the transport of all resources. Manure becomes 'waste' which causes environmental pollution. The use of fossil fuels is a part of food and trade-related pollution, needed for growing and transporting feed, live animals and poultry products and to heat or cool poultry barns. Since poultry can be raised under many different local circumstances, it may be unnecessary to involve so much transport when meeting the need for human protein supplies (Vaarst et al., 2015). Poultry houses are climate-controlled and are energy-intensive; however, sources of renewable energy, such as solar heating, and natural ventilation can reduce energy from fossil fuels. Providing outdoor access can be an enrichment for birds, provide increased space, and range, particularly fresh forage, can contribute to nutrients.

Breed diversity: There are about 8000 registered domesticated livestock breeds, including hundreds of poultry breeds. Most of these breeds are kept by smallholder farmers throughout the world, who produce a large part of the world's poultry meat and eggs, and at the same time conserve many of the world's livestock breeds. Many of these breeds are locally adapted, multipurpose, and indigenous, and suited to various farming enterprises. However, the current breeds in industrial systems are almost exclusively single purpose breeds, either broilers or layers. Very few genotypes exist in today's broiler production, which are selected for growth and FCR, but are more vulnerable in terms of immune competences (Rauw et al., 1998). Eradicating breeds means not only losing natural capital or genetic 'wealth' but also eradicates the possibilities for adapting to different environments and situations, e.g., climate changes (FAO, 2015). Thus, bird genetics should be adapted to the production system. A variety of hybrids and breeds, including standard breeds, help increase biodiversity and can be bred to be adapted to local conditions.

Working conditions in industrial poultry farms: Production sustainability is related to working conditions of workers in industrial poultry farms. Poultry farmers and their family members are at higher risk of being infected with antibiotic-resistant bacteria as well as of developing allergies to antibiotics. Other risks include health problems due to dust exposure from litter and feathers. Many workers in the food producing industry are exploited unfairly and employed under poor conditions to keep the direct costs of production low.

Benefits of poultry farming

1. The poultry farming does not require high capital for starting. Most of the poultry birds are not costly enough to start raising. It requires less investment compared to rearing other livestock. Thus, small and marginal farmers and even agricultural landless households can start poultry farming on a small scale. It can be well suited to many other components in an integrated farming system.
2. Poultry farming does not require a big space unless it is started commercially. Some birds can easily be raised on backyard with one or more coops or cages.
3. Poultry farming ensures high return of investment within a very short period. Some poultry birds like broiler chickens take shorter duration of time to mature and generating profit. Chicken start laying eggs in 10 to 12 weeks and broilers can be sold for meat between 6 and 10 weeks. The interval between generations is very small and hence, production can be phenomenally increased within a short period.
4. Poultry is capable of utilising large quantities of farm by-products like bran, substandard grains, vegetables, fruits, etc. which normally go as waste.

Broilers' intake of feed is comparatively very low while it produces maximum possible amount of food for humans.

5. Poultry farming is a continuous source of income. It is not seasonal and can produce income for the entire year. Meat, eggs and manure of poultry birds are all saleable and income generating.
6. Poultry farming requires very little water for both drinking and cleaning. One litre of water is sufficient for 5 birds for a day.
7. Poultry farm structures do not require high maintenance. Diseases and illness in poultry can be minimised by following proper hygiene and care. Diseases are less in some poultry birds like quails, turkeys, etc.
8. Poultry droppings are rich in nitrogen and organic material and hence, are considered valuable as fertilisers.
9. Poultry offers full time or part-time employment opportunity to farmers. Unemployed educated youth can easily create an income and employment opportunity for them by raising poultry. Women and students can also do this business along with their daily activities.
10. Poultry provides fresh and nutritious food and has a huge global demand. Poultry products like egg and meat have high nutritional value. Among all edible meat, broiler meat has the least fat content. Poultry meat contains more protein and essential amino acids than other meats and is low in its cholesterol content. Poultry products are not much expensive and most of the people can afford those.
11. Poultry feathers are also used for making pillows and fancy articles.
12. Marketing poultry products is very easy. There is an established market for poultry products in both urban and rural areas. So poultry products can easily be sold in the nearest local market.

Chicken farming

Besides subsidiary income generation, rural backyard chicken provides nutrition supplementation in the form of valuable animal protein and empowers women. Backyard chicken farming can suitably be integrated to any farming system.

Backyard chicken rearing requires hardly any infrastructure setup. Backyard rearing is mostly seen in villages but some houses in the urban areas also rear chicken in the backyard. The birds are provided with lots of kitchen waste and food leftovers. They scavenge and eat termites, ants, earthworm and insects. During the dry seasons, the birds may additionally be provided with rice bran or wheat bran mixed with the rice starch and other vegetables. It has a very positive impact to improve the socioeconomic factors of the socioeconomically backward people. Backyard chicken production can cater to the nutritional requirements of the family as well as cater to a niche market, thus providing a source of subsistence income. Backyard chicken can be taken up by every household as a subsidiary occupation, as a source of additional income or to cater to the egg and meat requirements of the family by taking up coloured bird units ranging from 10-20 birds per family in their backyards. Such units require very little hand feeding and can give a fairly handsome return with bare minimum night shelter.

Semi-intensive system may be adopted where the amount of free space available is limited, but it is necessary to allow 10-20 m^2 area per bird of outside run. In the semi-intensive system of rearing, feed and water are placed in the outside run or yard. The floor in the run can be cemented with addition of litter over the cemented floor or it could just be a sand floor. The run is fenced by using wire, bamboo or any other local material. The birds are driven to the pen during dusk, and they are well protected during the night. At daytime the birds are let out into the yard or run and stay there till the evening. The yard and the pen must be cleaned regularly once the birds are out of them. Under the semi-intensive system of rearing birds are provided with feed and clean potable water. The feed could be a well balance mash or pellet, or it could be the remains of household waste, vegetable waste mixed with some cheap grains and by products of grains.

Native chicken have undergone natural selection and are best adapted for the local climatic conditions, they can survive on kitchen and agricultural wastes and survive well on scavenging and leftover feed. They are hardy and resistant to many diseases. They survive well from predators. Native hens are good brooders and have good mothering ability. But native chicken are low on productivity, so to increase productivity improved varieties of chicken need to be reared.

There is a wide variety of chicken breeds, developed for egg production, meat production, and/or good looks. While many breeds are adaptable to a backyard setting, certain breeds are better than others for backyard conditions. Improved varieties like, Cauvery, Kalinga Brown, Nirbheek, Gramapriya, Blackrock and Vanaraja due to their moderate bodyweight and high egg yielding capacity are more preferred under semi-intensive farm conditions in rural areas. The feather colour of all the rural type birds are multiple coloured and mostly brown eggers. They have the better immune capacity and thereby strength of survivability is maximum. Because of moderate bodyweight the birds can escape from predators.

Advantages of rearing indigenous birds over exotic birds

1. Acceptability of the coloured *desi* bird by the landless labourers or marginal farmers.

2. Use of broodiness for hatching the chicks.
3. Capability of self-defence from predators due to its alertness, light bodyweight longer shank length, camouflagic characters and aggressiveness.
4. Can thrive well under adverse environments like poor housing, poor management and poor feeding.
5. Indigenous birds are comparatively disease resistant to protozoon and ecto-parasites.
6. Have better adaptability to extreme climatic conditions prevailing in the country.
7. They are comparatively hardier and need less health care than exotic birds.
8. The meat from native fowl has significantly higher amino acid contents (arginine and lysine) than meat from exotic birds and is widely preferred especially because of their pigmentation, taste, leanness and suitability for special dishes and often fetches higher prices.
9. The brown-shelled eggs of native fowl are rich in threonine and valine than farm eggs, have good flavour and fetch premium price.

Chicken housing and management

Good and suitable housing play a vital role in raising chickens. Birds can be raised in both free range and indoor production systems. In case indoor production system, it is very crucial to manage the environment. Chickens need accurate management and environment for better production. Sufficient space should always be kept in the houses so that birds can live, grow and produce happily. Poultry house should never be overcrowded. A minimum 3-5 square feet floor space is required per bird, including outdoor space. The house should be well ventilated and sufficient flow of fresh air and light inside the house should be ensured. The house and equipment are to be cleaned on a regular basis and the house should be sterilised before bringing new chicks. Care must be taken for suitable drainage system inside the house for cleaning it properly. Predators and harmful animals should be kept away from the shelter. The main predators are rats, owls, hawks, stray dogs and cats.

A quality coop is essential to backyard chicken production. Layers need nest boxes, one per 4-5 birds. Coops must provide protection from the weather. There should be a well-insulated area with a light bulb or heat lamp for the winter months as well as ventilation for fresh air. An enclosed space for them to stay at night is essential to their protection. The housing system must have to be weather proof. Weather proof housing system keeps the birds safe from the adverse weather conditions such as cold, rain, wind, hot sun, etc. The chicken housing system should have the facilities of providing warmth, particularly during brooding period and winter season.

Well ventilation inside poultry housing system plays a very vital role to control the indoor environment. It removes moisture, heat, gas, etc. and brings fresh air inside the chicken house. The chicken houses should be provided with wide windows and side curtains to ensure the entrance of fresh air to make the house cold during summer season. A pair of vent in the roof of the house will help to remove the hot air. A roof of at least six feet height from the floor is suitable for this purpose.

Light is also a very important element for chicken farming as birds are very sensitive to light. Light helps the birds to be productive, finding food and simulating them for reproduction. Besides light, the poultry birds also need dark period for keeping good health and producing melatonin hormone, which is very important for immune function. Birds require 8 hours of darkness period and 16 hours presence of light. Darkness is helpful for some fast growing broiler species, and helps them for reducing leg disorders and builds their body frame. However, young chicks require 24 hours light daily after hatching for finding food and water pot.

For sustainable chicken farming and getting desired production, litter management is very important. Usually the litter is used for covering the floor of housing system which may be made of concrete, wood or earth. Litter absorbs the moisture of closet and dilutes the manure. It also works as the bed for the poultry birds. Rice hulls, hay and straw and soft wood shavings are the common materials used as litter in chicken house floors. A good litter contain about 20 to 30% moisture and depth about 5 to 8 cm. High moisture in litter is very harmful for chicken health. Wet litter causes some problem like sores, and blister on birds and produce ammonia gas which hampers the respiratory system of birds. Aluminium sulfate or hydrated lime may be used to reduce ammonia gas from the litter. After disposal of all birds the used litter from house floor should be removed. This litter can be used as good manure in the agricultural land for crop production. Dry poultry manure contains about 3.5% nitrogen, 2.0% phosphorus and 1.5% potassium.

Chicken feed management

Rural chicks need brooding care during the initial 6 weeks of age. After 6 weeks, they can be let free for scavenging in the backyard. For healthy life, proper growth, maximum production and suitable breeding, all types of poultry birds need some essential nutrients. Most of these essential nutrients come from the natural sources. Chickens are omnivorous. They eat grains, fruits, and vegetables as well as insects. Chickens should typically be fed a prepared feed that is balanced for vitamins, minerals, and protein. A healthy laying hen diet should also contain crushed oyster shell for egg production, and grit for digestion. Birds also love fruit and vegetable scraps from the kitchen and garden, as well as bread. Feed consumption may increase in the winter, and decrease in the heat of the summer. A critical part of a chicken's diet is continual access to clean, fresh water. This is especially true in the summer as they cool themselves by panting.

During the first day, feed may be sprinkled or provided in the trays for encouraging the new born chicks to pick up feed. From day two onwards feed is provided in trough type of feeders. As the chicks grow bigger suitable feeders are used. The feeders should be at proper height for the birds to eat properly. As chicks grow the feeder should be lifted up by adjusting their height to the back level of the bird. The level of feed in feeder has a direct correlation with feed wastage. As a thumb rule, 10 percent feed is wasted if the feeders are two thirds full compared to 3 percent wastage if they are half full and only one percent if they are one third full. Therefore, feed should be offered more frequently with small quantity at each time and helps to gain weight more uniformly. During first week, frequent feeding of small quantity should be practised to stimulate feed consumption.

Clean and fresh water should be provided to the chicks. Water may be provided using troughs, bell shaped drinkers and caps. With these drinkers 2 cm of water space per bird is recommended. Water should be provided before the chicks are released under the brooders. Height of the drinkers needs to be adjusted according to the chick height. One chick drinker is enough for 100 chicks up to 2 weeks of age and regular bell drinker is sufficient for 50 birds from third week onwards.

The most expensive poultry feed element is protein. Animal protein is a great source of adding required protein in poultry feed. This type of protein is very helpful for proper growth and better egg production, than vegetable protein. Carbohydrate is essential for a starved flock. This poultry feed materials can be found in grains and grainy products. Carbohydrate provides energy to the birds and keeps them productive. Proper ratio of fat in poultry feed is also very essential. Calcium carbonate helps to formulate the egg shell. Calcium and phosphorous make the major part of poultry bird's bone. But excess amount of phosphorous in poultry feed lead to crooked bones and slipped tendon in chickens and poultry birds. Salt is also a reliable source of essential minerals.

Any vitamin deficiency in poultry feed rations can hinder the natural growth of young poultry birds. Vitamin A comes from green feed ingredients, yellow corn and fish oils. It is essential to protect the poultry birds against colds and infections. Vitamin D helps to prevent leg weakness and rickets. It is found in synthetic products and also available in sea fishes. Vitamin D is a must added elements in poultry feed. Vitamin B-complex is available in milk, green feed, liver, synthetic riboflavin, etc. It helps to increase the growth of chickens. It also helps to prevent curled-toe paralysis in young chickens. While preparing poultry feed, adequate amount of vitamin B-complex should be added in the poultry feed mixture.

Sanitation and chicken health

An important element to bird health is sanitation. In order to maintain a clean, healthy environment, the coop and outdoor area should be cleaned out weekly or as needed to control manure and odour build-up. Feeders and waterers should be regularly cleaned and disinfected. It is important that at least once a year, a thorough cleaning is done on the coop and yard. Also cleaning before introducing new birds to the area will limit the spread of disease. Inhalation of dried chicken manure can be harmful to humans.

Healthy birds will be active and alert with bright eyes. Chickens raised in backyard settings generally stay healthy and are not easily susceptible to diseases. The birds must be vaccinated against Marek's and ranikhet diseases. Birds should be dewormed starting from 7th week and repeated at 3-4 months intervals so as to give a total of 4 deworming.

Duck farming

Ducks form about 10% of the total poultry population and contribute about 6-7% of total eggs produced in the country. Ducks are mostly concentrated in the eastern and southern states of the country. Among the egg laying breeds, 'Khaki Campbell' is the best producer. A duck lays about 300 eggs per year. Khaki Campbell ducks weigh about 2 to 2.2 kg, and drakes 2.2 to 2.4 kg. Egg size varies from 65 to 75 g. 'White Pekin' is the most popular duck for meat purpose. It is fast growing and has low feed consumption with fine quality of meat. It attains about 2.2 to 2.5 kg of bodyweight in 42 days of age, with a feed conversion ratio of 1:2.3 to 1:2.7.

Benefits of duck farming

1. Ducks lay more egg per bird per year than chicken.
2. The size of the duck egg is larger than hen egg by about 15 to 20 g.
3. Ducks require lesser attention and thrive well in scavenging conditions.
4. Ducks supplements their feed by foraging. They eat fallen grains in paddy fields, insects, snails, earthworms, small fishes and other aquatic materials.
5. From commercial point of view, ducks have a longer profitable life. They lay well even in second year.

6. Ducks do not require any elaborate houses like chicken
7. Ducks are quite hardy, more easily brooded and more resistant to common avian diseases.
8. Marshy river side, wetland and barren moors upon which chicken or no other type of stock will flourish, are excellent quarters for duck farming.
9. Ducks lay 95-98% of their eggs in the morning before 9.00 AM. Thus, it saves lot of time and labour of the grower for collecting eggs.
10. Ducks are suitable for integrated farming systems such as duck-cum-fish farming, duck farming with rice cultivation. In duck-cum-fish farming the droppings of ducks serve as feed for the fishes and no other feed or manuring of the pond is necessary for fishes (200-300 ducks per hectare of water area). Under integrated duck farming with rice cultivation, the ducks perform four essential functions, viz., intertillage as they search for food, their bills loosen up the soil around the rice plants, insect control and manuring.
11. Ducks are good exterminators of potato beetles, grasshoppers, snails and slugs. In areas plagued liver flukes, ducks can help correct the problem. Ducks can be used to free the waterbodies from mosquito pupae and larvae.
12. Ducks are quite intelligent, can be tamed easily, and trained to go to ponds and come back in the evening of their own.

Duck housing, feeding and management

Ducks do not require elaborate houses. Usually adult ducks do not require housing. But it would be better if they are given some protection. They will lay eggs better and damage less food. Ducks generally love darkness. So, it is not important to lightening the house. However, the house should be well ventilated, dry and rat proof. The roof may be of shed type, gable or half round. The floor of the duck house may be built with dirt, concrete, wood or bamboo. Straw can be used inside the house as bed. It is very easy to clean and make dry of the floor, made with straw bed. Under semi-intensive system the house should have easy access to outside run as the ducks prefer to be outdoors during the daytime and even during winter or rains. The run should gently slope away from the houses to provide drainage.

Though duck is a water fowl and very fond of water, water for swimming is not essential at any stage of duck rearing. However, water in drinkers should be sufficiently deep to allow the immersion of their heads and not themselves. If they cannot do this, their eyes seem to get scaly and crusty and in extreme cases, blindness may follow. In addition, they also like to clean their bills periodically and wash them to clear off the feed.

Ducks may be grown on dry mash, a combination of dry and wet mash or pellets. Ducks prefer wet mash due to difficulties in swallowing dry mash. The pellet feeding, though slightly costly, has distinct advantages such as saving in amount of feed, minimum wastages, saving in labour, convenience and improvement in sanitary conditions. Ducks are good foragers. The use of range, pond or supplementary green feed, reduces the feed cost. Ducks should never have access to feed without water. During the first eight weeks, birds should always have access to feed, but later on they may be fed twice a day, i.e., first in the morning and then late afternoon. A 'Khaki Campbell' duck consumes about 12.5 kg of feed up to 20 weeks of age. Afterwards the consumption varies from 120 g and above per bird per day and depending upon the rate of production and availability of greens.

Ducks are resistant to common avian diseases. Duck diseases are similar to those of chicken and some are common for both, but the course of disease may vary. Since certain infections of chicken may be transmitted to ducks, it is essential that there is strict segregation of different species.

Quail farming

Quail farming is very profitable like other poultry birds. Almost all types of weather conditions are suitable for quail farming. Meat and eggs of quail are very tasty and nutritious. Quail eggs are very nutritious than other poultry eggs. Because quail eggs contain comparatively more protein, phosphorus, iron, vitamin A, B1 and B2. Quail farming can play a vital role to meet up the demand of food and nutrition. It requires small capital and labour. Quails can be raised along with other poultry birds for meat or eggs production.

Quails are very small sized bird. An adult quail weighs between 150 to 200 g and an egg weighs around 7 to 15 g. Female quails start laying eggs within their 6 to 7 weeks of age and continuously lay one egg daily. They lay about 300 eggs in their first year of life. After that they produce about 150 to 175 eggs in second year. Egg production gradually decreases after their first year of laying period. Quail egg is very suitable for human health. It contains 2.47% less fat than chicken egg. Many people believe that quail eggs help to prevent blood pressure and diabetic. Fat is very low in their meat. So quail meat is very suitable for blood pressure patients. Eggs are very beautiful with multiple colours. Quails do not incubate their eggs. So an incubator or brooder chickens are required for hatching their eggs. For successful breeding purpose one male quail is reared with every five female quails.

A number of quail breeds are available, which are very suitable for profitable quail farming business. Some of these breeds are famous for egg production and some are popular for meat production. Popular layer quail breeds include

Tuxedo, Pharaoh, British Range, English White and Manchurian Golden. Popular broiler quail breeds are Bobwhite and White Breasted. Besides these exotic breeds a few breeds like CARI Uttam, CARI Ujjawal, CARI Sweta and CARI Pearl have been developed by Central Avian Research Institute, Izatnagar which are very popular among the farmers.

Benefits of quail farming

1. Quails are smaller sized bird, so they can be raised within small place. 6 to 7 quails can be raised in the same place that is required for one chicken.
2. Feeding cost of quails is comparatively lower than chickens or other poultry birds.
3. Diseases are less in quails, and they are very hardy.
4. Quails grow very fast and gain maturity faster than any other poultry birds.
5. They start laying eggs within their 6 to 7 weeks of age.
6. It takes about 16 to 18 days to hatch their eggs.
7. Meat and eggs of quail are very tasty, delicious and nutritious. So it is a great source of food and nutrition.
8. Quail farming needs small capital, and labour cost is very low.
9. Quail meat contains less fat. So, it is suitable for high blood pressure patients.
10. Their food to meat or eggs converting efficiency is satisfactory. They can produce one kg meat or eggs by consuming three kg food.
11. As the quail eggs are smaller in size, so the price is also lower than other poultry egg. So marketing is not a problem.

Quail housing, feeding and management

Housing is very important for quail farming. Quail can be raised in both litter and cage systems. But quail farming in cage system is more suitable than raising them in deep litter system. In cage system, management is very easy and diseases or other problems are less. There should be proper ventilation system and proper flow of air and light inside quail shelter. 50 quails can be raised in a cage measuring 120 cm length, 60 cm wide and 25 cm height. Wire net is used for making the cages. Measurement of the net would be 5 mm × 5 mm for adult quails. Plastic cages are most convenient for quail farming business. The house must have to be out of the reach of wild animals.

For keeping the quails healthy, proper growing and highly productive, they should be provided with balanced feed regularly. An adult quail consumes about 20 to 25 g of food daily. Chick feed should contain 27% and adult feed 22-24% of protein. The feed is comprised with broken wheat, sesame oilcake, rice bran, broken oyster shell, salt and mineral mix.

Presence of adequate light is highly recommended for desired egg production. Artificial light and heat can be provided by using electric bulb or heater. Demand of light and heat varies depending on the season. For successful breeding and hatching eggs, one male kept with every five female quails. For getting better egg production, highly productive breeds are selected and their house is always kept dry and clean. Egg production also depends on temperature, feeding, water, care and management.

Diseases are less in quails, compared to other poultry birds. But good care and management is a must for profitable quail farming business. Generally they are not provided any disease preventive vaccines. Quail chicks cannot tolerate weather change, and sudden temperature change. So they get affected by disease, if they experience sudden temperature or weather changes.

Guinea fowl farming

Compared to chickens, guinea fowl are low-cost and low-maintenance, and do a standout job as chemical-free pest control. Thus guinea fowl has the potential for landless households, and marginal and small farmers of arid and semiarid region due to its hardy nature as a tool to lift their socioeconomic status. Three available varieties are Kadmbari, Chitambari and Swetambari. Weight of the fowl at 8 weeks becomes 500-550 g and at 12 weeks 900-1000 g. The female fowl lays egg at 230-250 days age. Average egg weight is 38-40 g. Egg production in one laying cycle from March to September is 100-120 eggs. A guinea is never caught by its legs, as in a chicken. The correct way to catch a guinea is to clap both hands against its wings.

Benefits of guinea fowl farming

1. Guinea fowl are highly free moving birds and possess excellent foraging capabilities with unique ornamental value.
2. They act as bioinsecticides and bioherbicides. Working as a team, guineas will eat any pest they can get their beaks on, but unlike chickens, do so without tearing and scratching the garden plants. Since the birds are free-range, these will hunt ticks (or beetles, fleas, grasshoppers, crickets, snakes) all around. These are a more natural option to control the insect pest population than pesticides.
3. Guinea fowl are hardy and resistant to many common diseases of other poultry birds.

4. Guinea fowls are used as a watch dog in protecting the farm flock from intruders and predators by its loud, harsh, cry and its pugnacious disposition.
5. Guinea fowls are extremely good runners which help them to escape from predators.
6. Hard egg shell provides minimum breakage and long keeping quality.
7. Tolerate weather extremes fairly well after they are fully feathered and can be suitable to any agroclimatic conditions.
8. Input requirements are low. Consumes all non-conventional feed stuffs, usually not used in chicken feeding. Birds are more tolerant to mycotoxin and aflatoxin. No requirement of elaborate and expensive housing.
9. Guinea fowl meat is rich in vitamins and low in cholesterol. Guinea meat is darker and richer than chicken, with less fat and lower cholesterol.
10. The nutrient-rich dropping of the birds can be composted and used in the garden.
11. The feathers of guinea fowl are used in arts and crafts such as hat and head gears.

Guinea fowl housing, feeding and management

Domestic guinea fowl are not fully domesticated and are still a bit wild. They are not used to staying inside a house. They should be kept on free range during the day and kept inside a house at night to be protected against predators and thieves. Guinea fowl will get sick and die if they are kept in a wet or dirty house. Therefore their house should be spacious and well ventilated. This guinea fowl house of 3 m × 3 m is large enough to keep as many as twelve adult guinea fowl comfortably. The house should have a well maintained grass thatch roof, covered with plastic sheeting to ensure that it does not leak. It should be well ventilated with two opposite windows, each 0.75 m × 0.25 m. In order to protect the birds from predators, the windows of the house and the play area should be covered with chicken wire mesh or closely spaced bamboos. The birds need perches. At least one bamboo perch should be faced to the walls of the house at about 1 metre from the floor. The floor of both the house and the play area should be covered with a thick layer of sand. Sand is the best bedding material for guinea fowl as it provides them with a source of grit. Droppings should be swept out once a week using a broom. The sand should be changed once a month.

Adult guinea fowl are one of the most self-reliant poultry birds around. Domestic guinea fowl thrive and breed well when they are allowed on free range during the day. They usually move together in groups and scratch for food such as insects, snails, seeds and berries with their feet or their bill. Although they do pick up food while on free range, they should still be fed each day. By providing food and water inside the guinea fowl house, the guinea fowl are trained to return to their house each evening. Guinea fowl need to be fed a well-balanced diet in order to gain weight and produce more eggs. A balanced diet consists of a mixture of different types of food which contain protein, fats, carbohydrates, minerals, vitamins and water.

As most keets are not safe to graze in the wild they should be given plenty of feed while they are inside the brooder. As for feeding the keets, a 21 to 23% protein ration consisting of either commercial turkey starter or mashed hard-boiled eggs mixed with cottage cheese and a little cornmeal or oatmeal is given. After four weeks, 18% grower ration, or a mixture of cracked corn and whole wheat supplemented with young grass clippings, chopped lettuce, and other fresh greens are given. Water is always provided to the birds.

REFERENCES

Chakraborty, M and Gupta, D. 2017. Small ruminant value chain in India: opportunities and constraints. *Indian Journal of Economics and Development* **5** (7): 1-5.

De Vries, M. and De Boer, I. J. M. 2010. Comparing environmental impacts for livestock products: A review of life cycle assessments. *Livestock Science* **128**: 1-11.

FAO and IDF. 2011. Guide to good dairy farming practice. Animal Production and Health Guidelines. No. 8. Food and Agriculture Organisation of the United Nations and the International Dairy Federation, Rome.

FAO. 2015. Coping with climate change - the roles of genetic resources for food and agriculture. Commision on genetic resources for food and agriculture. Food and Agriculture Organisation, Rome. pp. 110.

Gerbens-Leenes, P. W., Mekonnen, M. M. and Hoekstra, A. Y. 2013. The water footprint of poultry, pork and beef: A comparative study in different countries and production systems. *Water Resources and Industry* **1-2**: 25-36.

Gerber, P. J. and Steinfeld, H. 2008. Global environmental consequences of the livestock sector's growth. *Outlook on Agriculture* **37**(1): 7-13.

http://dahd.nic.in/about-us/divisions/cattle-and-dairy-development

https://globalagriculturalproductivity.org/sustainably-meeting-dairy-demand-in-india/

https://sd.appstate.edu/community/sustainable-poultry-production

https://static.vikaspedia.in/media/files_en/agriculture/livestock/commercial-goatery.pdf

https://static.vikaspedia.in/media/files_en/agriculture/livestock/sheep-and-goat-farming/sheep-farming.pdf

https://sustainableagriculture.net/our-work/issues/animal-agriculture/

https://www.government.nl/topics/livestock-farming/sustainable-livestock-production

https://www.livestocking.net/livestock-farming-overview

https://www.thehappychickencoop.com/guinea-hen

https://www.worldanimalprotection.org.in/blogs/sustainable-dairy-business

Kumar, S. and Roy, M. M. 2013. Small Ruminant's Role in Sustaining Rural Livelihoods in Arid and Semiarid Regions and their Potential for Commercialisation. In: Prasad, S., et al. (eds.). New Paradigms in livestock production from traditional to commercial farming and beyond. Agrotech Publishing Academy, Udaipur. pp. 57-80.

Quazi, W. 2016. Dairy farming in India: A sustainable path. https://www.ecoideaz.com/expert-corner/dairy-farming-in-india

Rauw, W. M., Kanis, E., Noordhuzen-Stassen, E. N. and Grommers, F. J. 1998. Undesirable side effects of selection for high production efficiency in farm animals: a review. *Livestock Production Science* **56**: 15-33.

Steenfeldt, S., Afrose, S., Hammershoj, M. and Horsted, K. 2013. How to safeguard adequate nutrition in organic poultry production? Proceedings of the 19th European Symposium on Poultry Nutrition, pp 60-66. Potsdam, Germany, August 26-29, 2013.

Steinfeld, H., Gerber, P., Wassenaar, T., Castel, V., Rosales, M. and De Haan, C. 2006. Livestock's long shadow - environmental issues and options. FAO, Rome, Italy, pp. 390.

Vaarst, M., Steenfeldt, S. and Horsted. 2015. Sustainable development perspectives of poultry production. *World's Poultry Science Journal* **71**: 609-620.

Williams, A. G., Audsley, E. and Sandars, D. L. 2009. A lifecycle approach to reducing the environmental impacts of poultry production. Proceedings and Abstracts, 17th European Symposium on Poultry Nutrition. pp. 7.

16

Agricultural Waste Management

The residues from production and processing of raw agricultural products such as fruits, vegetables, meat, poultry, dairy products and crops are called agricultural wastes. The by-products of agricultural activities are also referred to as agricultural waste. These wastes may contain material that can benefit man but whose economic values are less than the cost of collection, transportation and processing for beneficial use. According to the United States Environmental Protection Agency, agricultural waste is the by-products generated by the rearing of animals and the production and harvest of crops or trees. Organisation for Economic Cooperation and Development defines agricultural waste as waste produced as a result of various agricultural operations including manure and other wastes from farms, poultry houses and slaughterhouses; harvest waste; fertiliser runoff from fields; pesticides that enter into water, air or soils; and salt and silt drained from fields. Agricultural waste can be in the form of liquids, slurries, or solids.

Expanding agricultural production has resulted in increased quantities of livestock waste, agricultural crop residues and agroindustrial by-products. A large number of agriculture wastes have been produced each year in recent years around the world. It is estimated that the annual increase of agricultural wastes is at 5-10% (Wang et al., 2015), mainly due to intensification of farming systems. Agricultural wastes are widely available, renewable and mostly free, hence they can be an important resource. These can be a valuable source of energy as well as raw materials for animal feed, manures, etc. However, most of the agricultural wastes are still underutilised. The utilisation of agricultural waste by farmers for agricultural production also reduces the environmental pollution. Thus, a greater awareness of the public and farmers is needed for proper management and utilisation of organic wastes in agriculture to diminish fears and preconceived notions of nuisance problems, decrease in land values as well as environmental degradation (Westerman and Bicudo, 2005).

AGRICULTURAL WASTE GENERATION

Although the quantity of wastes produced by the agricultural sector is significantly low compared with wastes generated by various industries, the pollution potential of agricultural wastes is high on a long-term basis. Majority of the agricultural wastes are readily biodegradable. They also contain significant quantities of plant nutrients, and human and animal pathogens. The quantity and quality of waste generated depends on the type of agricultural activities carried out.

Waste from cultivation activities

Intensive farming in tropics causes the development of insects, diseases and weeds. This creates a high demand for pesticides and agrochemicals to control insects, diseases and weeds. The pest control practices very often lead to the abuse of pesticides by farmers. After using pesticides, most of the bottles and packages holding these pesticides are thrown into fields or ponds. These wastes have the potential to cause unpredictable environmental consequences such as food poisoning, unsafe food hygiene and contaminated farmland due to their potentially lasting and toxic chemicals. In addition to this, existing stagnant or unused pesticides and pesticide packages with residue from the original contents poses serious environmental consequence in that they could be stored or buried in the wrong way which may leak or enter the environment through osmosis and thereby affecting the environment (Obi et al., 2016).

Inorganic fertilisers maintain or enhance the agricultural productivity. However, often farmers use more fertiliser in their fields than the amount needed by the crop plants. The serious consequence of such an excessive application of fertiliser is that it is used to the point of abuse in order to increase the annual agricultural output. The rate of absorption of such fertiliser compounds varies depending on the land characteristics, plant types and method of fertilisation (Thao, 2003). A portion of the excess fertiliser is retained in the soil, a portion enters ponds and other waterbodies through either surface runoff or the irrigation system adopted, which results in the pollution of surface water; a portion enters the groundwater, and a portion released to atmosphere causing air pollution (Obi et al., 2016).

Agriculture wastes also include crop residues which can be divided into field residues and process residues. Field residues are residues that present in the field after the process of crop harvesting. The field residues consist of leaves, stalks, straw, pods, stems, etc. The process residues are residues present even after the crop is processed into alternate valuable resource. These include molasses, husks, bagasse, seeds, leaves, stem, straw, stalk, pulp, stubble, roots, etc. which are used for animal feed, soil improvement, manures, etc. Huge amount of field residues are generated and most of them are underutilised. Agricultural residues are differentiated on the basis of their availability as well as characteristics that can be different from other solid fuels like charcoal, wood, and char briquette (Zafar, 2014).

Waste from livestock production

Waste from livestock activities include solid waste such as manure and organic materials in the slaughterhouse; wastewater such as urine, cage wash water, wastewater from the bathing of animals and from maintaining sanitation in slaughterhouses; air pollutants such as H_2S and CH_4; and odours. Air pollution includes odours emanating from cages resulting from the digestion process of livestock wastes; the putrefaction process of organic matter in manure; animal urine and from redundant foods. The proportion of NH_3, H_2S, and CH_4 varies with the stages of the digestion process. This untreated and non-reusable waste source can generate greenhouse gases and may also cause water pollution. In livestock waste, water accounts for 75-95% by volume. The rest includes organic and inorganic matter, microorganisms and parasite eggs. These germs and substances can spread diseases to humans and cause many negative effects on the environment (Obi et al., 2016).

Waste from aquaculture

Feed is an extremely important factor of production in aquaculture and its level of importance depends on the type of culture technique (Biswas et al., 2006; Dauda et al., 2017). Feed is the major source of waste in aquaculture systems (Akinwole et al., 2016; Martins et al., 2010). Solid wastes are primarily derived from the uneaten feed and faecal droppings of cultured fish (Akinwole et al., 2016). They occasionally include those fish that do not survive the culture process. In a properly managed farm, approximately 30% of the feed used will become solid waste. Increase in temperature results in increased feeding which gives rise to increased generated waste (Dauda et al., 2019). Water flow patterns in production units are important for waste management because a proper flow will minimise the fragmentation of fish faeces and allow for rapid settling and concentration of the solids (Obi et al., 2016). Solid wastes have been classified as the most dangerous waste in fish culture systems and should be effectively removed as quickly as possible as they can clog the fish gills and lead to death (Akinwole et al., 2016).

Waste from agroindustries

A huge amount of organic residues and related effluents are produced every year through the food processing industries like juice, chips, meat, confectionary, and fruit industries. These organic residues can be utilised for different energy sources. Approximately, 20% of the production of fruits and vegetables in India are going waste every year (Rudra et al., 2015). Similarly, the waste produced from food industries contains high value of chemical oxygen demand (COD) and biochemical oxygen demand (BOD), and other suspended solids. Most of these wastes are left unutilised or untreated, which cause adverse effect on environment as well as human and animal health but the composition of these wastes contains a large number of organic compound that produce a variety of value-added products and also reduce the cost of production (Sadh et al., 2018). Especially in oil industries, huge amount of processed residue, known as oilcakes are produced after oil extraction from the seeds.

AGRICULTURAL WASTE AND THE ENVIRONMENT

Agricultural waste is now considered as an important source of environmental pollution. Huge quantities agricultural wastes from the livestock farms and agroindustries are entering into the environment (Wang et al., 2015). The impact of agricultural waste on the environment depends not only on the quantity generated but also on the disposal practices followed. Some of the disposal methods pollute the environment (Sabiiti et al., 2004; Tumuhairwe et al., 2009). Cereal straw burning is a common practice in the developing countries which generates a lot of harmful gas, smoke and dust (Varma et al., 2015; Karak et al., 2015). It is a major source of atmospheric pollution by releasing pollutants such as carbon monoxide, nitrous oxide, nitrogen dioxide and smoke carbon (Ezcurra et al., 2001). These pollutants are accompanied by the formation of ozone and nitric acid (Hegg et al., 1987) and contribute to acid deposition (Lacaux et al., 1992) which poses risk to human health and ecological environment. Agricultural residues are also directly discharged into water, leading to serious contamination of aquatic environment (Wanga et al., 2016).

Animal wastes such as faeces, urine, and respiration and fermentation gases cause an acute and serious environmental problem in countries with high concentrations of animals on a limited land base for manure disposal. Animal wastes are excreted in solid, liquid, and gaseous forms. Respiration and fermentation gases are lost to the environment soon after being produced by the animal. After excretion, solid and liquid animal waste is subjected to microbial conversion, mainly through anaerobic process, which converts organic substrates into microbial biomass and soluble and gaseous products. Some of these products have an impact on the environment, as well as water quality, soil deterioration, and air pollution (Sabiiti, 2011). Animal manure also contains many pathogens, parasite eggs, heavy metals etc. Application of excessive animal wastes on land as fertiliser and soil conditioner is subject to surface runoff and leaching that may contaminate ground or surface waters. Nitrate leaching is considered a major nitrogen pollution concern on livestock farms (Mackie et al., 1998). When phosphorus enters the surface waters from land application of excessive animal manure it can stimulate the growth of algae and other aquatic plants. Their subsequent decomposition results in an increased oxygen demand that interferes with the fish and other aquatic animals. Manure decomposition can be a major source of methane, ammonia and nitrogen oxides, which contribute to accumulation of greenhouse gases. Volatilisation of ammonia causes acid deposition, which contributes to acid precipitation (Likens et al., 1996). Odour pollution is also reported due to improper disposal of livestock wastes.

All kinds of agricultural wastes, especially poultry, animal faeces and crops straw, have a very high nutrition potential and can also improve soils for sustainable production ability (Ibrahim, 2015). Therefore, the effective transformation of agricultural waste recycling and utilisation was important in the controlling of environment pollution. Besides, facing with the problem, it can also address the serious energy crisis. The recycling and utilisation of agricultural wastes are considered to be the important step in environmental protection, energy structure and agricultural development (Wanga et al., 2016).

RECYCLING AND UTILISATION OF AGRICULTURAL WASTES

Agricultural wastes cause environmental pollution, which imperil sustainable development. Improper disposition of agricultural wastes also cause wastage of a lot of valuable biomass resources. Thus, recycling of agricultural waste is a necessity for the practice of sustainable agriculture and pollution free clean environment. The recycling and utilisation of agricultural wastes are considered to be the important step in environmental protection, energy structure and agricultural development (Wanga et al., 2016). Usually, agricultural wastes are utilised in the traditional ways with the low utilisation efficiency. The disposal and utilisation of agricultural wastes depend upon the availability of viable technology and agricultural automation processes, government policies, rules and regulations, and social service systems available in agricultural waste management (Nguyen et al., 2014). There are several factors such as the cost, technology innovation, local natural geographical environment, incentive policy, population density, industry and agriculture development (Gabhane et al., 2014; Kung, et al., 2014) that affect the recycling and utilisation process of agricultural wastes.

The pollutants from agriculture wastes have four characteristics, including huge quantities, bad qualities, low price and excessive danger (Rangabhashiyam et al., 2013). However, the target should be to make agricultural wastes a resource that can be utilised and not just discarded. Agricultural wastes can be used to enhance food security mainly through their use as biofertiliser and soil amendment, use as animal feed, and energy production. They contain large amounts of organic matter, and many of them can be directly added to the soil without any risk (Sabiiti, 2011).

Agricultural waste utilisation technology must either use the residues rapidly, or store the residues under conditions that do not cause spoilage or render the residues unsuitable for processing to the desired end-product. There are a number of applications to which these wastes can be used (Obi et al., 2016).

Organic fertilisers

The utilisation of animal manures for plant nutrients has a definite impact on input energy requirements at the farm level (Timbers and Downing, 1977). Converting crop residues and animal manures into organic fertilisers through composting is one of the waste treatment technologies that make it possible to use organic waste as plant nutrients. Composting reduces the volume of the waste, hence solving serious environmental problems concerning disposal of large quantities of waste, kills pathogens that may be present, decreases the germination of weeds in agricultural fields, and reduces odour (Jakobsen, 1995). Poultry manure contains high phosphorus which has positive effect on the growth and productivity of crops. It is also effective when combined with mineral phosphorus fertiliser for farm use. Adding organic manure to soil increases its fertility because it increases the nutrient retention capacity, improves the physical condition, the water holding capacity and the soil structure stability (Hargreaves et al., 2008).

Animal feed

Both crop residues and animal waste can be used as animal feed. Crop residues are high in fibre content but low in protein, starch and fat content. To increase the livestock production the forage and pasture should be supplemented with grains and protein concentrate to meet the protein needs of the animals. Thus, use of the grain and protein for human food will compete with such use for animal feed. These problems may be overcome by utilising crop residues to feed the animals (Hussein and Sawan, 2010). The nutrient content of animal waste depends on the animal species, type of feed, and bedding material used (Mackie et al., 1998). The ruminants are useful in converting crop residues into food, hence contributing substantially to reducing potential pollutants. The rumen contains the microbial enzyme cellulase, which is the only enzyme to digest the most abundant plant product, cellulose. With ruminants, nutrients in crop by-products are utilised and do not become a waste-disposal problem (Oltjen and Beckett, 1996).

Anaerobic digestion

Methane gas can be produced from agricultural wastes particularly manures. The gas is best suited for heating purposes as in broiler operation, water heating, grain drying, etc. The anaerobic digestion of agricultural waste to form methane-rich gas is a two-step microbial fermentation. Initially, acid-forming bacteria break down the volatile solids to organic acids which are then utilised by methanogenic organisms to yield methane-rich gas (Fig. 16.1). The composition of the typical gas produced is methane, 50-70%; CO_2, 25-45%; N_2, 0.5-3%; H_2, 1-10% with traces of H_2S; and the heating value of the gas is in the range of 18-25 MJ/m^3 (Timbers and Downing, 1977). Some of the major disadvantages of the digestion system are the high capital costs and the explosive properties of the methane gas. However, the advantages far outweigh the disadvantages. Anaerobic digestion makes the treatment and disposal of large poultry, swine and diary waste feasible, minimising the odour problem. It stabilises the waste and the digestion sludge is relatively odour-free and yet retains the fertiliser value of the original waste (Obi et al., 2016).

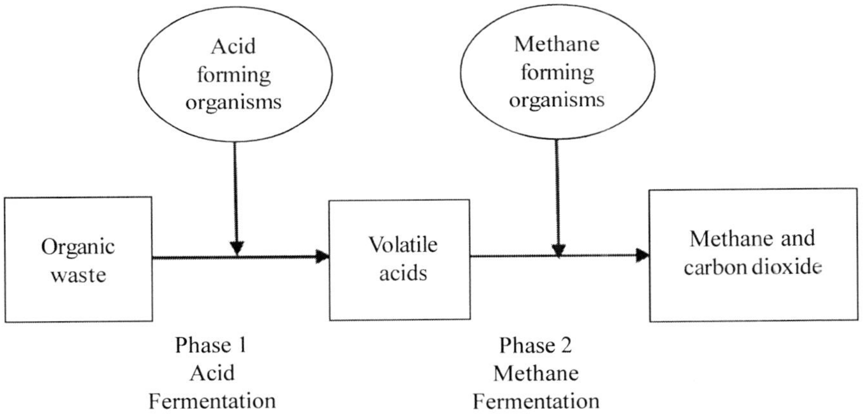

Fig. 16.1. Methane production by two-stage microbial fermentation

Source: Timbers and Downing, 1977

Adsorbents in the elimination of heavy metals

Excessive release of heavy metals into the environment due to industrialisation and urbanisation has posed a great problem worldwide. Unlike organic pollutants, the majority of which are susceptible to biological degradation, heavy metal ions such as copper, cadmium, mercury, zinc, chromium and lead ions do not degrade into harmless end-products (Gupta et al., 2001). The presence of heavy metal ions is a major concern due to their toxicity to many life forms. It is reported that use of agricultural wastes is a low cost alternative for the treatment of effluents containing heavy metals through the adsorption process (Chand et al., 1994). The low cost agricultural waste such as sugarcane bagasse (Mohan and Singh, 2002), rice husk (Ayub et al., 2002), coconut husk (Tan et al., 1993), sawdust

(Ajmal et al., 1996), neem bark (Ayub et al., 2001), etc. have the potential for the elimination of heavy metals from wastewater.

Pyrolysis

In pyrolysis systems, agricultural waste is heated up to a temperature of 400-600° C in the absence of oxygen to vapourise a portion of the material, leaving a char behind. This is considered to be a higher technology procedure for the utilisation of agricultural wastes. Others are hydro-gasification, and hydrolysis. They are used for the preparation of chemicals from agricultural waste as well as for energy recovery. Agricultural wastes can be utilised to prepare alcohols for fuel, ammonia for fertilisers, glucose for food and feed by pyrolysis. Pyrolysis of agricultural waste yields oil, char and low heating value gas (Obi et al., 2016).

Direct combustion

Complete combustion of agricultural waste consists of the rapid chemical reaction of biomass and oxygen, the release of energy, and the simultaneous formation of the ultimate oxidation products of organic matter, CO_2 and water (Klass, 2004). The energy released is usually in the form of radiant and thermal energy provided oxidation occurs at sufficient rate; the amount of which is a function of the enthalpy of combustion of the biomass. If agricultural waste is to be utilised efficiently through thermal conversion process, there is need to fabricate these biomass wastes into solid form. It is usually burnt for heating, cooking, charcoal production, and the generation of steam, mechanical and electric power applications (Klass, 2004).

AGRICULTURAL WASTE MANAGEMENT SYSTEM

The organic wastes, especially manure generated by animals, if improperly managed or left untreated can result in significant degradation of soil, water and air quality. Uncontrolled decomposition of organic wastes produces odorous gases as well as ammonia volatilisation, leading to acid rain (Wright, 1998). The following concerns have been raised due to the intensification of livestock farming (Fabian et al., 1993).

- Water quality resulting from higher nitrogen and phosphorous loadings
- Pathogens and antimicrobial compounds in the manure
- Foul odours and air quality from ammonia, methane and nitrous oxide emissions
- Soil quality because of potassium and phosphorous loading

Agricultural waste management (AWM) has now become an issue of concern for policy makers for maintaining the ecological balance and development of sustainable agriculture. The usual approach to agricultural waste management has been discharge to the environment with or without treatment. However, the wastes should be considered as potential resources rather than undesirable and unwanted, to avoid contamination of air, water, and land resources, and to avoid transmission of hazardous materials (Obi et al., 2016). So there is a need to change in the conventional approach of dumping the agricultural wastes. There should be appropriate technology and incentives for agricultural waste management; moreover, a change in philosophy and attitudes of the farmers in this regard is urgently required.

The United States Department of Agriculture (USDA) has defined agricultural waste management system (AWMS), as a planned system in which all necessary components are installed and managed to control and use by-products of agricultural production in a manner that sustains or enhances the quality of air, water, soil, plant, and animal resources (USDA, 2012). The agricultural waste management system is developed using total systems approach, i.e., it is designed to cater for all the waste associated with agricultural production to utilisation throughout the year. The total solids concentration of agricultural wastes is the main characteristic that determines the handling of the material. For example, the total solid concentration of animal manure depends on climate, type of animal, amount of water consumed by the animal, and the feed type. In most systems, the consistency of the waste can be anticipated or determined. The total solid concentration of the waste can be increased by adding beddings or other solid waste to the waste, decreased by adding water, and stabilised by protecting it from additional water. The total solid concentration is important because it affects the total volume of the waste to be handled. Liquid waste management systems are often easier to automate and manage than those for solid wastes; however, the initial cost of the liquid handling equipment may be greater than that for solid waste systems (USDA, 2012).

Agricultural waste management system consists of six basic functions such as production, collection, transfer, storage, treatment and utilisation (Fig. 16.2). For a specific system, these functions may be combined, repeated, eliminated or rearranged as per the requirement (USDA, 2012).

Production

Production is the function of the amount and nature of agricultural waste generated by an agricultural enterprise. The waste requires management if the quantity produced is sufficient enough to become a resource concern. A complete analysis of production includes the kind, consistency, volume, location, and timing of the waste produced. The waste management system may need to accommodate seasonal variations in the rate of production. The production of unnecessary waste should be kept to a minimum (USDA, 2012).

Collection

Collection refers to the initial capture and gathering of the waste from the point of origin or deposition to a collection point. The AWMS plan should identify the method of collection, location of the collection points, scheduling of the collection, labour requirements, necessary equipment or structural facilities, management and installation costs of the components, and the impact that collection has on the consistency of the waste (USDA, 2012).

Transfer

Transfer refers to the movement and transportation of the waste throughout the system. It includes the transfer of the waste from the collection point to the storage facility, to the treatment facility, and to the utilisation site. The waste may actually be transferred several times before utilisation (Fig. 16.2). For example, a liquid or slurry waste may be collected, transferred to a storage facility, and then to a solid/ liquid separator (treatment). From there, the solid portion may be transferred to another storage facility for additional treatment or reutilised as bedding, while the liquid portion is applied (transferred) to a crop field to supply plant nutrients (utilisation). The waste may require transfer as a solid, liquid, or slurry, depending on the total solid concentration. The system plan should include an analysis of the consistency of the waste to be moved, method of transportation, distance between transfer points, frequency and scheduling, necessary equipment, and installation and management costs of the transfer system (USDA, 2012).

Storage

The storage function has to do with the temporary containment or holding of the waste. The storage facility of a waste management system provides control over the scheduling and timing of the system functions such as the treatment and application or use of the waste which could be affected by weather or interfered with by other operations. The storage period should be determined by the utilisation schedule. The waste management system should identify the storage period; required storage volume; type, estimated size, location, and installation cost of the storage facility; management cost of the storage process; and impact of the storage on the consistency of the waste (USDA, 2012).

Treatment

Treatment is any function designed to reduce the pollution potential or modify the physical characteristics of the waste, such as moisture and total solid content, to facilitate more efficient and effective handling. Manure treatment is comprised of physical, biological, and chemical unit processes. It includes pre-treatment activities such as analysis of the characteristics of the waste before treatment; a determination of the desired characteristics of the waste following treatment; the selection of the type, estimated size, location, and the installation cost of the treatment facility; and the management cost of the treatment process (USDA, 2012).

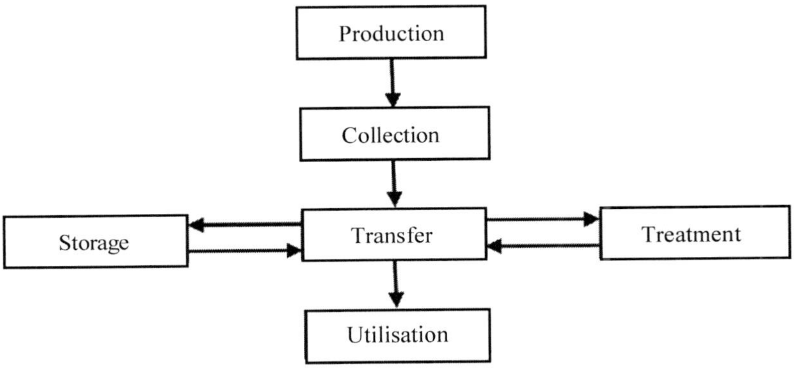

Fig. 16.2. Agricultural waste management functions

Source: USDA, 2012

Utilisation

Utilisation includes reusing and/or recycling of waste products. Agricultural wastes may be used as a source of energy, bedding, mulch, organic matter, or plant nutrients. A common practice is to recycle the nutrients in the waste through land application. A complete analysis of utilisation through land application includes selecting the fields; scheduling applications; designing the distribution system; selecting necessary equipment; and determining application rates and volumes, value of the recycled products, and installation and management costs associated with the utilisation process (USDA, 2012).

'3R' APPROACH TO AGRICULTURAL WASTE MANAGEMENT

The concept of minimising waste reduces the quantity and ill-effects of waste generation by reducing quantity of wastes, reusing the waste products with simple treatments and recycling the wastes by using it as resources to produce same or modified products. This is usually referred to as '3R'.

The principle of reducing waste, reusing and recycling resources and products (3Rs) aims at achieving efficient minimisation of waste generation by choosing to use items with care to reduce the amount of waste generated, repeated use of items or parts of items which still have usable aspects, and the use of waste itself as resources.

Reduction

The reduction concept refers to reducing the amount of unnecessary wastes by careful preparation of raw agricultural materials. Through reduction efforts, greenhouse gas emissions can be avoided or reduced in a number of sectors related to agricultural waste, including fertiliser and pesticide production, land use change, deforestation for agriculture, and energy harvesting (Hai and Tuyet, 2010).

Reuse

Reuse refers to activities in which something that cannot function in its original purposes comes to serve as a new type of product with little or no processing. Wastes from the farming process, animal husbandry, and agricultural services should be reused to the greatest possible extent, the using time of products should be extended, and energy efficiency and waste generation reduction should be given appropriate attention. Crop residues such as rice husks can be applied directly for soil mulching. This practice requires low labour input and low investment. In terms of climate change reduction, this reuse of plant residues can also avoid a certain amount of greenhouse gas emissions from the burning and composting processes.

Livestock waste contains many of plant nutrients. These nutrients are mainly in the form of complex organic compounds that are indigestible. These wastes, therefore, require pre-treatment before using as an organic fertiliser to the crop plants. After anaerobic digestion, organic fertiliser destroys weed seeds, insects and plant pathogens, because heat energy is generated during decomposition of organic substances. Livestock waste can also be used as a food source for fish after anaerobic digesting. When this nutrient waste discharges into a pond, a portion is used directly as feed by carps and other species of fish, while the rest is transformed to non-nutrient matter by microorganisms. Algae and aquatic plants in the pond can absorb these substances and transform them into nutrients. These algae and aquatic plants help development of phytoplankton in the pond (Hai and Tuyet, 2010).

Recycle

Recycle involves a complex set of activities in a process to recover resources from waste. There is a great potential in recycling agricultural residues for biogas production. Compared to the 'reduce' and 'reuse' approaches, 'recycling' requires greater inputs of time and labour for collecting waste materials and greater investment for processing them. The recycling process may also lead to greenhouse gas emissions. But these greenhouse gas emissions are often less than the amount that would occur as a result of landfill disposal. The major objectives of recycling of agricultural organic waste are to recover valuable plant nutrients and energy by means of composting and anaerobic digestion.

In terms of environmental benefits, a recycling approach can contribute to sustainable agriculture by reducing the use of agrochemicals. The cow manures can also be used to produce biogas by anaerobic digestion. This recycling approach has some advantages over composting such as requiring a shorter time and a smaller area for operation and also producing organic fertiliser that contains low quantities of heavy metals. The biogas can then be used for cooking and/or electricity generation (Hai and Tuyet, 2010).

The 3Rs individually or collectively saves fresh resources exploitation, add value to the already exploited resources and very importantly minimises the waste quantity and its ill-effects (USDA, 2012).

'3R' hierarchy in agricultural waste management

Waste minimisation can be achieved in an efficient way by focusing primarily in a hierarchical order; first, 'reduce', followed by 'reuse' and then 'recycle'. The waste hierarchy refers to the '3R', which classify waste management strategies according to their desirability. The 3Rs are meant to be a hierarchy, in order of importance. The waste hierarchy has taken many forms over the years, but the basic concept has remained the foundation of most waste minimisation strategies. The aim of the waste hierarchy is to extract the maximum practical benefits from products and to generate the minimum amount of waste (USDA, 2012).

The '3R' approach is conventionally expressed through a pyramid hierarchy in which increase in environmental benefits of each approach is placed from bottom to top (Fig. 16.3).

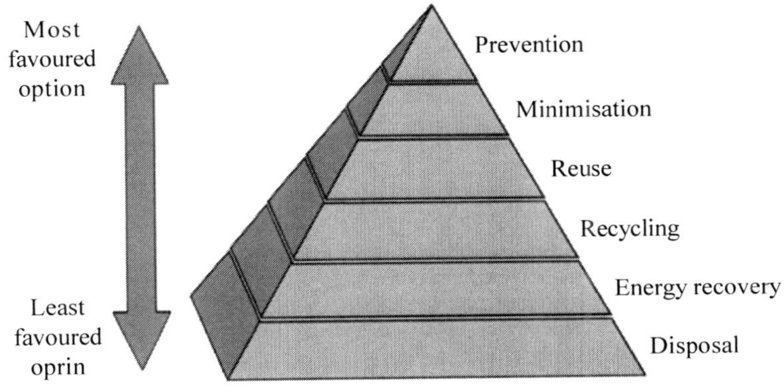

Fig. 16.3. '3R' hierarchy in agricultural waste management

Source: Government of Bangladesh, 2010

DAIRY WASTE MANAGEMENT SYSTEMS

Dairy farming requires high standards of sanitation and must prevent problems associated with flies. Small dairy operations are commonly managed on pastures in partial confinement. While animals are on pasture, their waste should not be a resource concern if stocking rates are not excessive, grazing is evenly distributed, manure from other sources is not applied, and grazing is not allowed during rainy periods when the soils are saturated. To prevent waste from accumulating in feeding, watering, and shade areas, the feeding facilities can be moved, the number of watering facilities can be increased, and the livestock can be rotated between pastures. Most medium and large dairy facilities are managed as total confinement. All of the manure and associated waste components must be included in the agricultural waste management systems. The United States Department of Agriculture has prescribed a guideline for the management of dairy waste based on the six basic functions of agricultural waste management systems (USDA, 2012).

Production

Waste associated with dairy farming includes manure, contaminated runoff, milking house waste, bedding, spilled feed, and silage leachate.

Collection

The collection methods for dairy waste vary depending on the management of the dairy operations. Dairy animals may be partly, totally, or seasonally confined. Manure accumulates in confinement areas and in areas where the dairy animals are concentrated before and after milking. Unroofed confinement areas must have a system for collecting and confining contaminated runoff. The volume of runoff can be reduced by limiting the size of the confinement area, and uncontaminated runoff can be diverted if a roof runoff management system and diversions are used. The manure and associated bedding accumulated in roofed confinement areas can be collected and stored as a solid. The manure can also be collected as a solid in unroofed lots in humid climates where the manure is removed daily. Manure can be removed by a flushing system. The volume of contaminated water produced by the system can be greatly reduced if provisions are made to recycle the flush water.

Transfer

The method used to transfer the waste depends largely on the consistency of the waste. Liquid and slurry wastes can be transferred through open channels, pipes, or in a portable liquid tank. Pumps may be used to transfer liquid waste. Solid and semisolid waste can be transferred by mechanical conveyance equipment, in solid manure spreaders, and by pushing them down curbed concrete alleys. Semisolid waste has been transferred in large pipes through the use of gravity, piston pumps, or air pressure.

Storage

Milking house waste and contaminated runoff must be stored as a liquid in a waste storage pond or structure. Manure may be stored as a slurry or liquid in a waste storage pond designed for that purpose or in a structural tank. It can be stored as a semisolid in an unroofed structure that allows for the drainage of excess water and runoff or as a solid in a dry stacking facility. In humid areas, the stacking facility should have a roof.

Treatment

Liquid waste can be treated in an aerobic lagoon (an artificial pool for the treatment of effluent), anaerobic lagoon, or other suitable liquid waste treatment facilities. Solid material can be separated from the waste stream through various solid/liquid separation techniques. Energy production and odour abatement can be accomplished through anaerobic digestion or thermo-chemical conversion. Following solid/liquid separation, solids from the waste can be composted or used as bedding.

Utilisation

Dairy waste is used as compost, as an energy source or through land application. Waste may be hauled and distributed over the land in a dry or liquid manure spreader. Liquid waste and slurries may be distributed through an irrigation system.

POULTRY WASTE MANAGEMENT SYSTEMS

The United States Department of Agriculture has prescribed a guideline for the management of poultry waste based on the six basic functions of agricultural waste management systems (USDA, 2012).

Production

Wastes associated with poultry farming include manure and dead poultry. Depending upon the system, waste can also include litter, wash-flush water, and waste feed.

Collection

The manure from poultry farming is allowed to accumulate on the floor where it is mixed with the litter. The manure litter pack forms a 'cake' that generally is removed between flocks. The litter pack can be removed frequently to prevent disease transfer between flocks. In layer houses, the manure that drops below the cage is collected in deep stacks or is removed frequently using either a shallow pit located beneath the cages for flushing or scraping or belt scrapers positioned directly beneath the cages.

Transfer

The method used to transfer the waste depends on the total solid content of the waste. Liquid waste can be transferred in pipes, or tank wagons, and dried litter can be scraped, loaded, hauled as a solid and transported using a truck.

Storage

Litter from poultry operations is stored on the floor of the housing facility or outside the housing facility. When it is removed, it can be transported directly to the field for land application. In some areas the litter may be compacted in a pile and stored in the open for a limited time; however, it is better to cover the manure with a plastic or other waterproof cover until the litter can be used. If it is needed to be stored for a long time, the litter should be stored in a roofed facility. If the manure from layer operations is reasonably dry, it can be stored in a roofed facility. If it is wet, it should be stored in a structural tank or an earthen storage pond.

Treatment

Poultry litter can be composted. This stabilises the litter into a relatively odourless mass and helps to kill disease organisms so that the litter can be reused as bedding or supplemental feed to livestock. The litter can also be dried and burned directly as a fuel. Liquid manure may be placed into an aerobic digester to produce methane gas.

Utilisation

The waste from poultry facilities can be used for agricultural land application or sold because of the high nutrient value of the litter. Poultry waste can also be used for the production of methane gas, buried directly as a fuel, reused as bedding, or used as a feed supplement to livestock.

Dead poultry disposal

Because of the large numbers of mortalities associated with large poultry farms, the disposal of dead birds is a resource concern. Poultry facilities must have adequate means for disposal of dead birds in a sanitary manner. To prevent spread of disease, the dead birds are often collected daily by hand. Disposal alternatives include incineration, rendering, burial, or composting. The dead birds are mixed with litter and straw, composted, and the finished material is stored until it can be applied to the land.

FOOD PROCESSING WASTE MANAGEMENT

Food processing facilities produce large amounts of waste, some of which are suitable for land application. Food processing waste can be solid, slurry or liquid. The chemical properties of the waste must be determined before a waste handling system can be designed. If the waste is biological in nature, it can be treated and handled much the same as livestock waste. Waste treatment lagoons can be used for some food processing waste. The material must be analysed for its volatile solids content or its biochemical oxygen demand concentration so that volumetric or areal loading rates can be determined. Because some canneries are seasonal, lagoons may need to be oversized to accept anticipated periodic heavy organic loading. Co-mixing food waste with animal manure can significantly increase methane production in an anaerobic digester (USDA, 2012).

REFERENCES

Ajmal, M., Rao, R. A. K. and Siddiqui, B. A. 1996. Studies on removal and recovery of Cr from electroplating wastes. *Water Research* **30**(6): 1478-1482.

Akinwole, A. O., Dauda, A. B. and Ololade, A. O. 2016. Haematological response of *Clarias gariepinus* juveniles reared in treated wastewater after waste solids removal using alum or *Moringa oleifera* seed powder. *International Journal of Acarology* **6**(11): 1–8.

Ayub, S., Ali, S. I. and Khan, N. A. 2001. Efficiency evaluation of neem (*Azadirachta indica*) bark in treatment of industrial wastewater. *Environmental Pollution Control Journal* **4**(4): 34-38.

Ayub, S., Ali, S. I. and Khan, N. A. 2002. Adsorption studies on the low cost adsorbent for the removal of Cr from electroplating wastewater. *Environmental Pollution Control Journal* **5**(6): 10-20.

Biswas, G., Jena, J. K., Singh, S. K. and Muduli, H. K. 2006. Effect of feeding frequency on growth, survival and feed utilisation in fingerlings of *Catla catla*, *Labeo rohia* and *Cirrhinus* in outdoor rearing systems. *Aquaculture Research* **37**: 410-414.

Chand, S., Aggarwal, V. K. and Kumar, P. 1994. Removal of hexavalent chromium from the wastewater by adsorption. *Indian Journal of Environmental Health* **36**(3): 151-158.

Dauda, A. B., Ajadi, A., Tola-Fabunmi, A. S. and Akinwole, A. O. 2019. Waste production in aquaculture: Sources, components and managements in different culture systems. *Aquaculture and Fisheries* **4**: 81-88.

Dauda, A. B., Ibrahim, H. I., Bichi, A. H. and Tola-Fabunmi, A. S. 2017. Assessment of fish farming practices, operations, water resource management and profitability in Katsina state, Nigeria. *Journal of Northeast Agricultural University* **24**(4): 89-96.

Ezcurra A. I., Ortiz, deZarate, Dhin, P. V. and Lacaux, J. P. 2001. Cereal waste burning pollution observed in the town of Vitoria, northern Spain. *Atmospheric Environment* **35**: 1377-1386.

Fabian, E. E., Richard, T. K. D., Allee, D. and Regenstein, J. 1993. Agricultural composting: A feasibility study for New York farms. https://www.cfe.cornell.edu/

Gabhane, J., Prince William, S. P. M., Gadhe, A., Rath, R., Vaidya, A. N. and Wate, S. 2014. Pre-treatment of banana agricultural waste for bio-ethanol production: Individual and interactive effects of acid and alkali pre-treatments with autoclaving, microwave heating and ultrasonication. *Waste Management* **34**: 498-503.

Government of Bangladesh. 2010. National 3R strategy for waste management. Department of Environment, Ministry of Environment and Forests, Government of the People's Republic of Bangladesh. pp. 3-4.

Gupta, V. K., Gupta, M. and Sharma, S. 2001. Process development for the removal of lead and chromium from aqueous solution using red mud - an aluminium industry waste. *Water Research* **35**(5): 1125-1134.

Hai, H. T. and Tuyet, N. T. A. 2010. Benefits of the 3R approach for agricultural waste management in Vietnam. Institute for Environment Science and Technology, Hanoi University of Technology Hanoi, Viet Nam. pp. 27-29.

Hargreaves, J. C., Adl, M. S. and Warman, P. R. 2008. A review of the use of composted municipal solid waste in agriculture. *Agriculture, Ecosystems & Environment* **123**(1-3): 1-14.

Hegg, D. A., Radke, L. F., Hobbs, P. V., Brock, C. A. and Riggan, P. J. 1987. Nitrogen and sulphur emissions from the burning of forest products near large urban areas. *Journal. of Geophysical Research* **92**: 14701-14709.

https://www.progressivegardening.com/agricultural-wastes

Hussein, S. D. A. and Sawan, O. M. 2010. The utilisation of agricultural waste as one of the environmental issues in Egypt. *Journal of Applied Sciences Research* **6**(8): 1116-1124.

Ibrahim, R. A. 2015. Tribological performance of polyester composites reinforced by agricultural wastes. *Tribology International* **90**: 463-466.

Jakobsen, S. 1995. Aerobic decomposition of organic wastes 2. Value of compost as fertiliser. *Resources, Conservation and Recycling* **13**: 57-71.

Karak, T., Sonar, I., Nath, J. R., Paul, R. K., et al. 2015. Struvite for composting of agricultural wastes with termite mound: Utilising the unutilised. *Bioresource Technology* **187**: 49-59.

Klass, D. L. 2004. Biomass for renewable energy and fuels. In: Cleveland, C. J. (ed.) Encyclopedia of Energy, Vol. 1. Elsevier, San Diego. pp. 193-212.

Kung, C. C., McCarl, B. A., Chen, C. C. and Chen, L. J. 2014. Environmental impact and bioenergy potential: Evaluation of agricultural commodity and animal waste based biochar application on taiwanese set-aside land. *Energy Procedia* **61**: 679-682.

Lacaux, J. P., Loemba-Ndemb, J., Lefeivre, B., Cros, B. and Delmas, R. 1992. Biogenic emissions and biomass burning influences on the chemistry of the fog water and stratiform precipitations in the African equatorial forest. *Atmospheric Environment* **26**: 541-551.

Likens, G. E., Driscoll, C. T. and Buso, D. C. 1996. Long-term effects of acid rain: Response and recovery of a forest ecosystem. *Science* **272**: 244-245.

Mackie, R. I., Stroot, P. G. and Varel, V. H. 1998. Biochemical identification and biological origin of key odour components in livestock waste. *Journal of Animal Science* **76**: 1331-1342.

Martins, C. I. M., Eding, E. H., Verdegem, M. C. J., et al. 2010. New developments in recirculating aquaculture systems in Europe: A perspective on environmental sustainability. *Aquacultural Engineering* **43**(3): 83–93.

Mohan, D. and Singh, K. P. 2002. Single and multi-component adsorption of cadmium and zinc using activated carbon derived from bagasse - an agricultural waste. *Water Research* **36**: 2304-2318.

Nguyen, T. A. H., Ngo, H. H., Guo, W. S., Zhang, J., Liang, S., Lee, D. J., Nguyen, P. D. and Bui, X. T. 2014. Modification of agricultural waste/by-products for enhanced phosphate removal and recovery: Potential and obstacles. Bioresource Technology **169**: 750-762.

Obi, F. O., Ugwuishiwu, B. O. and Nwakaire, J. N. 2016. Agricultural waste concept, generation, utilisation and management. *Nigerian Journal of Technology* **35**(4): 957-964.

Oltjen, J. W. and Beckett, J. L. 1996. Role of ruminant livestock in sustainable agricultural systems. *Journal of Animal Science* **74**:1406-1409.

Rangabhashiyam, S., Anu, N. and Selvaraju, N. 2013. Sequestration of dye from textile industry wastewater using agricultural waste products as adsorbents. *Journal of Environmental Chemical Engineering* **1**: 629-641.

Rudra, S. G., Nishad, J., Jakhar, N. and Kaur, C. 2015. Food industry waste: mine of nutraceuticals. *International Journal of Environmental Science and Technology* **4**(1): 205-229.

Sabiiti, E. N. 2011. Utilising agricultural waste to enhance food security and conserve the environment. *African Journal of Agriculture, Food, Nutrition and Development* **11**(6): 1-9.

Sabiiti, E. N., Bareeba, F., Sporndly, E., Tenywa, J. S., etc. 2004. Urban market garbage: A hidden resource for sustainable urban/peri-urban agriculture and the environment in Uganda. *The Uganda Journal* **50**: 102-109.

Sadh, P. K., Duhan, S. and Duha, J. S. 2018. Agroindustrial wastes and their utilisation using solid state fermentation: a review. *Bioresources and Bioprocessing* **5**: 1-15.

Tan, W. T., Ooi, S. T. and Lee, C. K. 1993. Removal of Chromium from solution by coconut husk and palm pressed fibre. *Environmental Technology* **14**: 277-282.

Thao, L. T. H. 2003. Nitrogen and phosphorus in the environment. *Journal of Survey Research* **15**(3): 56-62.

Timbers, G. E. and Downing, C. G. E. 1977. Agricultural biomass wastes: utilisation routes. *Canadian Agricultural Engineering* **19**(2): 84-87.

Tumuhairwe, J. B., Tenywa, J. S., Otabbong, E. and Ledin, S. 2009. Comparison of four low-technology composting methods for market crop wastes. *Waste Management* **29**: 2274-2281.

USDA. 2012. Agricultural Waste Management Field Handbook. Chapter 9, Agricultural Waste Management Systems. United States Department of Agriculture, Soil conservation Service, USA.

Varma, V. S., Yadav, J., Das, S. and Kalamdhad, A. S. 2015. Potential of waste carbide sludge addition on earthworm growth and organic matter degradation during vermicomposting of agricultural wastes. *Ecological Engineering* **83**: 90-95.

Wang, H. Y., Liu, S., Zhai, L. M., et al. 2015. Preparation and utilisation of phosphate biofertilisers using agricultural waste. *Journal of Integrative Agriculture* **14**: 158-167.

Wang, S. R., Ru, B., Dai, G. X., Sun, W. X., Qiu, K. Z., Zhou, J. S. 2015. Pyrolysis mechanism study of minimally damaged hemicellulose polymers isolated from agricultural waste straw samples. *Bioresource Technology* **190**: 211-218.

Wanga, B., Donga, F., Chena, M., et al. 2016. Advances in recycling and utilisation of agricultural wastes in China: Based on environmental risk, crucial pathways, influencing factors, policy mechanism. *Procedia Environmental Sciences* **31**: 12-17.

Westerman, P. W. and Bicudo, J. R. 2005. Management considerations for organic waste use in agriculture. *Bioresource Technology* **96:** 215-221.

Wright, R. J. 1998. https://www.ars.usda.gov/is/np/agby-products/agbyexecs ummary.pdf

Zafar, S. 2014. Waste management, waste-to-energy. https://www.bioenergy consult.com/ tag/waste-to-energy

Subject Index

D

E

F

G

H

I

K

L

M

N

O

P

Q

R

S